Electronic Classics

Electronic Classics

Collecting, Restoration and Repair

Andrew Emmerson

Newnes

OXFORD BOSTON JOHANNESBURG MELBOURNE NEW DELHI SINGAPORE

Newnes
An imprint of Butterworth-Heinemann
Linacre House, Jordan Hill, Oxford OX2 8DP
225 Wildwood Avenue, Woburn, MA 01801-2041
A division of Reed Educational and Professional Publishing Ltd

 A member of the Reed Elsevier plc group

First published 1998

British Library Cataloguing in Publication Data
A catalogue record for this book is available from the British Library

ISBN 07506 3788 9

Library of Congress Cataloguing in Publication Data
A catalogue record for this book is available from the Library of Congress

Typeset by Scribe Design, Gillingham, Kent
Printed and bound in Great Britain by
Biddles Ltd, Guildford and King's Lynn

Contents

Acknowledgements

No man is an island and the author is the first to acknowledge the inspired ideas and suggestions made by other enthusiasts in various publications and on the Internet. Plaudits are accordingly due to

Martin Ackroyd
Arden Allen
Australasian Telephone Collectors Society
Bernard Babani's handbooks of the 1940s
Bobbi Barmore
Alan Betz
Ray Bintliff
Jeffrey Borinsky
David Boynes
Ben Bradley
James Binkley
Malcolm Burrell
Gordon Bussey
Lenox Carruth Jr
Dale Davenport
William Donzelli
Tony Duell
Dexter Francis
George Gonzalez
Dave Grant
Richard Hager
Ben Hall
Ron Ham
Mike Hanz
Steve Harris
Jeffrey Herman
David Higginson
Jonathan Hill
Ian Hopley
Dicky Howett
Mike Izycky
Jeremy Jago
Al Klase
Mike Knudsen
Jim Lockwood
Larry Lovell
Chas. E. Miller
Andy Mitz

Ray Mote
Rob Murrell
Paul Nelson
Geoff Dixon-Nuttall
Old Timer's Bulletin
Barry Ornitz
A. Padgett-Peterson
Brian Pethers
Gary Pewitt
Larry Rau
Tudor Gwilliam Rees
Rétro-Phonia
Scott Robinson
Daniel Schoo
Ludwell Sibley
Bob Smallbone
Gerard Tel
Dube Todd
Hank van Cleef
John Wakely
Gerry Wells
Linley Wilson
Karl Zuk
and many other members of the Boatanchors group too many to mention.

Before you start . . .

SERIOUS WARNING!

Throughout this book you will read about chemicals which may be noxious and electrical voltages which may be lethal. The author assumes you know what you are doing and will take all necessary precautions. In no way will he be responsible for the actions of you, the reader. You have been (politely) warned!

For some strange reason people find safety boring or at least something they can put off until later. We don't agree but so as not to detain your reading now, the full section on safety in the workshop appears at the end of the book. Do read it before starting the practical work – it could save your life!

LESS SERIOUS WARNING

This is an unconventional book and it will take some getting used to. It is a mixture of indisputable fact and good advice, mixed up with highly charged opinions and insights. You probably will not be able or inclined to read it all at one go – never mind, try dipping in now and again until you feel comfortable with it. I genuinely believe it will repay you richly to study it, and you may even end up enjoying it and recommending it to your friends. Whatever you do, don't lend it to them if you want to see it again! By the way, if you encounter any unfamiliar terms (such as *boatanchor*), you'll find them all explained at the back of the book. Of course, they do say if you have to ask what a boatanchor is, you won't understand the answer! On the other hand, if you have ever carried one back to your car, you will!

A WORD TO THE WISE

This book is written in British English (more or less) and for people who are familiar with the British electrical environment. There's nothing chauvinistic in this and we would like this book to appeal to the widest possible readership. So please allow for a few unfamiliar expressions if, for example. your preference is the word *tube* rather than *valve*. Some of the other expressions which vary according to which side of the Pond you are will be found in the Reference Section at the back of the book. We have assumed throughout a mains electricity supply voltage of around 230 volts AC, 50 hertz, so if you live in an area with 60 Hz or a lower voltage, please make the necessary mental adjustment. Thanks!

FOR AMERICAN READERS

This book is for you too just so long as you remember that we are two great nations separated by a common language!

for	*read*
anode	plate
earth	ground
heater	filament
HT	B+
mains	line
Marconi-Osram	Genalex
picture tube	kinescope
resin flux	rosin flux
socket	receptacle
Tandy shop	Radio Shack store
valve	tube

International telephone numbers quoted in this book include the 00 international dialling code prefix used in most countries of the world. North American readers should dial the prefix 011 for overseas calls and omit the prefix altogether for numbers within the USA and Canada.

CHAPTER 1

Introduction (Don't skip reading this!)

Have you noticed how American books can be more fun to read than British ones? Somehow their authors manage to make them livelier than ours, yet still putting over all the information you need to know. They also make learning enjoyable and more digestible. Well, this is not an American book but I still hope you'll enjoy reading it because I've certainly had a lot of fun writing it. If you like it tell the publisher, tell me. If you don't like it, I still want to know!

FIRST THINGS FIRST

I don't want anyone feeling cheated so . . . this is not a complete book. That's not an admission, it's an assertion. This book does not waste time telling you things you already know or which you can find easily elsewhere. That may make it look unbalanced but its main purpose is to fill in the gaps with information that's missing elsewhere.

WHY I WROTE THIS BOOK

The reason why I wrote this book is simple and no, it wasn't money. It was because someone just had to write it. For twenty, or perhaps more like thirty years, I've been fooling with electrical hobbies and most of the knowledge I've gained I picked up the hard way. It wasn't in books and nobody made it easy for me. All the same, I had a lot of fun out of it and now it's time to put back something into the hobby.

IS IT NEW?

You bet! Although the subject is old (by definition), at least 90 per cent of this book is new, and even what isn't may well not be familiar. I make the claim that never has so much handy material on restoring old sets been presented together in one compilation. And whilst the information has been gathered from here, there and everywhere, this book really is my own creation!

IS IT DIFFERENT?

I really do think so. For a start, it 'takes the lid off' the subject and lets you in on all kinds of secrets I have never found in any other book. It also tells you not just what to do but also how and why you should go about it this way. Many books leave out the explanations, leaving you to wonder just how reliable their advice is.

NOT JUST RADIO IN THIS BOOK

This book naturally reflects my own personal experience and interests, and I enjoy telephones, telephone exchanges and telegraph instruments just as much as I like wireless and television. I have some treasured items of classic hi-fi as well. Many people also find a common interest across all these fields and for this reason I have found it convenient to treat them all together. There are no exclusion zones in this book, however, and you will almost certainly find things to interest you throughout the text.

You may feel that radio techniques dominate this book, but in truth, most of the information and techniques are equally valid to all collecting and restoration interests in the field of vintage electronics. Don't feel ignored, therefore, if you happen not to share all these interests; please remember this book was still written for you. Just spit out and leave the bits you don't savour. And do yourself a favour; do read the chapters that don't look relevant – you may well discover a technique which can be applied to your particular passion. Call this technology transfer or lateral thinking or whatever you like. By the way, if you are wondering why I didn't arrange the chapters to cover radio, television, telephones, audio and so on, that's easy. It would have involved much repetition; it's the techniques you're interested in, after all, and these hold good regardless of the particular object in question.

A BIT BITTY

I have deliberately made the information in this book 'bite size'. You have to plough through some books to find the part you need and I wanted to make this book easy to use and to assimilate. You don't have to read it all at once (although I'd be flattered if you did). What's also relevant is that some of the source material for this book was in magazines that I have edited, and magazines are often written in short chunks, probably to assist the people who don't buy them but read them in W.H. Smith's library (and prevent me from getting anywhere near the shelves when I actually want to buy a magazine or paper!).

INFORMATION GAP

If you ask people who have been in this hobby (or employed in the trade) for years, they'll tell you that you pick up all kinds of knowledge gradually over the years. That's fine for them but what about newcomers? They need to know now! Also, for people used to low voltages in solid-state equipment it is vital to be aware of safety issues. It is true that all sorts of useful articles have appeared here and there but never together in one book. Most people don't have access to all these articles nor the time to sift through so much reading matter.

SOMETHING MISSING

More important, though, is that all the books I have read on the subject seemed to leave out information that I feel other enthusiasts would have wanted to read, which is why I eventually felt I had to write this book. Most of the books already published on vintage radio just show you how to mend non-working sets. But this book tells you how to restore sets and that goes beyond just mending them. It means making them look smart, finding the authentic parts, right down to using the right flex and mains plug. None of those other books takes you to that stage!

CLOSED SHOP?

I also wanted to demystify the restoration of radios, TVs, amplifiers and telephones in order to make it more accessible. Some people like to make out that restoring old sets is a black art, for experts only. It's not. Some people think it's also very expensive. It shouldn't be. Furthermore, most of the books you can buy talk only of repairing, not restoring, and then they tend to concentrate only on wireless. I happen to be interested in hi-fi, television, test gear and telephones as well, and it's not always obvious how far radio techniques apply to these other fields.

EUREKA!

I believe this is the first book that tells you how to find the old sets, how to buy them, where to go for parts and how to avoid being stung in the process. Read through these pages and I'll tell you quite a bit about the business side of this hobby as well. It may annoy you, it may make you want to join in, but it certainly should keep you amused and better informed.

SAVING MONEY

There's another reason for reading this book. You must have noticed the rising price of old radios and other electrical devices. In a way this is satisfying, because it shows that people now value these items as antiques, but it does mean that an increasing number of enthusiasts can no longer afford to indulge their hobby to the extent they'd like. Or at least, they can no longer afford to buy decent, working, second-hand items, and most of the ones they can afford are faulty, cobbled-up mongrel sets or plain fakes.

But there are plenty of untested, 'sold as seen' or definitely defective items around which are still in everyone's price bracket. Yes, these will take far more work to put right but I guarantee that you will get more pleasure and satisfaction from this policy because each radio, television or whatever you bring back to life will contain a little bit of you. You might even use your new-found skills to make a bit of pocket money, restoring sets for antique dealers or for other hapless souls who can't do it for themselves.

SPENDING MONEY

This book is not for cheapskates. It is aimed at serious hobbyists who want to enhance the value of their treasures and suggests professional methods of achieving this. Some techniques will force you to put your hand in your pockets and spend money, but I guarantee the results will be worthwhile. If you are not interested in using proper tools and materials and if you are satisfied with sloppy consequences, you'll probably not find much benefit from this book. On the other hand, if you'd like an insight into the way the professionals go about their business, read on!

SECRETS BIBLE

This is not a book for all-round experts, rather its aim is to help you become one – or at least something approaching one. Most seasoned hobbyists and collectors are proficient in one or more fields of their particular interest, but few of them profess to be all-round experts. Perhaps you are a wizard at electrical repairs but have to concede defeat when it comes to professional-looking cabinet restoration. Equally you may envy other collectors who seem to possess an uncanny knack of sounding-out rarities (or finding the bargains first every time). That's why I call this book a secrets bible – it will let you in on their techniques so that you too can become as successful all-round as the experts.

By the way, I don't count myself as an expert practitioner nor do I have all the answers. But I do keep my eyes and ears open, meaning I have gleaned a goodly number of ideas over the years, and since nobody else has offered a similar compilation, I felt it might as well be me who did.

THANKS

Whilst this compilation is entirely my own, I cannot take the credit for all the contents. Ours is a mutual, co-operative hobby and I willingly acknowledge the contributions made by friends. These are all credited in the text.

One more thought: each of the books I have written has been to fill a knowledge gap where no comparable book existed. In effect I ended up producing the book I hoped was already written. I trust this book turns out to be the one you were hoping to find.

Thanks for your patience: you now have my permission to read the rest of this book. Enjoy it with my best wishes and my thanks for buying it.

PART 1

A Great Hobby and More

CHAPTER 2

Approaches to collecting

I'm assuming you are a collector – are you? I suppose you may be a dealer or museum conservator, and in this case your interest is purely professional. But most of you reading this book will be hobbyists and enthusiasts, using your own time and money to build up your collections, and this means you can break all the rules that professionals must follow.

Professionals are expected to follow specific remits on what they buy in and make sure that what they purchase is good value for money. Hobbyists don't have to: if there's something really special they can bend the rules! What's more, they don't face the sack if they make an ill-considered purchase (although there may be other, worse, sanctions!).

Ask five collectors how they go about their hobby and you'll probably get five different answers. One may collect sets which have a personal connection or a special meaning, another may settle for a single-manufacturer compilation. Some people try to make their collection as representative as possible (with each set illustrating some special point), whilst others will go for one specific type of set (perhaps early transistors or novelty radios). Finally, some collectors just amass everything they see, particularly if it can be acquired at low cost. What the heck? It's your hobby, so you set the rules.

That said, a systematic collection will almost certainly give more pleasure and satisfaction than a random affair. Some people go as far as calling their collection a museum and style themselves as curators; if this is for a private collection with no access to the public, this sounds a bit pretentious to me (but it's your life, your choice).

WHERE DO YOU START?

Well, we all have to start somewhere – it may be with a gift or an inheritance or perhaps a chance find at an antiques fair or a car boot sale. Whatever, once the bug has bitten, there's no stopping, and what started as a casual interest can soon become serious! It's up to you: you can give house room to one or two conversation pieces or you can let them take over your home and life. With a bit of luck they will accumulate in value too; of

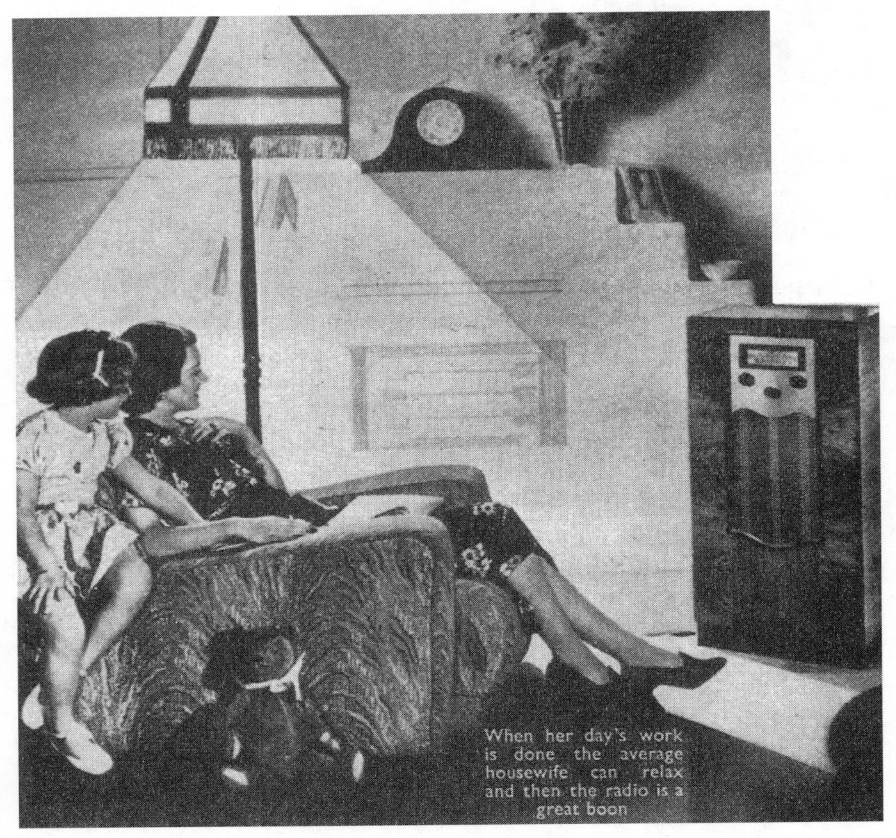

Figure 2.1 It would be foolish to deny the nostalgic appeal of old wireless sets; they have a marvellous ability to conjure up the cosy, warm comfortable feeling of days gone by, as symbolized by this *His Master's Voice* publicity illustration of the 1930s. Whether summers were truly sunnier in those days and people really were happier is irrelevant – today everyone warms to old radios either as conversation pieces or as collectors' items.

course everyone drops the odd clanger early on, but on balance your collection should end up more valuable, especially if you 'add value' to your treasures by following the advice in this book. It's probably not a good idea to treat every acquisition as an investment (you can fool your 'other half' in this way if you must!), but over time, values never fall.

Collecting old radios, televisions, telephones and hi-fi equipment is not a difficult or expensive hobby – everyone can play. You can explore relatives' lofts, visit jumble sales and boot fairs or attend household effects auctions; these are the low-cost entries to the hobby. Or you can visit antique and collectors fairs, big auctions, specialist dealers' shops and collectors' swap-meets; here you'll probably be spending rather more.

Junk shops should not be ignored either; only a few years ago an enthusiast spotted thirteen television discs in a Hastings junk shop. These had been brought in by someone who had cleared the house of television pioneer J. L. Baird following his death at Bexhill in 1946. They were put up for auction in December 1996 and fetched £14 490! Pre-war television sets sell consistently for over £2000 at auction but good examples have been found lately in an antique shop (£75) and at a car boot sale (just £25) – you just need to keep your eyes open! Tell friends and relatives about your hobby too.

Newspapers are another good source of bargains. Read the local paid and free newspapers and the specialist advertisement papers such as *Loot* and *Exchange and Mart*. Many of these offer free want ads too, so it must be worth giving these a try – those who have done so report excellent results.

When you get bored with your main hobby you can extend the boundaries. Radio collectors can start looking for advertising novelties such as china 'Nipper' dogs (the *His Master's Voice* trade mark), old copies of the *Radio Times* and sales brochures. Telephone collectors can look for pre-war telephone directories, manufacturers' catalogues and so on, and you'll soon be discovering new delights. Morse keys, cigarette cards, test equipment, enamel shop signs . . . the list goes on for ever!

If you really wish to start something novel, what about a virtual collection? It won't take up much space and all you'll need is a computer and a source of illustrations! You can add new items at little cost and nobody will ever steal your collection. On the other hand, you can share it with others by putting it on the World Wide Web, as in fact a number of spirited collectors are already doing.

TRENDS IN COLLECTING

It is an undeniable fact that at any given moment some items are fashionable and others are not. Many collectors feel they are missing out if they fail

Figure 2.2 Although transmissions on the old 405-line television system ceased in 1985, an enterprising band of enthusiasts keep their treasures alive with electronic standards converters, modulators and other gadgetry available from specialist suppliers. Period programmes (on videotape) and vintage technology make a powerful combination which always arouses interest when demonstrations are held at vintage radio exhibitions and other events.

to have the trendy items in their collection, both pushing up prices and leading to misery if the holy grails are out of reach. Speaking for Britain, the trends in telephones are coloured plastic phones (particularly the red, green and ivory 200- and 300-type instruments, which were never common even when they were new) and in radios any attractive bakelite sets, particularly the Bush DAC90 and DAC90A. Proof that fashion is capricious is found in the fact that the similarly styled Bush DAC10 sets are not rated at all!

In hi-fi the Quad 2 and Leak Point One amplifiers are among the most sought after. Across the Atlantic it appears that most art-deco-styled radios are fashionable, particularly those made in catalin plastic or with mirrors incorporated into the cabinet, whilst the TV to have is the Philco Predicta in one of its many variants. In Continental Europe art-deco sets are popular.

There is not a lot of logic in these fads but they cannot be ignored if your chief goal is a collection that will increase in value. On the other hand, if

Figure 2.3 Would you like one of these monsters in your living room? A growing number of enthusiasts find their fascination in the 'heavy metal' equipment of broadcasting. Having rescued cameras and other studio equipment from broadcasters, training colleges and scrap-yards, they now find their restored treasures in demand as props by television and film companies. The picture is also a useful reminder that shooting television pictures was still an energetic occupation thirty years ago. *Above:* The camera as found before restoration, complete with unwanted detritus. Renovation techniques for professional equipment such as this can be the same as for old automobiles.

you go deliberately for the unfashionable items, you can build up an excellent collection at modest prices, especially if the items you choose represent a progression or illustrate a particular point and are not just a random selection.

Another trick is to recognize trends at the outset, before prices start to rocket. Among British telephones, the original 700-type sets are worth seeking out, particularly those in the less commonly found colours. With radio the first transistor sets are already highly sought after, although the original sets using integrated circuits can still be found at quite reasonable prices. Novelty sets, modelled on cartoon characters or on cola bottles, hamburgers and the like, are not expensive (yet) either. Another good theme would be pre-war 'people's sets'; whilst not cheap, these can still be found at fair prices.

In television, pre-war sets have such a cachet as to be out of reach of many pockets but the receivers made immediately after the war are almost as attractive and a good deal easier to find. Among the post-war sets, any Bush moulded bakelite set is a must-have if you're following trends. Micro-TVs are already sought after and will become more so, as are the 'Sputnik', 'Videosphere' and 'space helmet' round-cabinet sets of the 1970s. First-generation colour sets are bulky but they were never plentiful even when new; in future they will become a lot rarer.

Finally, if people tell you that unless it has valves in it, it's not worth collecting, don't listen – they are just expressing their ignorance!

THE DRIVING FORCE

This is not the place for deep psychological investigation but it's interesting to note that a recent *Modern Maturity* survey of the collecting hobby in the USA revealed:

- 38% started collecting before age 25, 30% after age 40;
- 49% collect for love, 28% for nostalgia, 11% for money, 12% for posterity;
- 10% have 25 or fewer items, 64% have hundreds and more;
- 6% regularly swap or trade items, 31% occasionally, the rest never;
- 42% say displaying the collection gives the most pleasure;
- 51% don't care what anyone else thinks of their collecting habits.

Collector Bill Hawkins observes that the article touches on 'obsession' with a few examples, but that wasn't a question in the survey. He also notes that the fact that 63 per cent never part with an item once acquired is what keeps the estate-sale dealers in business, and what drives the heirs to distraction.

MONEY AND FASHIONS ARE UNIMPORTANT

One last point, and perhaps the most important one: vintage electronics is not a hobby that you can just 'throw money' at. If you approach the hobby purely from a financial angle, you'll end up very disappointed. For a start, it's unlikely you will be able to afford all you see and desire and, in any case, there are many items that money cannot buy in the short term. Demand outstrips supply and waving any amount of money in the air will not bring certain items; you'll just have to wait until they come on to the market. You don't want to be a sheep anyway, merely buying what other people say is collectable; far better that you learn the techniques of restoring the items other people have overlooked or discarded and acquire the skill of finding things that others don't know where or how to find. This book will teach you.

CHAPTER 3

Enhancing your collection and keeping track of it

Is this going to be a warning about collectomania? Not really. I could lecture you on the risks of never knowing when to stop . . . in fact I know of at least two people who were forced to buy the house next door in order to house their burgeoning collections (honestly!). I know of one collector who neglected his wife to the extent that she got another lover, who shot dead the husband at point-blank range in his own garden when he returned early one morning and disturbed them. But this is not the place to discuss matrimony, except to say that some of the best partnerships I know are where both partners appreciate each other's interests and support one another. In several cases each partner has a passionate collecting hobby, which tends to balance out any complaints! But this is dangerous ground.

THEME FOR A DREAM

How do you choose a theme for collecting? And should you? Most collections start by chance and some grow and grow, as the collector amasses more and more artefacts. This can be fun but not very satisfying, since the collection ends up not making much sense. The most successful collectors identify a theme, be it Eddystone communications receivers, early magneto telephones, 1960s transistor TVs or whatever. Some even set an upper price limit, e.g. radios I bought for £20 or less (and then turned into highly desirable restorations).

By adopting a theme, you can learn a lot and derive even more interest from your hobby. You may well become an expert or at least a specialist in your chosen field and, if you make the right noises, people will even come to offer you first choice on items they think will interest you. It's worth considering.

Once you have established your own theme, you will be less prone to fashionability fads, where suddenly all the material has gone sky-high in price, as has happened to catalin radios, bakelite telephones, round Ekco radios and so on. You can smile pityingly as prices for these sets go through the roof while your own collection of 1970s novelty radios grows at modest cost (and then suddenly everyone goes mad over novelty radios and your collection is already nearly complete). Try and anticipate the next trend; currently 1950s mains sets, particularly continental ones with lots of high-gloss veneer, ivory knobs and gold trim, are underpriced. They may not remain that way. Sound-recording equipment of the 1940s and 1950s is getting scarcer every minute too.

One theme, if you can call it that, is for professional equipment: marine and military radios, and also broadcast equipment. The beauty of this stuff is that it is relatively unloved and thus not too expensive. The other factor in its favour is that it is better made and thus far more easily restorable. Metal cabinets are easy to repaint. All components are numbered and easy to replace (generally) and professional equipment can virtually always be brought back to its original working condition. It may be big and ugly but it will serve you well. Added to which, there's a real satisfaction in acquiring for a few score pounds something which cost its original owners a fortune.

WRITTEN RECORDS

Maintaining a written record of a collection is one of the activities which distinguishes the true collector from the mere hoarder. It not only helps you get to know your collectables better, it adds to your enjoyment of your collection as a whole.

The practical advantages are many and include:

- You know what you have, where you acquired it and how much you paid for it.
- It is a great help in insurance arrangements, particularly if you have to make a claim.
- If through some misfortune a relative or friend has to arrange disposal of your collection, then well-maintained records can be an enormous help with a difficult task.
- If you donate your collection as a whole to a museum, detailed records save the recipient time in research and enhance the value of the gift.

Although museums and individuals use different methods for recording their holdings, depending on size and type, the information itself tends to follow a general pattern. In museums this is usually done in a number of stages, but for most collectors a simple, single-stage catalogue is more satisfactory.

The following suggestions may help collectors who want to keep a simple catalogue.

KEEPING TRACK OF YOUR TREASURES

Firstly, each item should be given an accession number. You could simply use sequential numbers beginning with 0001 and allocate numbers in order as the collection grows. The system used by most museums is a hyphenated two-part number. The first part is the last two digits of the current year and the second part is the sequential number in that year. For example 95-12 would be the twelfth item registered in 1995. Because the accession number becomes a unique identification, you may decide to allocate numbers also to the more valuable component parts (e.g. tubes) or any items you acquire separately.

The accession number should be marked on the item. A small adhesive label on the back or base is usually satisfactory, although in this day and age some kind of marking in indelible ink and possibly also in 'invisible' security ink may be advisable. Whatever you do, you should endeavour to avoid defacing the object or detracting from its value. Tie-on labels may be suitable for some items. Each entry in the catalogue begins with the accession number. This is followed by a description of the item, how acquired, source, date and price.

The amount of detail given in the description will vary with the item. For mass-produced items something like 'TELEVISION RECEIVER, Bush type TV22, as-new condition' may be quite adequate. For your rare and precious items, do not stint on details. Include manufacturer, overall dimensions and descriptions of components and their materials and of any distinctive features such as decoration and manufacturer's numbers. Comments on condition, including damage and missing parts, should also be made. A photograph of the item adds greatly to the written description and to the interest of the catalogue.

The entries on how acquired and source are an opportunity to not only state whether purchase, swap or gift but also to give details such as price or item swapped and name of supplier. This is not only useful information but also adds to that personal factor which is such a part of collecting.

There is no limit to the other information you can include. Details of references to the type of item in books and journals, such as *Radio! Radio!* or *BVWS Bulletin*, can be added, including photocopies. Particulars of restoration work done on the item should also be recorded, plus those of parts added.

Catalogues can be in the form of card files, bound books or loose-leaf binders. Keeping a catalogue on a computer is also a possibility. Make your choice but remember to design it in a flexible way, since you may want to add information to individual entries at a later date. If you are a computer user and have not yet set up a data base, there is an excellent program available you should consider. It is freeware (in other words provided at no cost) and is remarkably sophisticated. The name is the Antique Radio Database and it will suit many cataloguing pursuits, not only radio collectors. It will help you keep an inventory of a collection and the program stores information such as maker, model, year, repair notes, pictures and other details. Here is a list of the several ftp locations on the Internet where you can download this excellent program (it's a 32-bit application written for Windows 95).

ftp://ftp.cdrom.com/pub/simtelnet/win95/database/radio.zip
ftp://ftp.simtel.net/pub/simtelnet/win95/database/radio.zip
ftp://ftp.digital.com/pub/micro/pc/simtelnet/win95/database/radio.zip
ftp://ftp.lib.sonoma.edu/pub/simtelnet/win95/database/radio.zip
ftp://ftp.bu.edu/pub/mirrors/simtelnet/win95/database/radio.zip
ftp://oak.oakland.edu/pub/simtelnet/win95/database/radio.zip
ftp://ftp.rge.com/pub/systems/simtelnet/win95/database/radio.zip
ftp://ftp.ou.edu/pub/simtelnet/win95/database/radio.zip

The message should now be clear: if you have not already recorded your collection, make a start now and discover the advantages.

KEEPING HOLD OF YOUR COLLECTION

Possessing a detailed record of your collection may come in useful if anything unpleasant occurs. Sadly the increasing desirability of old artefacts and the perceived antique market value for our collectables means there is a ready market for period radios, telephones and so on, and there are plenty of unscrupulous people who are prepared to go to extreme lengths to get hold of them. There are even people who will steal to order.

Collections have been taken from museums and private homes alike, and only strenuous security precautions will keep burglars out. Even the opportunist burglar with no specialist knowledge may take a fancy to an obviously prized exhibit. Whether they are professionals stealing to order or just casual sneak-thieves who strike lucky is irrelevant; they will get your collection if it is unprotected and some of the tales I have heard are heartbreaking. Insurance is not an entire answer; quite likely the possibility of finding replacements for your rarities is pretty slim. In this connection, don't assume your collection is automatically covered by your household contents insurance. These policies often exclude cover for collections of valuable items.

BEATING THE BANDITS

Sturdy door – and window – locks and a powerful alarm (preferably one linked to a central monitoring station) are the only deterrent; whatever the cost, it will be less than what your treasures are worth. Indeed, after I myself was burgled a few years ago, the most lasting impression (apart from bitterness and frustration) was that my lost earnings during the time spent dealing with police and insurers afterwards would have more than paid for a decent alarm system. Thieves tend to take the line of least resistance and if they see a bell box bearing the Telecom Security or ADT name – or window locks – they will probably target another house nearby. Of course you will also qualify for reduced insurance premiums if you take these additional precautions. It's a sad fact that most householders install security measures only *after* they have been broken into (that's my own story too), but the consequent loss is always greater than the cost of installing security.

Don't just put up a dummy bell box, though; the burglars recognize the design of the dummy covers you can buy in the shops. A bell box bearing the name of a national or local security company is a more powerful deterrent and the name of a phone line-connected alarm generally means that they won't give you another glance. If you don't have a household alarm system, get one now before you are burgled: installing one afterwards will already be too late. Consider, too, a system which dials a distress call automatically when the alarm is triggered; the cost is not excessive.

One last thought – you can also make burglary less likely by your discretion. Do your friends and neighbours know what you collect? Do you place want ads in magazines? Have you had your collection featured in the local paper? It's highly unlikely that you are the only person who knows about your collection.

DISPLAY YOUR COLLECTION...

Dust is another enemy and for smaller items a glass case is the answer. The larger out-of-town furniture stores carry small wall cabinets with glass doors, and these are ideal for displaying unusual valves and tubes, advertising novelties and other small items. A range of cabinets with sliding glass doors are also made by specialist firms, primarily for model collectors to show off their die-cast cars or model trains; these are advertised every month in *Collectors' Gazette* magazine.

... AND USE IT!

Nearly all of our treasures benefit from regular use; they were all made originally to fulfil a function and they deserve to be used and enjoyed. Periodic use will keep the item in good order and prevent the kind of problems that occur with storage in less than perfect locations. Mechanical parts are less likely to seize up, whilst electrolytic capacitors benefit particularly from regular use, since they otherwise need reforming if left unused (unexercised) for periods. Any faults come to light more quickly in this way and switch shafts will be less likely to freeze up and volume controls will not get so dirty.

There is no need to overdo this, but experts suggest that sets be used for at least one hour every month or so. One hour's operation or more allows them to warm up properly, which for the valves is better than shorter periods of operation. There is no need to treat old sets with kid gloves either; they were expected to be used regularly and were designed as such. What's more, if they have survived fifty years or more, they are unlikely to fail now!

CHAPTER 4

How old is it and what's it worth?

The essential appeal of ancient treasures is their age, so it's only natural that we want to know how old they are. Collectors insist on knowing in which year a design was introduced (and perhaps when it was last made), as well as when this particular example was made.

ESTABLISHING A DATE

For many radios, televisions and telephones the year of introduction can be found in textbooks, reference books, trade yearbooks, catalogues or collectors' guides. You may not find the information immediately, so you will just have to be diligent. Dealers and museum staff may be able to help, whilst a letter to the manufacturer may also bring results (although many firms have minimal archive resources these days).

As regards date of manufacture, you may be lucky enough to find the year marked on the case, possibly in code. British telephones, for instance, normally showed the last two digits of the year, prefixed by the code of the manufacturer (e.g. H49, S56). With consumer electronics (radios, televisions and so on) the date may be less obvious or possibly absent altogether; this protected dealers with slow-moving stock, who might not want their customers to know they had just bought last year's model rather than the latest thing.

Detective work may be needed to decipher the vital date. You need to examine the components inside and the best clue is normally given by electrolytic capacitors; since these age and need to be reformed if used a long time after being manufactured, they normally carry a date. Many electronic components carry a four digit number indicating the year and week of manufacture, so an integrated circuit marked 7436 would have been made in the 36th week of 1974.

Finally, you may find clues in the way patents are asserted and the country of origin is marked. The German abbreviation for 'registered design' was DRGM before 1945 but DBGM in post-war West Germany. 'Made in West Germany' or 'Made in US Zone' are clearly post-war indications too.

DATING RADIOS

An excellent list giving the introduction date for 6000 radio receivers (mainly European) can be found on the Rétro-Phonia Web site:

http://members.aol.com/retrophoni/bienvenue.htm

Similar information can be found in many of the various European and American collectors books and price guides listed in Chapter 26, 'Recommended reading'.

An alternative approach, giving the approximate date of manufacture, is to examine clues given by the stations marked on the tuning dial or by the circuitry. Let's check out the station names first; these offer a strong clue. If certain stations in the country where the radio was produced are missing, this is a clear indication that the set was not produced when the station was operating. This does not always hold good for stations in other countries, however.

Wavelengths are allocated by international convention, with frequency allocation conferences being held at intervals. One of the earliest such plans dates from 1929 (the Prague Plan dealing with long and medium wave broadcasting), and each successive plan has caused some significant changes to the arrangement of station names. Major rearrangement occurred following the Lucerne Plan of 1934 and the Copenhagen Plan of 1948 (still in force but revised at the 1975 Geneva Conference). VHF/FM radio and television services are regulated by the Stockholm Plan of 1961.

Unfortunately this subject has not been documented in great detail. The first edition of Jonathan Hill's book *Radio! Radio!* contained a date-at-a-glance chart covering wavelength changes for British stations, whilst a more comprehensive list covering radio stations across Europe can be found on Gerard Tel's Web site at the address:

http://www.cs.ruu.nl/people/gerard/radios/dabydi.html

In summary, all British regional and national services were suspended with the outbreak of war in 1939 and replaced by the Home Service. The Light

Figure 4.1 What makes an item collectable depends on a number of factors, but the opulence of an object and a high original purchase price are no guarantee that it will be sought-after today. Bulky radio-gramophones such as this find little favour in the compact homes of today and seldom fetch high prices.

Programme was introduced in 1945 and the Third Programme in 1946. Radios 1, 2, 3 and 4 were introduced in 1967 after the suppression of nearly all of the offshore stations. The name Daventry is found before 1934 and Droitwich from that year.

Radio Luxembourg was found on 1190 m from 1932 to 1935, on 1293 m from 1935 to 1940 and from 1946 to the present. Its post-war medium wave outlet on 208 m lasted from 1950 to 1991.

The era of the offshore pirate stations (such as Veronica, Caroline, London and Noordzee) was the 1960s. Transmitter sites which have changed nationality are another pointer. Breslau, Danzig, Gleiwitz and Stettin were ceded by Germany to Poland, Königsberg to the USSR and Lwow and Vilna/Wilno by Poland to the USSR, all in 1945.

Some sets had a bandspread facility to simplify the tuning-in of weak stations in the crowded medium wave. British sets of the early 1960s often provided this for Radio Luxembourg (208 m) and around 1964/65 for Radio Caroline (199 m). British radios made before the Second World War sometimes have a tuning position for 'television sound' or 'USW' (ultra-short waves); this indicates that the sets were made between 1936 and 1939. British and American receivers may also have band markings for aircraft, amateurs and police but these are not very meaningful.

Sets made in the USA in the height of the Cold War era carried Civil Defence markings, which were required from 1953 to 1962. Most radios made after 1964 do not show them whilst many radios were made without CD marks during that time period. In short, CD marks are a guide but they cannot be relied on as absolute proof that a radio was or was not made in the Cold War period.

There are of course traps for the unwary. Changes made under the Montreux Plan of 1938 were not in fact implemented, thanks to the Second World War, but this did not prevent some receivers being made with dial glasses showing the new arrangements. These sets are highly collectable of course. In addition, many new sets were produced in the first few years after the war and their owners were not pleased when the implementation of the Copenhagen Plan made their tuning scales obsolete. Many manufacturers were forced to make replacement dial glasses for these nearly new sets, proving that station names are not a faultless guide to a radio's age.

When it comes to dating sets by their VHF stations, there is similar confusion. American sets with a VHF waveband date from 1941 onwards, following the start of commercial FM broadcasting the previous year. The frequency band allocated was 42.1–49.9 MHz, with alphanumeric call signs indicating the frequency and city. The FM band was redefined as 88–108 MHz in 1945, making that year the turning point for dating American radios having an FM band (some FM receivers such as Zenith covered both the 42–50 and 88–108 MHz bands).

In Europe the situation is slightly more complex. The VHF band was defined in 1948 as ranging from 87.5 to 108 MHz but this band was not fully used at the outset. Initially only the lower part, from 87.5 to 100 MHz, was used for broadcasting, and the upper half (100–108) was reserved for mobile services. The first European FM radios appeared around 1950 and their FM coverage was to 100 MHz; this limit was universal until around 1964. In the mid-sixties it became clear that the upper half of the band would be necessary also; hence manufacturers started to extend the tuning range of their FM radios to 104 MHz. Finally, the WARC Conference of 1979 assigned the entire band for broadcasting.

Until the early 1980s some sets produced for domestic coverage (for instance by Philips) offered only the 87.5–104 MHz FM tuning range, whereas sets produced for international export already covered the full range with 108 MHz as the upper limit. Sets produced later than the early eighties all covered the full FM band from 87.5 to 108 MHz. None of this applies to the communist countries of eastern Europe, by the way. Their FM band ran from 65 to 73 MHz; stations are transferring gradually to the 87.5–108 MHz band.

Summarizing then, if your European set has no FM, no conclusions can be drawn. FM was introduced in 1950, but many sets were produced without FM after 1950. However, these were mainly cheap sets, portables or DIY construction kits. A domestic set without FM is likely to be older than 1955. If the upper limit is 100 MHz, the date is roughly between 1950 and 1965, and if it is 104 or 105 MHz, it will date roughly between 1964 and 1980. If the upper limit is 108 MHz, the set probably dates from after about 1980.

Figure 4.2 In terms of functionality there is probably little to choose between these two radios of the 1930s. Both were cheaply made, yet today for the round Ekco bakelite set you would pay ten times the price of the wooden radio. Two factors are responsible for this state of affairs; the innovative styling by Wells Coates made the Ekco set a design classic and its fragile bakelite cabinet means that many sets never survived to the present day.

Television receivers can also be dated from their tuners. A British set with no tuner or the channels 1–5 only must be pre-1955, whereas one with VHF channels 6 and above will date from 1955, the year in which independent television was introduced. The first UHF channels came into use in 1964.

Similar knowledge can be applied to sets made in other countries. A German television set marked I, II and III will be pre-war, whilst an American set with a channel 1 is almost certainly pre-1946.

Dating older radios by their valve complement or circuit features is a technique reserved for experts but there is a very detailed and informative guide on the Internet entitled the *Rec.antiques.radio+phono frequently answered questions (FAQ) list*, to be found at the location:

http://www.accessone.com/~philn/radfaq.htm

WHAT'S IT WORTH?

A true collector would never ask this question, of course. If something appeals, you have it, and that's all there is to it. This must be true, because no genuine collector starts out with an eye to cash values; a speculator might do, but not a collector. All the same, the time will surely come when a loved one asks why on earth you clutter up the house with all this old junk and at times like this it can be convenient to claim those dusty objects are really valuable antiques.

All the same, it's all too easy to get bound up in cash values. Whilst it is useful to assign some kind of total value to your collection for insurance purpose, you'll probably end up disappointed if you become obsessed with valuing every item. For a start, there is no set price for our treasures and although for some objects you can buy price guide books (mainly from the USA), these represent no more than one person's opinion at one moment in time. Factors such as condition, temporary fashionability and whether you are selling or buying play a major role in determining an item's price. In the final analysis, it all boils down to what you're willing to pay and what the vendor is prepared to accept.

HOW TO ASSESS VALUE

Sotheby's, the auctioneers, offer nine guideposts for assessing the value of antiques. They are:

- Authenticity
- Condition
- Rarity
- Historic importance and provenance
- Size
- Medium
- Subject
- Fashionability (trends)
- Aesthetic quality

These categories are really intended for works of art, so we can eliminate medium, subject and aesthetic quality, but otherwise they fit our kind of treasures very well. Authenticity and rarity are pretty obvious, although it's worth mentioning that some quite rare items are not highly valued on the collector circuit, either because they are not fashionable or else because of their size (large antiques are not welcome in most small modern homes). Historic importance and provenance can be important, although perversely some very utilitarian items can hold high appeal, such as the catalin and bakelite radios intended for distinctly low-income households which could not afford the higher cost of sets with real wooden cabinets. How times change!

CONDITION

Condition has a major bearing on value and prices. Some items are extremely uncommon in first-class condition and command high prices with ease for this reason. Most collectors today are choosy and want only top-quality merchandise with no defects. Items that are scratched, dirty or with

Figure 4.3 Portable radios of the post-war era are particularly collectable; they are old enough to be considered antique, yet relatively modern in styling. This Ultra was a valve set but transistor portables are sought after too.

parts missing have very little appeal, even to newcomer collectors. This is something that some collectors – and vendors – find very hard to grasp. A cracked dial-glass, a missing knob or a single blemish can easily knock 50 per cent or more off the accepted value of an item. The sole exception to this rule is with very old or rare objects, where a totally untouched and unrestored item is considered more desirable than a 'tarted-up' equivalent.

Over-restoration is to be avoided too – most collectors dream of finding sets that look as if they have stood in a living room for the last 50 years, dusted carefully once a week. Some restorations are better than others and with really rare items, restoration to any less than museum standard will actually reduce its value. Does the item have to be in working order and does that increase its value? No and yes for most people – many collectors are satisfied with static exhibits, although I feel a radio or telephone that doesn't talk or a TV set you cannot watch is a sham. Many collectors are unable to repair their sets but after reading this book perhaps you'll have a go and enhance the value of your sets. These themes are discussed in more detail in the chapter on 'Repairing and restoration'.

PRICE OR VALUE?

A specialist will be able to tell you what an item might fetch at auction, and also what you might get from a private purchaser and from a dealer. All three will be different. A lot of people also confuse price with value but they are not the same thing. In truth, only the purchaser can determine the value of something and then only to him or herself.

The actual value of a particular object will depend on factors such as its condition, scarcity, general desirability and fashionability. Supply and demand factors as well as local economics may also have an influence. Even two established experts may come up with varying estimates of an item's worth.

Location is an important factor; anything will sell cheaply if there are few buyers interested. If you are the sole bidder at a country auction sale, you'll probably end up with a bargain. To us British collectors, prices in the USA are generally lower than here. That cannot be for lack of interest there so the reason must lie in a larger supply. In Continental Europe the supply is smaller and prices tend to be a lot higher. A far higher proportion of old artefacts was lost there during the last war and until recently outdated objects were generally discarded without sentiment. Attitudes have changed belatedly – almost too late!

Auction prices are notoriously high – they are generally paid by people who are desperate to buy something and don't know where else to look (or haven't the time to spare to chase around dealers' premises). Moreover, although auction records indicate the actual price at which an item was

Figure 4.4 House telephones are among the most commonly found types at antique fairs, turning up when old houses and offices are cleared. Their value is depressed because they were installed in very large numbers, often using the wiring of an older bell system used to call the servants. Despite their ornate design, they are not loved by serious enthusiasts – they cannot be used on normal telephone lines nor are their parts generally useful for restoring 'real' telephones. Their appeal is mainly to beginners, casual collectors and people looking for ornaments.

sold, the identity of the buyer and the circumstances under which the item was sold may have a profound effect on the price. As a seller at an auction, you will of course have to pay the auctioneer's premium, so you don't get all the sum realized. In particular, one-off auction prices do not establish the value of an item and until several similar items have been sold, no-one can even name a regular going price for the items.

In a nutshell, auction prices are not current values; they are the result of the excitement of the auction process, the skill of the auctioneer and the specific interests (and bank balances) of the participants. Nevertheless, auction prices serve as useful references and as one of the several elements in the value-determining process.

Figure 4.5 The 'skeleton' telephone (called the 'sewing machine' in France) dates from 1895 and was made by several manufacturers for nearly forty years afterwards. As well as its artistic looks, it presents a clever piece of industrial design since the curved legs form the magnets of the hand-generator and the induction coil for the speech circuit is concealed within the turret below the handset cradle. For many telephone collectors this is the 'holy grail' and such examples fetch several hundreds of pounds. The decorative transfers are not found on all examples and a fair number of these transfers – and other vulnerable components – are modern reproductions.

TRENDS

Auction prices do have a real value in that they make price movements easy to track. Trends are easy to follow in consequence. Of course, if you visit many swap-meets and dealers' shops you can compare their prices and check if they match the auction price trends. Prices always rise when an item becomes fashionable, perhaps as a result of media exposure. The moulded bakelite Bush TV22 television receiver is a case in point; black bakelite telephones are another. If a large number of people come on to the scene and all try to build up collections in a short time, then prices are bound to rise. If the fad is short-lived, then prices will fall again later. For a while, however, the total exposure brings a lot of material (mainly uninspired) on to the market and many of the vendors are disappointed with the prices they are offered, having been led to believe that *all* old items are valuable. Most are not but eventually prices do rise, simply because dealers cannot buy in further stock except at higher prices. Television programmes such as *Antiques Roadshow* and the like have a lot to answer for in this respect.

Figure 4.6 The internal arrangement of this turn of the century wall telephone is clearly visible in this view. Starting at the top we see the magnet emblem of the General Electric Company, a carbon and mica lightning arrester, the ringer and bell gongs, the switch hook with the handset hanging from it, the speech induction coil, the magneto generator and the battery holder. Most of the cabinet is made of wood, although the front of the battery box is of steel, tin-printed with a wood-grain finish. Telephones of this design were made by GEC and Ericsson, some remaining in use until the 1970s in little-used Irish post offices and some English electricity sub-stations. They are popular with collectors but not particularly easy to find now.

SUPPLY AND DEMAND

Every week more people are attracted to collecting 'our' kind of treasures, yet it's obvious that the total supply of old artefacts is fixed. Rising prices may lure a few items from long-established collections but, in general, few collectors part with their treasures before death. The trade is already finding it hard to locate stock and in some fields (phonographs for instance) there are no further stocks to discover; every item which comes on to the market is from an existing collection, not from someone's loft. Other fields have not reached this stage yet, but for really old radios, objects fresh from their long-term resting places are quite uncommon and often in poor condition.

With time, interest in certain items may decline, however. Dealer Steve Harris reports that the price of phonographs has not risen in line with radios or telephones, largely because the market has not expanded. Prices are stable and currency conditions mean that the export market is flat. Another reason for the relative unpopularity of phonographs, he states, is that they are less practical to use than gramophones and the repertoire of music for them is distinctly limited, both in range of music and in the number of cylinders in circulation. The long-term outlook is more positive, he asserts, because phonographs are significant historically and eventually the slack in prices will be taken up. Buy now while stocks last, is his suggestion!

BACK IN THE REAL WORLD

It's as well not to delude yourself as to the value of your collection. You may think you know what a particular item would fetch at a show or at auction but what if it had to be sold quickly to raise cash? What would 'ordinary people' give for it? Probably not much, so if you do need to raise cash in a hurry, how do you go about finding the person who will give you the best price? A collection can be a source of great satisfaction but justifying it as an investment is unwise. Peterson's Law says that if you are buying to collect, the only question is whether you can afford it. If you are buying to invest and later sell, then unless you can double the investment within six months, forget it – your money would be better off earning interest in a building society. Phil Nelson adds two more wise thoughts: 'Buy for love, not for investment' and 'Buy only what you like – don't be swayed by other people's tastes and opinions'.

OBTAINING A VALUATION

If an insurer asks for an independent valuation you will need to consult an expert. Remember that no-one can put a value on something without seeing it or having a very accurate photograph and/or description; with these a specialist can hopefully tell you its age, manufacturer and perhaps its sale value. Most dealers and auction houses will help in this way as well, but you need to beware of sharks at antique fairs and the like. Some museums will help on identifications, though most decline to give valuations. Don't expect valuers to work for nothing, either; a professional opinion is worth money. You would not expect a consultant surgeon or barrister to work for nothing, would you, so why should a valuer?

PRICES IN THE REAL WORLD

By now we know that value and price are not the same thing, but what drives prices? Can we ignore them and just think about values? Hardly.

Prices vary widely depending on where you live and upon supply and demand. As mentioned above, 'our' kind of treasures generally cost less in the USA than in Europe. Fads also have a role to play and these can drive up prices. Of course, you are not obliged to follow these trends in collecting, but if you do, then you will pay the price. Many collectors have short-sighted ideas on prices and see the price they pay purely in terms of the object they are purchasing. In most cases, however, the truth is that they are paying also for the time and expertise that went into its restoration, plus a guarantee of its future performance and an assurance of its provenance. Items bought from professional dealers or auctioneers are bound to cost more than similar goods found at swap-meets, even if they are similar.

CHAPTER 5

Buying and selling

Is this chapter really necessary? Surely everyone knows how to buy and sell? Yes and again yes. People think they know how to get the best deals but many tricks of the trade are revealed here, not to mention traps for the unwary. So do read on, your time will not be wasted!

WAYS OF MAKING A SALE

If you have something to dispose of, you're probably aware that the highest prices are realized at auctions, although this may mean waiting a while until the next sale takes place, and even then you stand the risk of people not taking a shine to your item. There will be a commission to pay the auctioneer as well.

Selling privately is an alternative but it also means waiting. You could go to an enthusiasts' swap-meet or an antiques fair and take a stall (more expense!), or else put an advertisement in *Exchange and Mart*, *Loot*, or the appropriate collectors' magazine. You can expect to receive a fair price but you will obviously have to wait until your ad appears (or the next swap-meet is held, and so on).

The third option is taking the object to a dealer. He will offer you a low price relative to what he will sell it for (perhaps half or a third of the asking price), which is disappointing. On the other hand, you will at least get cash in hand, straight away, and he then takes all the risk.

Finally, you may decide to keep the object yourself or donate it to a museum (now there's a nice thought).

RULES OF ENGAGEMENT

The same routes to market apply to buying but this is usually a more immediate process. If you're anything like me, the first thing you do when a new issue of a collectors' magazine or newsletter arrives is to dash to the small advertisements at the back. That's where you'll see the latest treasures and rarities on offer and hopefully find a bargain as well. Bargains don't hang about and many is the time that having rung the advertiser, you must make an instant decision and decide to buy – or decline – unseen. You're relying on someone else's description, and whether the deal has a happy ending or not depends on their honesty and on your common sense. Some of these deals end in grief so here are a few tips for saving disappointment.

IF YOU'RE A VENDOR

Once you strike a deal, it is totally immoral to cancel it when someone else offers you more; a deal is a deal. It's also a verbal contract and an aggrieved party could sue you successfully for breach of contract. Just because 'it's only a hobby' doesn't mean you can be a rat.

If you are waiting for the best offer, tell potential customers the cut-off time after which no more bids will be accepted – and stick to your word.

IF YOU'RE A BUYER

Having made a deal, realize you are now committed. You cannot back out unless the goods are substantially different from what was described. If you are negotiating unseen, make clear what you expect to receive, and if the vendor says something is in good condition, establish exactly what he or she means. It may be good for its age but rusty or worm-ridden nonetheless. Non-collectors are notoriously vague when it comes to estimating condition, size and weight.

Try to collect the goods in person; if you have to rely on a carrier, get a quote on costs before finalizing the deal. Many people underestimate postage and shipping costs hopelessly, so do some research yourself, otherwise the vendor may come back to you for more money to cover carriage. There's nothing unfair in this; he is not a professional shipper so your own investigation is part of 'doing business'.

BUYING UNSEEN

Buying unseen is a fraught business, simply because descriptions of the condition made over the telephone or in writing can be extremely vague. An item described by a vendor as very good may turn out to be a grave disappointment, especially if he or she is unskilled in these matters – or dishonest. Professional dealers ascribe clear, unambiguous meanings to words such as 'good' and 'excellent' and enthusiasts ought to adopt the same code to avoid misunderstanding.

As new can only be used when the item is in the same immaculate condition as when bought. *Very good* can describe an item which shows small signs of wear – but no damage. *Good* is the term used for the used and worn item which is in average condition for its age but has no parts missing. *Fair* is a disputable term, so anything less than good is *poor* and will either need substantial restoration or is suitable only as a source of spares. *Original* means untouched and entirely authentic, whereas *restored* and *repaired* mean what they say.

What happens if your goods arrive damaged or not as described? A lot depends on the deal you struck. If the goods were purchased *FOB shipping point* (wherever that was), it is the purchaser's responsibility from the moment it is handed over to the Post Office or carrier. If it was purchased *FOB your address* (wherever you are), it is the shipper's responsibility until delivery is completed. If these terms are not used (or not explained) things get more complicated. Many sellers use these terms without explaining them, and they are quite legal – if *you* don't understand them, it's your job to find out. FOB means free on board and has an accepted legal connotation. If you buy from a trader, the normal Trades Description conditions apply but this is not the case in dealings with private individuals.

PACKING AND TRANSPORT

This is a classic area in which you get what you pay for; the lowest price is not always the best deal, especially for rare or delicate goods. Sadly the normal parcel post in the UK has an appalling reputation for smashing and losing items, and should be ruled out for carrying anything of value. Whilst compensation is generally paid, this is no help if irreplaceable items are concerned.

Many people have sadly inadequate ideas on packing, probably because they have never seen how parcels are actually handled in transit. Assume your package will be dropped six feet and plan accordingly! A good plan is to pack the item once, then suspend the package inside a much larger carton. This double-skinned packing gives your treasure a 'second skin' and double the protection.

Many of the items we send are scarce, expensive or both, so they deserve decent packing. Balls of newspaper and plastic worms (plastic peanuts) are

UPS guidelines for good packaging

At UPS our objective is to get every package safely to its designation. *You* can help us achieve this goal by properly packing and labelling your shipments. The outer carton, the packaging material, the closure and the address label are four key elements in a properly prepared package. By following the simple suggestions below, you help ensure that your UPS-shipped merchandise arrives in good condition.

The carton

A corrugated carton is the ideal way to ship most items and provides the best protection for your package contents. We recommend that you select a carton with a minimum 200-pound test rating as described on the carton manufacturer's seal printed on the bottom of the carton. Always use cartons in good, rigid condition with all flaps intact. Cartons that have punctures, tears, rips or corner damage should not be used. Make sure the carton is large enough to allow room for adequate cushioning material on all sides of the product.

The cushioning

A minimum of two inches of packing material should completely surround each item in the shipment. Crumpled Kraft paper makes a good, inexpensive cushioning material, as does bubble-wrap. (Caution: 'foam peanut' cushioning material can settle or shift position in a carton during transit, thereby reducing its effectiveness as a cushioning material.)

Begin by packing several inches of the cushioning material you have chosen in the bottom of the carton. Next, wrap individual items separately and place them toward the centre of the carton at least two inches apart. Finally, stuff cushioning material firmly around, over and between the items.

Use enough cushioning material so that the contents cannot move easily when you shake the carton. Several inches of cushioning material all the way around your merchandise should do it. But remember, the further you keep fragile articles away from the corners and sides of the container, the less chance there is of damage. Extremely delicate items should be double-boxed for added protection. Carefully pack the delicate items in the first container with cushioning, and then place the sealed package in a second cushioned container.

The closure

Proper closure of a package is as important as adequate cushioning. To close a carton securely, use a heavy-duty tape designed for shipping, at least two inches in width. Pressure-sensitive plastic tape and water-activated paper tape, with or without fibre reinforcement, are all acceptable closures.

The 'six-strip method' for package sealing provides maximum protection against cartons popping open. First, tape the bottom centre seam, leaving a two-inch overlap on each end. Next, seal both bottom edge seams. Pack the contents, using the cushioning guidelines above. Now tape the centre seam and both edge seams of the box top, exactly as you taped the bottom.

Never use masking or cellophane tapes. These tapes do not provide the strength necessary for proper package sealing. In addition, avoid using string or cord on your package. These can catch on automatic sorting equipment. Finally, don't paper-wrap your package. If the paper tears, your address label may become separated from the package.

totally useless; newspaper loses its spring and compresses, whilst heavy items tend to sink their way down in plastic worms until they sit on the hard bottom of the carton. In fact the best low-cost packing material is the bubble-wrap with the big one-inch bubbles; you can find this at packing specialists and at some large out-of-town office equipment warehouses, and for what it does, it's not expensive. Three wraps in one direction and tape, a wrap or two in the other direction and tape, and in the box it goes. It's so easy to use; you just cut it off the roll as you need it. Cartons are particularly vulnerable to corner drop hits, so it is worth protecting contents by layering corrugated cardboard on the inside of the shipping box. It's worth building up the walls of the box to about half an inch for objects weighing more than 50 lb.

But what about large or really heavy articles? The answer is to do as the professionals do . . .

1. Start with a special pallet (solid top deck, not slotted), sized to match (ask the local motor cycle dealer for old crates or spend a modest amount at the local lumberyard).
2. Lift the object on to the pallet.
3. Nail blocks down on all four sides to hold it in place – a total of eight blocks, two at each corner. Put some soft foam such as upholstery foam (fully compressed) between the blocks and the item.
4. All the way around the top, lay hard-foam angle stock (a common packaging material made of a foamed plastic which is very rigid and has a waxy feel).

5. Run banding or load-binders ('motor cycle tie-down straps' from the local bike dealer) under the pallet deck and over the object. Cinch down firmly. With the rigid foam to protect the upper corners, the paint doesn't get damaged.
6. Build a five-sided, open-bottom box around it that reaches from the foam on top to the top of the pallet. Old appliance cartons and duct tape will work well here.
7. More banding to hold the carton on.
8. 'This side up' stickers, the more the better.

This way your treasure will arrive undamaged. Other methods cannot be relied upon.

Finally the following information is supplied by United Parcel Service; it is their official statement on the subject and should be treated as the minimum requirement for packing.

PURCHASING ABROAD

If you have exotic tastes in collecting, you may well find some tempting items on holiday abroad. Whether in antique shops, specialist dealers or flea markets, I have found that only cash talks. Travellers cheques and plastic are of little negotiating value. In Moscow, dollars (and they must be clean ones, not well worn examples) are the only universal currency.

Tracking down the kind of plunder we like is not the easiest task abroad; you can hardly look in the *Yellow Pages*, although some tourist information bureaux can be remarkably helpful if you ask them how to find the flea market or the antique shops. Locations like this are not easy to list in any comprehensive way; I can only tell you of the locations I have found. All the same, other places must have their special areas as well . . . you have just got to seek them out!

The list includes flea markets, dealers' shops and proper organized swap-meets. Do bear in mind that details may change and you'll need some fluency in the language of the country concerned. For society swap-meets, you will need to examine the bulletin or magazine concerned; it might be worth joining the relevant club in advance.

If anyone else would like to offer some suggestions, I'll be pleased to use them in subsequent editions of this book.

BAGGAGE RESTRAINTS

Beware of getting carried away on your travels! It's easy to be tempted by a bargain but consider also how you are going to bring it home. If you are travelling by car it's not so much of a problem but if you are flying or using other means of public transport, you may have problems. Problems are only opportunities in disguise of course, so here are some thoughts.

Everything gets heavier after five minutes of carrying. This obvious fact is not known to academic physics but it's as valid a law of science as Ohm's law. What you think you can carry for a little while may feel extremely weighty after a time. Small items, such as telephones, table radios and even tiny portable televisions, are best carried as hand luggage, especially on planes. You'll have some explaining to do at the X-ray machine but at least you can control how this one piece of baggage is handled. Anything left to the mercy of baggage handlers may not survive.

One purchaser of a large radio travelled to and from Paris by coach and hovercraft. His smile vanished when he realized the enormity of getting the lump home. The solution was to plan a careful course of bus routes and interchanges to bring the radio back to the hovercraft terminal (a taxi was considered too expensive and climbing up and down the stairways of the Métro was out of the question). Back at the terminal a radio purchase of a huge, low-price holdall provided an 'outer skin' for the radio, which was further padded with items of clothing. This protected the precious set once it was handed over to the baggage handlers on the coach and hovercraft.

CUSTOMS RESTRICTIONS

Buying a treasure in western Europe does not pose any customs problems to British residents now that we have a customs union. Items purchased outside the EC may be problematical, however. The USA is a popular source of old radios and telephones and, whilst they are technically liable to import duty, you'd be unlucky to be made to pay. Some of the countries of central and eastern Europe have heritage laws which effectively ban the export of any object made before 1945. I successfully brought some items out of Poland without a problem but that was only because no-one inspected the contents of the car; I was ready to argue that the things were German-made and hence not part of Poland's heritage but this reasoning might not have carried much weight. A bribe in dollars might have helped, but would you want to try this?

Books are both plentiful and cheap in the USA and I have found vintage technology items in just about every second-hand bookshop I have visited there. If, like me, you buy in quantity, don't try and bring them back as hand luggage. Books are incredibly heavy and you won't be able to lift your suitcase. Excess luggage charges on airlines are substantial and you may not be able to smile or bribe your way out of them (although I'm told twenty dollars to a baggage handler can work wonders – personally I wouldn't know). Fortunately the US Mail has an incredibly low rate for sending books by fourth-class surface mail; go to any post office there and ask for an M-Bag. They'll give you a substantial sack and a cardboard label; fill the sack with books (preferably well wrapped in stout paper bags from any supermarket there) and fill in your name and address on the label. Return these to the post office and expect to pay 72 cents (something like 47p) per pound weight of postage. In my experience, these M-Bags arrive in under a month, which is pretty good.

ABBREVIATIONS USED IN CLASSIFIED ADS

BO	best offer
ca.	circa (approx. date)
ea.	each
consignment	In USA this means commission, as in 'to sell on commission'
FOB	free on board (meaning that the vendor will pack the item and take it to a shipping agent or whatever, but after that the purchaser pays all charges
IB	instruction book
IMO	international money order
LSASE	Long SASE (send one which will take the full width of A4 paper, size 10 in the USA)
MIB	mint in box
MO	money order
Mod	model, modification
NIB	new in box (as new condition, in the original box, but its sales history is unknown)
NOS	new old stock, i.e. old stock which is unused and unsold, an original, unused vintage item
NT	no tubes
OB	in original box
OBO	or best offer
OIRO	offers in region of
ONO	or near offer
P&H	postage and handling
P&P	postage and packing
ppd	postage paid (shipping included)
PU	pick up only (will not send by carrier)
SAE	self-addressed envelope
SASE	self-addressed stamped envelope
S&H, s/h	shipping and handling (i.e. postage and packing)
UPS	United Parcel Service, in the USA generally held in higher regard than the US Postal Service

w/, WI	with
WIO, w/o	without
WT	with tubes

BOOKS USED AS REFERENCES BY AMERICAN COLLECTORS

Blast = *Radio – A Blast From The Past*
Bunis 1, 2, 3 or 4 = *Collector's Guide to Antique Radios*, 1st, 2nd, 3rd or 4th editions
Collins = *Radios – The Golden Age*
Douglas V1, V2 or V3 = *Radio Manufacturers of the 1920s*, Volumes 1, 2 or 3
E.O.R. 1 or 2 = *Evolution of the Radio*, Volumes 1 or 2
F.O.S. = *Flick of the Switch*
Grinder/Fathauer = *Radio Collector's Directory and Price Guide*
G.O.R. = *Guide to Old Radios*
Johnson's A.R. = *Antique Radios – Restoration & Price Guide*
M.A.J.A. = *Machine Age to Jet Age*
O.T.B. = *Old Timer's Bulletin* published by Antique Wireless Association (AWA)
Sidell = *Classic Plastic Radios*
V.R. = *Vintage Radio*
(V.R., p. 30) means *Vintage Radio*, page 30 and the same system can be used by European collectors and others elsewhere.

CONDITION

exc	excellent
f	fair
p	poor
g	good
vg, vgc	very good

(If good or better, mention non-original parts)

The above lists of abbreviations were written with a tip of the cap to John Terrey of *Antique Radio Classified*.

Radio amateurs have their additional abbreviations of their own such as FB (fine business, good), but these are not widely recognized outside ham circles. For technical abbreviations and expressions, see the chapter devoted to this subject.

CHAPTER 6

Patronizing dealers

After I had chosen this title I realized that you might read it two ways and I must admit, there are indeed a few patronizing types among the dealer fraternity. But what I mean is dealing with traders. Traders are an essential part of our hobby, not the parasites that some people portray them to be. Many of us bought our first set from a dealer and quite probably pestered him for advice, spare parts, service data or whatever. Many of these dealers are nice guys and so they should be, after all their task is to make the extraction of your money as painless as possible.

In fact you could take this a step further and argue that were it not for dealers, most of the items in hobbyists' collections would not exist. It is dealers who have the time, the money, the knowledge, the dedication and the expertise to seek these items out and make them available to everyone else. Unlike the casual hobbyists, dealers do not obtain their merchandise by way of the occasional relaxed Saturday afternoon trip to an antique shop or Sunday morning boot fair. It calls for an infinite amount of hard work and dedication.

SYMPATHY FOR THE DEVIL

Are dealers rip-off merchants then? Not in the main. Yes, the prices they charge are higher than you will find at swap-meets, collector's fairs and charity shops but you would expect this, wouldn't you? Dealers have to lay out real money for months, sometimes years, before the time comes (if it does) when they eventually make a sale. That's money that might well have earned more interest in the bank! Traders also have significant overheads which private sellers do not: rentals on premises, general business rate and so on. If you want a particular, hard-to-find treasure in a hurry it's to a dealer you must go and, in my opinion, dealers perform a very useful service to collectors and enthusiasts (albeit at a price). If you prefer not to patronize them, that's your choice.

Dealers are not spread evenly across the country nor in every district, so occasionally you may have to travel quite a way to see one. Ring first is my advice – make a firm appointment and don't assume that because he normally opens on Saturdays he'll be there this weekend. Dealers do occasionally have to close their premises at short notice (e.g. illness if he's a one-man band). On the other hand, there is no excuse for breaking appointments or shutting early, even in the face of customers who had just arrived, as one cavalier trader in the West of England was prone to do.

BOUNDERS IN OUR MIDST?

What about rogue suppliers? A colourful trader once told me anyone who tries to make a business out of supplying hobbyists is either a rogue or a fool or both. Thankfully this is not entirely true and there are many businesses run on ethical lines who give their customers excellent service. There are also a (thankfully small) number of businesses who give lousy service; they bank your cheque or debit your plastic card and then take months (if ever) to do what they promised. They purport to be hobbyists but in fact form the scum of our hobby. That's why you won't see any mention in this book of certain traders.

How do you spot a scoundrel? Certainly not by looking for shifty types wearing striped T-shirts or too much Brilliantine on their hair. Rogues are generally very plausible and I myself have been taken in. Impressive advertisements are no guarantee of a reliable business and my best advice is to ask around.

Never pay large money up front for goods to be supplied in future and if you have any suspicions try to pay large sums by credit card (which gives you some safety net and the added muscle of the credit card company). For repairs, get a written receipt for your goods and do clearly understand the difference between an estimate (an informed guess) and a quotation (a firm offer of price). Many people don't know the difference and end up annoyed.

If you are buying something expensive, insist on a written invoice stating what you are buying, from whom and what guarantee you are being offered. Ensure this has the vendor's name and address, then get him to receipt it. With most traders this elaborate performance is totally unnecessary but there are a number of mighty smart operators in our midst.

SO YOU'VE BEEN HAD

One of these fly boys is selling faked-up coloured 'bakelite' telephones at the moment. They have British metalwork (with GPO markings) but the coloured cases are later productions imported from India; in some cases modern replacement parts are used to repair easily broken parts. The 'real thing' is hard to find today and there is a steady demand for old-style phones in attractive colours. Accordingly the price of the genuine article is high.

Not everyone wants an original; they may be perfectly happy with a mongrel or 'bitsa' telephone (made of 'bitsa' this, 'bitsa' that) or even with a complete reproduction. Concoctions like these should not sell at prices as high as the genuine article, nor is there any justification for mixing genuine GPO parts with imported cases, but that is no hindrance for this trade. A newcomer collector bought one of these concoctions and after realizing his mistake, asked for his money back. The trader declined, saying he never actually claimed what he was selling was a genuine GPO telephone.

Technically the trader was right; this is a classic case of *caveat emptor*, 'let the buyer beware'. Imagine you are buying (or selling!) a second-hand car; is there any obligation on the vendor to disclose defects? The answer is no, especially as most purchasers consider themselves experts! So always assume the worst. What you are buying is sold as seen, with all defects; never assume anything. Can you afford to lose the price? Are the parts you can verify worth buying, even if you can only use them as spare parts or to restore a better example? Once you have handed over the cash you will likely as not have a job getting it back.

SETTLING A DISPUTE

If you do end up in dispute with a trader (let's hope you never do) and you are convinced you have grounds for recompense, apply to your local library or citizens' advice bureau for help. You will need to contact the trading standards office in the local government area where the trader operates and often they can put pressure on a recalcitrant trader and also tell you if he's a regular cause of complaint. Your ultimate weapon is the small claims court, again in the district where the defendant lives (not where you live), and although you have to pay a fee to start proceedings, you win this back as well as all your other costs if the District Judge decides in your favour. You will of course need to show adequate documentation of your claim, so always demand a receipt if you buy something expensive.

The foregoing might incline you to believe that professional dealers are the villains of the piece and private traders are the nice guys. You'd be wrong. Virtually all of the dealers and all of the private traders are nice people who are just as keen on the hobby as you are. All the same, a few of the private traders at swap-meets (not many, thank goodness) are as sharp as knives and if you don't ask the right questions, they certainly won't prompt you. Don't assume anything, therefore.

Always ask if a set is working, if it is complete, if what's inside is authentic. 'Sold untested' may mean just that but more often it means 'not working' or even 'unrepairable'. Fair enough, nobody told you it worked but don't assume it might. Of course, if it does you have a bargain! 'Sold as seen' and 'as found' mean exactly what they say. 'Breakages must be paid for' has no validity in the case of a genuine accident but try not to be careless with things you haven't bought. With expensive catalogues, books and other paperwork, check pages are not missing; sometimes even the vendor may be unaware of defects.

DEALERS' COSTS

Going back to an earlier point, how much does it cost to be a dealer? Well, one of the well-known dealers told me his shop premises cost him £200 a week. Add in the other inevitable overheads such as telephone, general business rate, food at lunchtimes from expensive snack-bars, and a part-time helper and you're talking about £300 a week. That's £300 just to break even and it means running a risk that a lot of enthusiast collectors would prefer not to take. Taking that into consideration, are those dealers who have the cheek to ask £75 for a restored DAC90A radio that you could have bought for £35 at a jumble sale really quite the rip-off merchants you thought they were?

CHAPTER 7

The big-time league – buying at auctions and antique fairs

Prices tend to be higher at these kind of events but there are still bargains to be had if you are in the right place at the right time and know exactly what you are buying. Auctions and antique fairs are also where fools and their money are soon parted, so look before you leap!

BUYING AT AUCTION

Auctions represent an interesting and often ignored source of collectable merchandise which need not be expensive. After all, many dealers buy in their stock at auctions and since they have to add a margin to the buying-in price to make a living, this means that items bought at auction are normally well below shop prices. So if you are bored with the same old dreary, over-priced merchandise you see at collectors' fairs and antique marts, why not patronize the auctions? Contact auctioneers too if you wish to sell an item; with sales being held regularly, they are a good way of offering your goods to a broad range of interested and motivated people.

Some specialist auctioneers are listed in this book. If you are searching for a particular item then you are most likely to find it at a sale organized by a specialist auctioneer – that's obvious. But don't ignore antique fairs or, for that matter, common or garden furniture sales, though. These auctions often contain all sorts of 'general' items and many treasures turn up here first of all. Because these sales are held on weekdays they are not patron-ized much by the general public which means that dealers get the best bargains and then sell them on for a profit. Of course, this is how they make their living and how many interesting items eventually reach a wider public, so it's not all bad.

Some people are scared of auctions. Remember those daft stories in childhood comics where someone attending an auction sale waves after recognizing a friend elsewhere in the saleroom and ends up committed to buying a house full of furniture? Well, forget it – auctions don't work that way and no-one can buy things 'accidentally'. Auctioneers have sufficient sense to recognize real bidders and are well able to disregard those folk who are just adjusting their spectacles or blowing their nose. Indeed, in quite a few auctions you need to obtain a numbered bidding card beforehand and you signal your bid by raising this card. So you won't get caught buying something by accident or by default.

Make sure you know whether a buyer's premium is to be paid and whether transactions are subject to the dreaded VAT. Some of the larger auction houses charge buyers 15 per cent to cover their fees (not unreasonable if the vendor gets to keep all of the hammer price) but add this and VAT to the hammer price and suddenly your £100 bargain is going to cost you over a third more (£135.12 to be precise).

Some words of warning now. There is no guarantee as to condition with auction goods; they are sold as seen and if you find it difficult to inspect items properly, then that's your problem. Even authenticity may be in question, and many is the set bought at auction which turns out to have the wrong chassis or missing valves.

BUYING BY POST

Many auctions also take postal bids these days and this is a fine way to buy items without the need to trek to another part of the country. Virtually all auctioneers issue catalogues, usually at a modest price and often illustrated (they can be a useful reference book as well!). All contain full and honest descriptions of the items under offer and you can safely bid unseen, since if an item turns out to be not as described, you are not committed to buying it. This kind of problem seldom happens, therefore. Instead, you peruse the catalogue before the sale and if something takes your fancy, you just fill in the bid form supplied, entering the lot number required and your name and address.

If your bid is successful you will be contacted; you then send a cheque for the amount required and the item is posted to you. Alternatively you can send an open cheque or give your credit card number. Either way, you set an upper limit for your bid and the auctioneer spends the minimum necessary to secure the item for you (unless you are outbid). Auctioneers don't abuse your trust and the system does work!

SELLING TOO

What's sauce for the goose is also sauce for the gander, which means that if auctions are a good place to buy cheap merchandise, you cannot expect to raise a fortune if you are selling. It's worth classifying three types of auction and taking a closer look at the way they work.

Regular specialist auctions organized by hobby societies or magazines are good places to buy, but not to sell. The 'punters' have a shrewd idea of the value of the merchandise and are unlikely to pay over the odds except for truly exceptional items. Most of the items are bread and butter goods and the sale represents a good way to buy items cheap for spares or to clear out some duplicate items. In fact it's quite remarkable how people expect to sell items at auction which wouldn't sell if laid out on a table . . . and a few get away with it too!

When specialist items turn up in general auction sales, anything can happen. You are dealing with customers who are not well versed in values and relative scarcity; they may well take a shine to a quite ordinary item which they haven't seen for years, resulting in high bids. A good place to sell, then, but not to buy.

Finally, there are the occasional 'special event' sales run by major London auction houses on their own ground or in the provinces. These sales are well advertised and attract a wide, often international, gathering. There are undoubtedly bargains to be had at these events but prices in the main will be high. Again, a good place to sell but don't expect to buy cheaply here.

Some people believe they can sell their goods for higher prices at auction than, say, privately or to a dealer. That may be, but if your item remains unsold, you may regret having gone down this route. Any damage or loss sustained by your goods before and during the sale is often at your risk and don't expect to reap all of the hammer price. Expect to pay up to 15 per cent commission (plus VAT) and also for any catalogue photos, transport and so on. If the auctioneer sets the reserve price, or if there is none, you run the risk of having your treasure knocked down for a ludicrously low price if bidding starts to flag. Selling – and buying – at auctions really is a gamble.

You may wonder who actually pays those monster prices at the big specialist auctions. Short of going there and watching the bidding, it's hard to say, although in the main, the buyers are not normal or ordinary collectors. Most people do not have that kind of money to spend on their hobby so the inevitable conclusion is that most buyers are commercial. Some of the sets are bought as theatrical props; after all, a couple of hiring fees may repay the item's purchase cost. The rest of the merchandise seems to be going, either direct or through dealers, to collectors in more affluent countries, such as Germany, Italy and Japan. These countries also have more lavishly produced collector magazines than ours, so you could argue that these people take their hobbies more seriously than we British do.

Indeed the meanness of hobbyists at most swap-meets is renowned; many of them hate putting their hands deep into their pockets!

PRACTICALITIES

Your first visit to an auction may be a bit daunting, especially if you have not come prepared. Here are some pointers condensed from an Internet file by Doug Jefferys and Steve Ozdemir.

It's important to remember that most auctions are in two parts, the preview period and the actual bidding process. With large sales, the preview may be held the previous day and to come fresh to the bidding session may be unwise. During the preview period, all the items are available for inspection, enabling potential buyers to inspect the merchandise and decide for themselves what they're willing to pay for it. With electrical items you may be allowed to power them up and check their functionality. Once the bidding has started, your opportunity to inspect the goods has passed and you are effectively bidding blind.

The fun begins when the auctioneer asks for a starting price (an 'opening bid'). If nobody accepts this price, he will lower the opening bid until someone accepts his offer. He will then slowly raise the price as other customers show interest in the item. As the price rises past the personal limits of the various bidders, the bidders stop bidding. The last person to make a bid (i.e. the one willing to pay the most for the item in question) 'wins' the bid and secures the item. Bidding then proceeds to the next item.

Most of the auctions concerning our kind of treasures are advertised in the specialist magazines and newsletters. The larger auction houses have catalogue subscription services to make sure you never miss one of their auctions. Some of these catalogues are beautifully illustrated and printed – they become almost reference books in their own account, particularly since subscribers are sent a list of prices realized after each sale.

What should you bring along to an auction? In a few cases a mains extension lead is useful, particularly when amusement machines are being sold. There may be over a hundred games available, and very few of them will be in reach of the cords dangling from the saleroom ceiling. They supply the power, but it's up to you to get the power to the machine you want.

Another useful item is a flashlight, for spotting signs of rough handling, poorly performed conversions and so on. Many salerooms are poorly lit, so a small torch can be invaluable when determining the condition of a machine.

The third most important thing to bring is cash. Some places accept credit cards and cheques but many don't. In many auctions bids are accepted only from visitors with bidder's numbers (you are given a numbered flag to wave when you offer a bid). You must register before the sale starts and some establishments charge a 'bidding deposit' (usually around £100 or so), which is either deducted from your purchases or returned at the end of the sale if you do not buy anything.

You should also bring some food; some sales take several hours. You may also spend a long time standing up. Make sure you bring a pen and paper as well; some people write their bids and winning prices in the catalogue, others prefer to keep these books in pristine condition.

Finally, remember you will need to remove your treasures, so if you intend buying anything large, make sure you have proper transportation. Most auction houses allow you a couple of days to collect your goods but some do not.

Nearly everything is sold 'as-is'; it's up to you to decide whether the item is what you want and authentic and/or functional. Only in cases of blatant misdescription or misrepresentation have you any claim against the auctioneer or vendor. Radios and TVs may well have the wrong internals but that's for you to spot.

Auctions are social occasions, a great place to 'network' with other collectors who may prove to be valuable contacts in the future. Treat them as a low-cost method of gaining experience, and remember you don't have to buy anything.

BUY NOW, REGRET LATER?

Lastly, here is a 'government health warning', issued by the Office of Fair Trading.

Take care if you buy at an auction because auctioneers, unlike other traders, can refuse to accept responsibility for the quality of the goods they auction. Look out for exclusion clauses and read any notices and catalogues carefully. Note any conditions of sale, such as buyer's premiums, terms and methods of payment, deposits and time limits for removal of goods. You cannot back out of a deal once the hammer has fallen.

ANTIQUE FAIRS

If you look in your local paper, *Exchange and Mart* or any of the more upmarket collecting magazines you will see there's an antique or collector's fair going on somewhere almost every weekend. They range from simple local affairs in civic halls to vast affairs like the Newark-on-Trent county showground fair which covers a number of acres and runs for several days (many people say they cannot cover the whole fair in a single day!). Some of these fairs also have two-tier pricing: to get in at the beginning of proceedings or on the 'trade day' you have to pay extra, whereas it's cheaper after lunchtime or from the second day onwards.

Are these 'mega' fairs worth visiting? Serious collectors say yes, definitely! They list two reasons: it's at these fairs where some of the most desirable merchandise turns up and there are some fantastic bargains to be had at the end of the show. But if it's rarities you're after, you must be there as soon as the fair opens – literally. Seasoned collectors know the score: they find the best items in the first 15 minutes when the stall-holders are still setting out their wares and need the boost of a quick, sizeable sale to revive their spirits. Soon afterwards they are off on their rounds to find bargains themselves (having left a loved-one to hold the fort) and by the time they come back, they will have spotted and bought any underpriced items and also will be less inclined to 'give away' items to make a quick sale.

It all boils down to psychology and you might as well know what you're up against. Very often the best bargains at these fairs are to be had from dealers who specialize in something else. In other words if you collect telephones you are unlikely to find a bargain with a dealer who specializes in these items but you may well come across a model train dealer who has, say, picked up an odd telephone and wants to move it on quickly. So don't be too inclined to ignore dealers who appear to hold no interest for you.

Another philosophical point. Some 'old hands' of the antique fair game will tell you never to act too eagerly; don't show the vendor how desperate you are to get the item. Play nonchalant, try to knock the price down. Well, you do this at your peril. Quite a few vendors take offence at anyone who questions their price levels and even if you afterwards offer the original asking price, they will quite often refuse to do business with you. Time to eat humble pie? Either that or get a friend to go and buy it for you!

Always look under tables. The big, heavy stuff is usually lurking there and perhaps other prize items the vendor is holding back to lay out on the table after the first load of tasty goodies have been sold. Under tables is also where you will find the grocery boxes full of old magazines that were too heavy to put on the table.

Books are generally underpriced by dealers who do not specialize in books, so this is worth noting. Sometimes of course a dealer is displaying a rare catalogue or piece of sales literature or other ephemera and the price is obviously too high. You can try negotiating a lower price but you are unlikely to be successful: either the dealer knows the price is high (but paid too much and is desperate to get his money back from a 'mug') or else he is truly mistaken (and thinks your protestations are a trick). Tough luck, you'll just have to swallow your pride and find something else to buy!

As mentioned earlier, stall-holders like to go home with an empty car or van (and a full wallet), so by the time the clock reaches 4 p.m. they will be keen to shift their wares at almost any price. But if you spot something tasty early on, don't bank on it still being there at closing time. Bitter experience says it is always worth paying full price to secure a rarity rather than hope to get a bargain later on!

CHAPTER 8

Collector's luck?

More secrets are about to be revealed! How is it that dealers manage to offer interesting material even when the market seems to have dried up? And how come they constantly turn up much tastier stock than you ever encounter? Do they have some secret system? You bet they do!

NO SUCH THING AS LUCK

Why is it that some collectors seem to have far more luck than you do in finding rarities? Do items just fall into their lap? Are they very fortunate people and, if so, did they have to sell their soul to the devil to have this luck? Nothing of the sort, luck doesn't come into it at all.

So what's the secret? In both cases it's a mixture of hard work, perseverance and determination. Don't underestimate the hard work bit either. As Abraham Lincoln said, 'Things may come to those who wait, but only the things left by those who hustle'. If you're serious about your hobby – and you must be if you're reading this – you're going to have to follow the rules if you want to make things come your way.

Dealers know this very well. They are not amateurs and what motivates them is the survival instinct; if they don't keep on turning up tasty merchandise they won't live. 'Big gun' collectors know this as well. For them it's a matter of pride, sometimes even a way of life. Chance plays very little part in finding new treasures for any of these people. Real life isn't that glamorous or capricious.

Chance, if it has any role at all, favours the prepared mind, as American author Don Lancaster says. Another favourite saying of mine runs: 'The reason why some people don't recognize opportunity is because it often comes disguised as hard work'.

In other words, you have to be systematic. Dealers and serious collectors work damn hard at tracking down the rarities and you can be absolutely sure that if they just sat at home and visited the occasional swap-meet, nothing exciting would ever come along! And now, hold on to your hat, their secrets are about to be revealed...

EXPERIENCE TELLS

One of the reasons dealers always seem to have stock is that, in the main, they have been at this game a long time. The stock they sell today may have been bought ten years ago (pause for accusations of hoarding and claims of 'Look at the value of the stuff today and I bet they bought it for a song then'). Actually this argument doesn't stand up to a moment's scrutiny. If every dealer bought his stock ten years ago he'd be bankrupt by now. The money laid out would have earned far more money in the bank, not to mention the cost of storing all this stock in the dry somewhere. No, what dealers sell today they probably bought last month and any other suggestion is plain stupidity.

'Ah, but they have a chain of contacts who sell them stuff cheap and then they ramp up the price to sell to collectors.' Really? Please point me in the direction of this army of people who deliberately sell quantities of merchandise at below its value. Certainly dealers sell at higher prices than they buy in, but wouldn't you? They too have mortgages and food bills to pay, and probably a lot of petrol bills as well for all the travelling they do buying stock. No, if dealers manage to keep merchandise rolling in they must have a system – and of course they do. But it's not really a secret. It's called hard work and involves looking in the not-so-obvious places.

LOOKING WHERE NO-ONE ELSE DOES

Provincial furniture sales have already been mentioned as a good source of nice sets at low, low prices. Charity shops and domestic boot sales are another. You may not have the time to visit these outlets but they do reap occasional rewards. Even smart antique shops may be pleased to off-load what they consider rubbish. A musical box shop in London was very pleased to sell a pre-war mirror-lid television for a couple of hundred pounds, whilst a lucky punter spotted a pre-war TV in an antique shop near Reading. The television, an HMV 5in-screen table set, was marked up at

Figure 8.1 Real treasures are still turning up for peanuts at car boot sales and the like. This unassuming looking contraption was recognized by its purchaser as an extremely rare mechanical television receiver of the Baird era. Despite its scruffy appearance, it is a unique and valuable relic of the earliest days of television and its restoration will not be difficult. All kinds of treasures are still waiting to be discovered – you just need eyes!

£100. 'You know that's not its true value,' said the punter. 'Oh well, go on, you can have it for £75' said the shopkeeper! Exit rapidly one very satisfied punter, having acquired a highly desirable piece of merchandise perhaps at a tenth of its market value.

THE ALMOST OBVIOUS

Are there any traditional radio and TV shops in your area, the independent dealer sort? If they have been in business for 30 years or more, it's quite possible they have an old-fashioned workshop with plenty of period test gear, components and service literature. Sooner or later the owner thinks of retiring and then has to sell off all the stock for a song. If approached properly, some of these people would willingly consider turning out their 'old junk' to an enthusiastic collector, especially if there's a penny or two to be made from it. These dealers may even have old show-cards or flags in the loft and you'll never find out unless you ask. Some unsold Baird televisors turned up in the basement of a Yorkshire dealer quite recently and I myself have been asked to help clear out a long established dealership when the premises were being sold. And yes, there were some forty-year-old sets in the attic.

ALMOST OBVIOUS AGAIN

Do you attend amateur radio rallies? If not, why not? Perhaps you have an aversion to the secret language of ham radio, its odd expressions (CQ, CQ) and apparently stuffy on-the-air parlance (73, Old Man and all that).

If so, more fool you – this is just a smoke-screen to keep out the CB radio users (sorry, no offence intended if you're a CBer). Every weekend throughout the spring, summer and autumn there is one or more amateur radio rally held in Britain; most are on Sundays but a few are two-day affairs. All are advertised in the events calendars of magazines such as *Practical Wireless* and *Ham Radio Today*, and each one of these rallies has a number of traders dealing in all kinds of surplus electronics (not just radio-related) – including vintage parts, books and sometimes sets. Virtually all rallies have a flea market area and a bring-and-buy stand, and a few rallies are entirely devoted to car boot sales (Luton in May and Rugby Truckstop in July for instance). You can always pick up components at a fraction of shop prices at these rallies, and often old radios, TVs, hi-fi gear and telephones as well. What a treasure trove!

ALMOST OBVIOUS 3

Don't neglect 'normal' car boot sales either. In Summer 1994 a collector of television sets was visiting a Sunday morning boot sale and happened to see a pre-war Murphy television worth at least £1000. 'That looks interesting,' he said, which pleased the vendor no end, as he admitted this was the third week he had brought the set along. 'How much do you want for it?' was the next question and the reply was 'Would £20 be too much?'

Another collector decided to place a want ad in his local free newspaper; all it said was 'WANTED: very old television set'. Within two weeks he had been offered a pre-war Marconi 702 for £20 and a Pye V150 for £70.

All three of these sets must have been worth between £500 and £1500 so the two collectors certainly got a bargain. But were they lucky? Not really, they got out and looked for the sets and the beauty is that you can try their methods as well!

ALMOST OBVIOUS YET AGAIN

Several 'serious' collectors tell me they despair of finding anything new at the established swap-meets; virtually everything on offer has been picked-over previously and really amounts to the rejects of dealers and other collectors. They are also tired of seeing well-disguised items cobbled from bitsa this and bitsa that, not to mention downright fakes and traps for the unwary. In fact, some discerning collectors have virtually given up attending the 'mainstream' events laid on for specialist collectors.

What they really want is to tap truly undiscovered material at source, *before* the 'usual suspects' and wheeler-dealers have snapped up the choice items and added their mark-up on the rest. Their answer is to escape from the mainstream and hunt where the competition is less intense. A lot more leg-work is involved but the pickings can be rich. What we are talking about is antique fairs – not the little village hall affairs (although these have yielded treasures in their time) but the mammoth provincial affairs held over a weekend or bank holiday at country fairgrounds.

Many of these fairs are so large that they take several hours or even a whole day to walk around, and many 'big' collectors rate them highly. These fairs are where the trade buyers come, although they do their deals on the first day of the show, when anyone without a recognized trade card has to pay a premium price to enter. The best deals are done in the first hour . . . and also in the very last hour when you can drive tough bargains with traders who don't want to take all their stock home.

You can find out where and when these fairs are held by reading the specialist magazines, and the money you spend on these magazines is a good investment. These magazines have readers' sales and wants advertisements, which also represent a fascinating new approach to trawling the market. Here are some of them:

- *Antique Calendar International.* Monthly guide to antiques and collectables fairs across Europe, issued in conjunction with the publications marked ACI below.
- *Antique Dealer & Collector's Guide* (ACI). British monthly.
- *La Vie du Collectionneur.* Low-cost weekly newspaper circulating widely in France and Belgium, with comprehensive calendar of antique fairs, auctions and swap-meets, feature articles and small ads.
- *Le Carnet du Collectionneur* (ACI). Belgian guide.
- *Trödler und Sammeln* (ACI). Monthly magazine circulating in Germany, with feature articles, comprehensive events listings and small ads.
- *Verzamelkrant* (ACI). Monthly magazine circulating in Belgium and the Netherlands, with feature articles, comprehensive events listings and small ads.

These publications also list the vast number of collectables fairs held on the Continent. These fairs are larger and better organized than the average British swap-meet; they are also more commercial. Most capital cities on the Continent also have proper flea market precincts, which to many visitors are a complete revelation.

THE NOT-SO OBVIOUS

Autojumbles. Do you visit them? Do you even know what they are? You should do. These are the automobile equivalent of the vintage radio swapmeet or boot sale, and a surprisingly large amount of radio material turns up at them. Not just car radios either. Before the war many garages and cycle shops sold radios or handled accumulator acid and the folks who go round clearing out defunct garages and cycle shops are turning up nice enamel advertising signs and all sorts of other radio-related goodies. Remember where you read this tip and thank me for it next time you get lucky!

Council rubbish dumps, or civic amenity facilities as we are supposed to call them these days. In this throwaway era a lot of very tasty merchandise gets dumped at these tips, which is why nearly all of them now have a resident 'totter' who pays the council for 'exclusive salvage rights'. This means anything dumped there is his and even if you spot something before he does, you still have to pay him. No matter though; if you want it, you'll probably buy it from him for far less than anywhere else. One dealer in the South of England confided that he obtains the bulk of his stock from rubbish dumps and he has standing contracts with the resident totters at each of them. That doesn't mean you cannot outbid him, however; just have some cards printed with your name, phone number and what you collect, then hand them out. The treasures will soon start rolling in!

PHOOEY!

Is that it? Were you hoping for something more dramatic? Perhaps you can't spare the time to attend furniture sales during the week or you can't be bothered to track down and attend all these autojumbles and ham radio jamborees. Oh dear, you probably weren't cut out to be a big-time collector. OK one more secret to come out.

GO FOR IT IF YOU'RE REALLY SERIOUS

You now know luck doesn't play any major part in finding gear – but you can still go out and beat the opposition at their own game if you're prepared to put as much effort into it as they do. This means being systematic, having spies everywhere. Tell your relatives, friends and neighbours that you collect old radios, televisions, telephones or whatever. Make a bore of yourself. Also, visit every last charity and antique shop in the district and tell them to give you a call if anything unusual comes in. Have some cards printed saying 'My name is Fred Bloggs and I collect old radios. Top prices paid, all phone calls returned. Commission paid for good leads'. Scatter the cards like confetti and see what turns up. What goes round comes round, they say.

Put a regular free ad in your local newspapers and in the papers for local towns. Be like a well-known collector in Nottingham and put display ads in papers and magazines with illustrations of the sets you are looking for and the prices you will pay. Of course, placing all those ads costs money but the system obviously works for him and what's more, it must pay for itself many times over.

Make yourself known as a collector. Write articles for county magazines, invite the local paper and radio station to interview you about your interesting hobby, establish yourself as a name in your hobby's specialist magazines.

All this will cost time and money, lots of it. It will definitely pay off, although not in the short term. But don't ever say other people get all the goodies if you're not prepared to put yourself out to the same extent as they do.

IT'S ALL SO EASY

Yes, it's amazing – or is it? Strange that you don't see many vintage radio (television/hi-fi/telephone) dealers driving around in Rolls-Royces. Honestly, if you want to get rich quick, forget this business – your money will earn better interest in the bank. The other thing is that, as a dealer, not only do you have to know the value of everything, you also have to know what it will sell for. The scarcest set in the world has no commercial value if nobody recognizes it for what it is (and your hyping it may not help!). These days fashionability plays a major role in the equation; art déco sets, bakelite, catalin and other plastics are the rage at present. In ten years' time we may be hallowing the 'Continental' look of the 1950s, with ivory knobs, gold trim and high-gloss lacquer woodwork – who knows?

Finally, never belittle the acquired experience of an expert; it counts for a lot. Surely it was only in jest when Jonathan Hill printed Michael's law of antiques?

i. Every mature antique dealer is surrounded by his mistakes.
ii. The quality of one's purchase is often inversely proportional to the distance travelled in order to acquire it.
iii. You have to be a considerable expert on a subject in order to realize just how little you know.

On the other hand, perhaps not.

CHAPTER 9

Fakes . . . and how to spot them

I shall stick my neck out and say there are not many fake sets in circulation. Not many. I have seen very effective faked telephones but never, I think, a convincing fake radio or television. How do you spot them? Not easily; I suppose instinct is your greatest asset. You need to be on your guard with any set that looks too fresh or is selling too cheap. Price is not a clear guide, however; some fakes are sold at outrageous prices. Don't be afraid to ask if the item is completely original and when you do buy it, insist on a detailed receipt with a description stating the item to be authentic. Make sure the seller's name, address and phone number is given too. If the vendor starts to get edgy (or refuses for any reason!) he is probably up to no good – if only because he is not declaring profits to the tax man! In a nutshell, use your judgement, ask trusted friends and, if in doubt, hang back or ask for a written statement of authenticity.

REVIVED FINISHES

It would be stretching a point to argue that a wooden cabinet which has been refinished sympathetically was a deliberate deception, whilst even the application of new transfers (new old stock or reproduction) does not turn a good restoration job into a deliberate fake. The same applies to metal items which had a japanned or stove-enamelled finish; if stripped down and refinished really well, they benefit from the operation, even to the extent where the restorer has then toned down or 'distressed' the finish to make it look less shiny.

The process becomes deliberate deception, however, when restoration techniques are used to turn a low-value item into something more valuable. This happens with certain telephones, for instance where a plain black BPO skeleton telephone is 'enhanced' with gold transfers to turn it into something much more attractive. In fact, examples made for the Post Office never carried transfers and any telephone with 'No. 16' markings and transfers as well is almost certainly not genuine. Fortunately for the serious collector, most reproduction transfers give themselves away by their bright, artificial look. The same applies to engraved brass name-plates and reproduction dial scales made for radios; the colouring and letter styles are always far too modern-looking.

MORE DODGY PHONES

There are currently a large number of fake telephones on the antique market scene at the moment. Some are merely antique-style instruments made during the 1970s and 1980s (and already fooling some collectors), whilst others are 'mongrels' or 'bitsa' phones made partly from original parts and partly from reproduction components. One such trick is to fit low-value plastic parts of Indian origin on to a British GPO base and sell this at a high price as a rare authentic British instrument.

Of course there's nothing intrinsically wrong with selling Indian instruments that look similar to older British phones, and these days it may be the only way open to traders for getting hold of a product for which there is clearly a keen demand. The appeal is not just to serious collectors of course and I have no doubt that the majority of the coloured 200- and 300-type phones on the market now are going to ordinary people who just appreciate a classic design. These folk are not fussed at all if their new treasure contains a mixture of old and new parts from more than one country.

ONE WORD, TWO INTERPRETATIONS?

For collectors, however, things are much more sharply defined. For them, authenticity is essential and in their eyes there can be no grounds for substituting some genuine British parts in an otherwise Indian phone unless the intention is to deceive. To them, a catalogue containing a photo of a GPO 232 and the description '200-type' can mean only one thing, the real thing. If the trader had written '200-style' (like the 'Cola-style' drinks you find at the supermarket) it would have been obvious that the phone was not a 'real' 200-type, but there you are, not everyone is precise with words. It all depends what you mean by the word 'type' and, equally likely, a seasoned collector would have known the difference. A newcomer to the hobby wouldn't.

COLOURED PLASTICS

With both telephones and radio sets the motto 'penny plain, tuppence coloured' certainly applies, at least where plastic cases are concerned. Most bakelite telephones came normally in black and radios in brown, and coloured examples of the same shape are much more highly prized. This has led to the appearance of replica cases fitted to original chassis and some buyers have been fooled. Reproductions can generally be distinguished in the following ways.

The modern cases have a shade and appearance totally different from original British-made parts. Pre-war cases, made of a urea-type material, have a very different 'look and feel' to those of post-war manufacture. The pre-war parts tend to have an eggshell sheen and a granular feel to the touch; the colours are generally quite muted too. More recent cases are moulded in different, more glossy materials and the colours are more vivid too. There are also differences in the quality of the mouldings; reproductions are seldom as clean and sharp as the originals.

'Impossible' colours are not always fakes, however. There are indeed telephone and radio cases which have been spray-painted in a way that almost defies detection. Look for areas of black bakelite under the edges of the case which have not been painted properly. You should also be aware that some pre-war British telephones and radios *were* spray-painted to achieve different colour schemes, and during the last war (and perhaps just afterwards) some parts were painted, probably owing to a shortage of coloured moulding powders.

UNUSUAL BRAND NAMES

Your suspicions may be raised when you encounter a familiar product with an unfamiliar name, sometimes with a minor design modification or altered specification. Some collectors treat these items as highly suspicious but more likely you have an export model or a 'clone'. Several British firms used a different name on equipment sold overseas, such as Genalex (General Electric export) and Berec (British Ever Ready Company). In both cases, this was because in some markets another, generally American, company had the rights to the more familiar name.

There are many instances of manufacturers selling identical (or very similar) products under two names – for instance HMV and Marconiphone,

Pye and Invicta, Philips and Stella. This was simply to allow the company to award two separate dealerships in the same town and hence sell twice as many sets without alienating the trade. The practice is extremely common in the automobile industry, where it is known as badge-engineering. Occasionally the manufacturer would make sets with no identification (for selling to cheap chain stores without damaging his high-class reputation) or with the branding of the dealer in question (such as the Westminster brand radios sold by Currys and the Defiant sets sold by the Co-op).

CHAPTER 10

Make a business out of your hobby

Well, why not? The idea is anathema to purist collectors but there are a number of successful traders who combine their hobby with a business. What's more, there must be room for a few more hobbyist-vendors of this kind, especially if they offer a superior service at a fair price. Some hobbyists are constantly mouthing off about how they could do a far better job than some of the traders, although we seldom see them turning words into action. But if you believe you could make a go of this business, why not put your money where your mouth is?

After all, you may have nothing to lose. Many people who thought they had a job for life have now learned otherwise and you may well find that being your own boss gives more security than an employer whose financial targets don't include you as a beneficiary. Self-employment has never been as popular as it is now, which means that there are plenty of magazines in W.H. Smith telling you how to start a business of your own. Your bank will have information packs and the local council will guide you towards any development grants or loans you may be entitled to.

Don't be profligate with money though, especially if it's a loan and not really your money. To begin with, your foray into the new world of commerce will be a bit of a gamble and one trader told me his motto was 'Never risk more than you can afford to lose altogether'. In other words, don't go in deeper than you (or your bankers) can afford to. You'll soon develop confidence and 'business sense' but don't blow everything at the outset. Fancy headed notepaper, a sign-written van and an executive notebook computer and mobile phone combo may have to wait till later.

Never underestimate your costs; you're doing this for a business now and you'll incur a lot more bills than just petrol costs. Stalls at swap-meets and antique fairs cost money and if you are travelling long distances there will be the cost of overnight accommodation and meals as well. Can you work without a partner? Unlikely, because you will need to disappear for meal breaks and personal needs, so bringing in an assistant will double your costs.

Buying in stock can be problematic, at least at the prices that will be economic for your business. You need stock but you don't want to be burdened with items which hang around, and finding space to store it will be expensive if you have to pay for such storage. Remember, unless you can double your investment within six months, forget it – your money would be better off earning interest in a building society. You may think that you can set your prices but in truth the market does this for you. This applies particularly to high-fashion sets, such as bakelite and catalin radios; they have clearly defined prices and if you are buying to resell, you must resist the temptation to pay too much, especially if repair or restoration is involved.

You'll have to price keenly in order to turn your stock over and the general rule is 'buy at one price, sell at twice that'. This applies almost universally in the sale of second-hand goods (radios, components, antiquarian books, antiques etc.); it covers the dealer's outlay and other business expenses and is recognized as a fair ratio by all well-informed people. Effectively this means that you now have to buy in at half the retail price and if your vendor knows the value of his treasure, you may find he won't sell as low as you'd like. You'll need to develop a whole new bunch of persuasive 'social engineering' skills! Never pressure buyers though; give them time to make up their minds and don't talk all the time, it just annoys them.

NEW ATTITUDE

Turning a hobby into a business will alter your attitude to the hobby. Time becomes money, simply because while you are doing one thing, you can't be doing another. Weekends suddenly become working hours and social phone calls that you used to enjoy may get in the way of earning your living.

Fellow hobbyists may even become enemies or at least a scourge. At swap-meets you will rapidly change your opinion of the punters that you hope will do business with you. Many will turn out to be 'tyre kickers'; total time wasters who will discuss the merits of your wares for three-quarters of an hour before walking away empty-handed, possibly screening out the customer who was going to buy that DAC 90 radio from you but has now spotted another one two stalls away. Oh what a rotten cynic I am! But life is often like that.

You may not get rich from your new business, particularly if you reckon it is to become your main source of income. In fact you'll need to resign yourself to a modest but steady living, but overall you may well derive a lot more enjoyment from this than some other nine-to-five activity. You will definitely need some capital and some money to tide you over when business is slack. You need to take good business advice from your bank and certainly don't go into it because I told you to; this chapter is not intended as a recommendation, merely as a guide. Speaking personally, I certainly wouldn't try and derive my income from vintage wireless but there are many others who are proving it can be done. Just bear in mind that as far as Britain is concerned (it's different in the USA) there are probably no more than 2000 serious collectors in the country and most of these spend no more than £1000 a year on their hobby. Whatever service you offer must be a good one to make them part with that money to give it to you!

That said, there are customers who desperately want good service. Several radio restorers I have spoken to tell me they have more work than they can handle, so there must be some opportunities out there. Combine it with some other activity in a different field of business (what economists call diversification) and you might have a nice little business.

GOOD BUSINESS

Snag! Yes, there's always a few snags. For a start, once you become a trader you acquire responsibilities. Most venues insist you display a sign giving your name, address and phone number. This is good business practice anyway, as is putting out cards with your name and address on (someone may pick one up and offer you some tasty merchandise a week later). Buy a duplicate invoice book and offer every customer a receipt (a rubber stamp with your name and address is handy for these).

Daft as it may sound, make sure any venue where you trade is licensed, especially for Sunday trading. Some smart Alec may come along and try to prosecute you! You may think it's the organizer's task to check out things like this. It isn't; it's yours. You are the businessman now and businessmen are professionals. Professionals don't get caught out; only amateurs do.

Also, it is one thing to restore electrical apparatus to a standard that you consider safe to use, another to sell such items to third parties. Technically it is an offence to sell an electrical item without a three-pin plug with sleeved pins. If you are restoring sets for customers or selling them on, you must either ensure that the apparatus conforms to current electrical safety regulations or else you should affix a label firmly to the mains cable indicating the item is sold as an antique, together with any precautions that must be taken. You are not allowed any excuses in this respect.

Similarly with old telephones, each item must carry a red triangle label and the words 'Prohibited from connection to a public telecommunication system', and certainly so if the phone is supplied with a modern plug-ended cord. If the item in question is 100 years old and clearly an antique, that's probably different but the Marking Order doesn't make any explicit exemption; technically it applies to all telephones offered for sale. The law is the law and who knows when the Trading Standards people are going to have a blitz on your trading venue?

DIFFERENTIATING YOUR BUSINESS BY OPENNESS

One or two traders try to hide their identity; if they even show a card giving their name at swap-meets and fairs, it either gives no address or a Post Office box number and the only phone number is mobile. This hardly inspires confidence. On the other hand, a business which distinguishes itself by openness sticks in everybody's mind and all the business books tell you that differentiation is what sets your sales operation apart from all the others today.

Guarantee

MY-BUSINESS-NAME undertakes to repair or replace at no charge for labour or materials any electrical or mechanical part found to be defective within twelve months of purchase provided that:

- No unauthorized repair or adjustment has been made.
- The apparatus has not been subject to misuse, accident, neglect or connection to an unsuitable electricity supply.
- The instrument has not been stored or used under conditions of excessive heat, cold, damp or vibration.

During the guarantee period, should repair be impracticable, MY-BUSINESS-NAME undertakes to refund in full the purchase price paid for the set. The cost of returning goods must be borne by the customer.

For safety reasons, it is imperative that the connections inside the 13-amp plug supplied with the set are not tampered with. Only a 3-amp replacement fuse should be fitted. The fitting of any other rating of fuse invalidates the guarantee.

MY-BUSINESS-NAME, address, phone number, fax, e-mail.

THIS COLLECTOR'S ITEM was made around the year 19xx and was intended to have a relatively short working life. The fact that it still works well is a credit to the quality standards of the original manufacturer. You can help prolong its active life by treating it with care and not leaving it switched on for extended periods.

Valve sets generate a fair amount of heat, so make sure there is adequate ventilation around the set. Allow time for the valves to warm up; there will be a delay of a minute or so before you hear sound. Slight drift on tuning after a while is normal. Sound quality may not be up to modern hi-fi standards. Do not use or store in damp atmospheres. If you need any advice, please call us.

MY-BUSINESS-NAME, address, phone number, fax, e-mail.

THIS FULLY-TESTED COLLECTOR'S ITEM was made around the year 19xx and was not intended to last this long! The fact that it still works well is a credit to the quality standards of the original manufacturer. You can help prolong its active life by treating it with care and not using or storing it in damp atmospheres.

It is technically illegal to connect this telephone to the public network in Britain, either directly or indirectly. Its sound quality, although adequate for the era in which it was first used, may not be up to today's standards. Your other telephone equipment may not be compatible with this instrument. If you need any advice, please call us.

MY-BUSINESS-NAME, address, phone number, fax, e-mail.

You should be proud of your name and reputation, so you will make a point of proclaiming who you are and where you come from. You will back this up with cards and leaflets, making it easier for people to contact you when they have something to sell or be repaired. You will also issue proper receipts from an account book stamped with your details and on all but the smallest items you will issue a guarantee. Guarantees are a sales aid, not a liability, and by wording the guarantee carefully you can filter out spurious claims. An example of the sort of thing you might care to consider is given.

Giving a guarantee on what may be a fifty-year-old product does have its down side, of course. Ideally you want to sell only to thoughtful, understanding customers who appreciate that running a fifty-year-old device eighteen hours a day may not be in its best interests, but these people are often less ready to spend their hard-earned cash than 'style freaks' who think it's 'cool' (or even 'kewl') to have a piece of old technology in their living room. For this reason it may be advisable to tie a label to the back of the device, with words along the lines shown in the upper box (they are intended for a radio set and you can adjust them to your needs). Telephones need to be marked as well; an example of suitable wording is given in the lower box.

Telephones can bring other problems too. Problems have been reported in Britain where people have tried to use antique 200- and 300-type telephones on cablecom telephone lines (Telewest was mentioned but other cable television companies may be similar). It appears that, whilst these exchanges can handle loop-disconnect telephones, they are much fussier over pulsing speed and make/break ratio than those of BT. It is also a moot point as to how much longer the public network will support loop-disconnect telephones and it might be worth advising your customers of this.

WHAT THAT GUARANTEE ENTAILS

Repairing radios and telephones for your own use and pleasure is one thing, making them 'bombproof' for the general public is another. Not only do you not know how they are going to treat your merchandise, you cannot even be sure they will understand how to use it. So be prepared for comebacks . .

'This dial telephone you sold me stops the other ones from ringing/sets off my answering machine when I dial out/sounds a lot fainter than my normal BT phone.'

'That valve radio you sold me – after an hour's use I find I have to retune it; surely that can't be right.'

These factors may be obvious to you but your customers may not be prepared to make allowances. What about the smell of burning dust as the valves warm up? You may find it entirely natural and quietly reassuring, but to someone who knows nothing about valve sets, it may be a cause of great alarm.

Are you sure the set is completely safe? Did you warn your telephone customer that it's illegal to connect antique phones to the network? None of these old radios and televisions was made to today's safety standards – just think of the live chassis, knobs with grub-screws which can come loose leaving a live volume-control spindle exposed, set-backs with great big holes inviting small children to poke a finger in . . . It's all very well for us enthusiasts because we know the risks, but your customers (who paid good money for the quaint old set) expect – and have – the full protection of consumer law. If you sell them something an independent trading standards examiner could find to be unsafe, you might find yourself in line for a hefty fine. In fact, if you intend to make a business of selling old sets, you ought to have a public liability insurance policy for half a million pounds. It's normally an integral part of a small business insurance package, which you will need to protect your workshop and your stock (remember, they will not be covered by your domestic policy), and this insurance will cost no more than a couple of hundred pounds a year. Again, it's all part of taking a professional attitude – when you join the professionals, you need to act like them.

THE TAX MAN COMETH

That's another part of professional business life. The tax man will be taking an interest in you from now on and even if you think he wouldn't be interested in a tuppenny-ha'penny part-time business like yours, who is to know that some disenchanted rival of yours isn't going to 'shop' you? It is an established fact that tax inspectors now make regular visits to Sunday car boot sales and note the faces they see there. Anyone they come to recognize as a 'regular' will be challenged to produce their books and, if they cannot, the tax man will determine your estimated turnover. It's then up to you to prove otherwise by showing the District Inspector your profit and loss accounts.

Actually, keeping accounts is not difficult, and for a small trader all you need is a day-book. This is a book (a diary will do) in which you show every item of expenditure and income: how much, on which day and with which person or firm. You will be surprised how rapidly the outgoings add up (petrol money, restoration materials, sets bought in etc.) and that your actual profit is quite modest and not the sort of thing that will put you into the super-tax bracket. But do declare it annually; I know one guy who carried on a lucrative spare-time job for five years before the tax man caught on. Not only did he have to stump up five years' back-tax, the Inland Revenue also informed his employer and he lost his day job. It's true.

So start off on the right footing; don't intend to 'go legal' later on when you have built up your business. Your local Citizens Advice Bureau and tax office will have leaflets on running a small business, whilst there are some good paperbacks in bookshops.

It's highly unlikely that you will pass the turnover threshold where you are obliged to register for Value Added Tax, although some people register voluntarily in order to claim back VAT on fuel and other input expenses. Modern computer accounts programs make calculating your VAT return child's play but collecting the extra 17.5 per cent will undoubtedly put your prices up. This is something you need to consider carefully.

After a while it may make sense to have a separate bank account for your 'business'. Your bank will help you on this and offer all manner of leaflets. They will also explain what's involved in opening a credit card account, which may be helpful when hard-up customers have to use their plastic and for overseas customers who cannot easily send sterling amounts.

SPENDING MONEY

Buying in stock shouldn't cause too many problems but you don't want to be stuck with stolen property, however innocently it is acquired. Make up a form which you can use when buying stock and photocopy sufficient copies. It should say the date and amount you paid together with the vendor's name and address. Include the words: 'I have received the sum of (---) pounds for these goods, which I warrant are my goods to sell'. Then always use these forms!

GOOD LUCK!

It should be fun (at least for a while) and certainly hard work. The worst result may be you lose some friends, as they will think you've sold out to commercialism. It's the old, old story. What is acceptable as a hobby is not acceptable (to some) as a business, and there is a feeling around that it is somehow not playing the game to make money out of a hobby. I'll not take sides on this except to say that if you run your business ethically and provide a service that people need at a price that is not unreasonable, few hobbyists will resent you. In fact they'll probably be delighted to do business with you. It would be unfair to name names but I can think of several businesses that are well regarded by enthusiasts – and of a few traders (only a couple) who are nothing more than parasites or scum. Good luck to you!

PRIVATE SALES...

What about if you don't feel like making a business from trading but would like to be a bit more successful when you take things for sale at swap-meets – or place small ads in magazines and newsletters?

Well of course . . . there are ways of securing success here as well. It's all down to cultivating a professional attitude and treating it the same as you would any other business transaction.

Printing some business cards with your name, address and phone number is good investment; they are an excellent sales aid and save time when you want someone to contact you. You can put an outline of what you collect or sell on the card as well.

A receiver accompanied by a bulging file of service information and sales pamphlets (originals or photocopies) will always look more interesting than one without. Be prepared to show the item working and give it a good dusting and polish before bringing it out (but don't tart it up too much). Label anything that plugs into the mains 'Antique/collector's item: do not plug in until tested'. Don't hide faults – it's dishonest and your reputation is worth more than that.

Writing classified advertisements that work is easy.

1. Start by mentioning the product or service you are selling (or want to buy). By doing so, you make it easier for the reader.
2. Always include the price. Research has shown that 52 per cent of people who read classified ads will not respond to ads that fail to mention a price. Don't ask for offers – people will think you're trying to get more than the item is worth! Don't say 'ONO' (for 'or near offer') either; it indicates all too clearly you are unsure of yourself and no-one will offer the full price.
3. Include your location; it will avoid wasted calls from people living far away.
4. Keep abbreviations and trade jargon to a minimum. Will the reader know what an NB207 is? If it's a 12in table-model TV from 1956, say so – and give the maker's name as well!
5. Be honest about the condition it's in.
6. Put yourself in the position of the reader. Is all the information included? The more information you provide, the more likely people are to call you.

If you make any statement about the condition or description of your item, make sure it is 100 per cent correct. Otherwise your purchaser might have grounds for action against you and the argument that you're not a professional vendor would hold no water. Finally, beware of accepting cheques from unknown purchasers; ask for a name and address as well as noting the cheque validation card number.

. . . AND PURCHASES

Do your homework and if you're unsure of an item's worth or authenticity, and try and get a second opinion. If something is claimed original but looks too good to be true, make a close examination. If the vendor claims an item is in working order, ask for a demonstration. The expression 'untested' may be true but it may also mean that the vendor just couldn't fix it! Always look for missing (or incorrect) valves and bring a small screwdriver. Ask if you may remove the back and if the vendor refuses, be very suspicious.

Unlike cars, it's not a criminal offence to sell a receiver or amplifier that's not up to specification. Unless the vendor guarantees an item is in working order and totally original, do not assume this – it's up to you to decide if it's what you think it is.

If you enjoy haggling, never make the opening bid. Instead push the vendor with words such as 'What's your minimum?' and then say, 'Well, that's not unreasonable but I had a little less in mind'. Bring – and offer – cash; it concentrates the mind. Cheques from unknown purchasers are always a liability.

Should the price be above your limit, leave your name and phone number; the vendor might have second thoughts.

If you are buying from a private individual you have very little legal protection; the Trades Descriptions Act does not apply and your deal is no more than a private bargain between two people. You have no redress if the goods turn out to be defective, although if any written statement you were given turns out to be untrue, you would have a legal case (whether it would be cost-effective to pursue it is another matter, though).

Finally, always remember this golden rule, attributed to the Victorian writer John Ruskin:

'It's unwise to pay too much but it's also unwise to pay too little. When you pay too much, you lose a little money, that is all. When you pay too little, you sometimes lose everything because the thing you bought was incapable of doing the thing you bought it to do.'

'The common law of business balance prohibits paying a little and getting a lot. It can't be done. If you deal with the lowest bidder, it's well to set aside something for the risk you run. And if you do that, you will have enough to pay for something better.'

By the way, the legal responsibilities described above relate to Great Britain. They may be different in other countries.

PART 2

Means of Making Good

CHAPTER 11

Repairing and restoring, cleaning and lubricating

What's the difference between repairing and restoring? Well, quite a lot actually. Repairing is servicing or mending, getting something which has stopped working going again. Most of the so-called restoration books concentrate exclusively or almost exclusively on servicing, and that's only half the story.

Restoring an old artefact is more than just servicing, it means making a set complete, making it functional and, most of all, making it look cared for. Much of this book is about restoration in its fullest sense.

This is not to decry the need for servicing – I think an object that doesn't work is only a shadow of itself. After all, radios, televisions, amplifiers or whatever, were sold as functional items, designed to do something useful. I know some collectors claim they are interested purely in the aesthetics of their treasures and having them working is unimportant. Really? Is that true or just a cop-out? I suppose it's a valid viewpoint for antique shotguns but it wouldn't wash with a vintage car enthusiast. I don't think it's a valid viewpoint for vintage sound and vision enthusiasts.

Of course some sets catch the eye in a beguiling sort of way and it's easy to see why many folk find the design, the very look and feel of these old items pleasing and somehow comforting. But that said, it seems a shame not to put the sets to their proper use. A live dog is more interesting than a stuffed one. The only genuine reason for not repairing a set is if you cannot do it yourself and you cannot afford to have it mended professionally. There is no shortage of people who will take on the task of servicing old radios, although I hope this book will encourage you to consider trying to learn to do this yourself.

And if you are going to make a set work again, why not make it look the part as well? There is no need to go the whole hog. It is normally impossible to put items that are fifty years old or more back into showroom condition and you run a severe risk if you try to. Yes, it's true, there are a few conservators or refinishers who actually can get a set back to showroom condition. They are the people with ultrasonic cleaning machines (the sort normally used for watches) which can remove dust and discoloration from bakelite knobs and brass terminals. They have the means to strip and re-cellulose cabinets, polish bakelite until it gleams and rub white paint into engraved panels, all to bring back that brand-new look. But if you don't have these skills and appliances, don't even think about it. The experienced French restorer Guy Biraud sums it up nicely when he says: 'Be restrained in your restoration work and may your objects keep as much as possible of their original condition and patina of age.'

On the other hand, there's no harm in making your pride and joy look cared for. Gordon Bussey has a good way of putting it; he says a set should look as if it has stood in the same place in the living room for fifty years and been dusted once a week by a little old lady. It's easy to remove built-up layers of old furniture polish and use a bit of foam cleanser to make a well-used set look well looked after.

AUTHENTICITY OR FUNCTIONALITY?

Restoration carried too far can easily spoil an item. A few collectors do not allow any restoration – they want equipment in 'as found' condition, and if cables are frayed or parts missing this adds to the authenticity of the thing. Of course, if you are not prepared to replace electronic components you probably won't be able to operate the device any more. Some collectors go to the trouble of hiding new components (such as electrolytic capacitors) inside the cans of the old ones, having first scraped them out, but that's too much like hard work for most folk.

I take the attitude that my cherished items were not in a dirty 'as found' condition when they were new or in loving use, nor would they be displayed in 'as found' condition in any self-respecting museum. We may not be able to restore items to museum standards but we can jolly well try to. Nor do I worry about using new components inside an old set. Working condition and electrical safety are far more important to me (and nobody else knows what's inside my sets). In any case, during a set's life, when it failed, the local dealer repaired it with entirely new components (probably from Radiospares!), so why not continue to do this now? If the set was last repaired thirty years ago, the components installed then are already antiques. If you expect your sets to be around in another thirty years from now (even if you aren't), what's wrong with installing the best components around today? You may even hold back the time when new components are next required!

However, this is a point argued about long and often. Most collectors are prepared to replace parts from other sources if these parts will enable the equipment to function properly, and will try to ensure all their treasures are capable of working. What's more, the sensible collectors endeavour to run their sets up every few months to exercise the components and avoid deterioration of capacitors and so on. Many old valve radios never give up if they are in daily use (keeping the electrolytics formed for instance). But allow those sets to go cold for two years and you'll almost certainly need to buy new 'smoothing bombs' as one of my friends so eloquently puts it.

What about unnecessary modifications? This seems particularly prevalent in the classic hi-fi world. People pay good money to have their Quad II amplifiers restored or 'upgraded' but to justify their often high charges, some of these 'restorers' change the power and signal connectors for modern designs and substitute alternative valves and valve holders to get around the current high price of KT66 valves. All I can say is it's not my idea of restoration, more like butchery.

WHEN TO STOP

Is it legitimate to remanufacture parts when originals are no longer available? There are many parts sold in the USA and here which are pattern parts made to look like the originals – but they are not actually old. The same applies to reproduction transfers, dial glasses and so on. You may even have to have some metal parts turned on a lathe if no originals are to be found. I have no problems with this really; in the classic car business it is normal to have reproduction and remade parts and nobody says they detract from the value of the vehicle. In fact, a good restoration job can easily transform and enhance the value of an old 'banger' and put it into the *concours d'élégance*. But if you are selling a set which contains a high proportion of non-original parts it is only fair to make this known.

Can you really ruin a radio with restoration? You bet you can! Here's a real-life example. A friend was lucky enough to buy a scarce mid-1920s Marconi set which had spent the last fifty years at least in a barn. It was complete right down to the original mains power pack but what a state it was in! The base of the set had rotted away with damp and battery acid, the cabinet had varnish flaking off, the copper wiring was green and yet the set was definitely restorable, in fact by some miracle even the original Marconi transfer was intact. My friend had the good sense to take advice from a 'guru' (not me!), one of the vintage wireless hobby's 'good guys'. And this is what he said.

Put the set in a tea chest, cover the top with a rug (not plastic sheet, let the set breathe) and leave it in the loft for twenty years. As it is, the set is already worth a damn sight more than the £150 paid for it but in twenty years' time it will be worth a small fortune. Restoring the set now to museum standard is beyond any amateur's capability; there are probably only six people in the country who could renovate the set without ruining it and these people all have waiting lists as long as your arm (and charges to match). With a set as scarce as this, any other standard of restoration would destroy any value it has, yet full professional restoration would put its value into the four-figure price bracket. And as my friend can't afford to have it restored to this standard, he has reluctantly put it in the loft. What would you do? Would you dare disagree with one of this country's experts on vintage wireless? Something to think about.

RESTORATION VS. RESTIFICATION – THE DEBATE RUMBLES ON FOREVER

During 1993 and 1994 there was a flurry of letters in the American magazine *Antique Radio Classified* on the subject of restoration and so-called improvement. It started with someone who had chrome-plated a radio (it happened to be an Arvin set but that doesn't matter), and the question was: is an enthusiast entitled to chrome-plate a radio that would never have received that kind of finish originally? On the other hand, was applying a process such as chrome plating – which could be removed later – less heinous than drilling a wooden cabinet to fit some important but non-original fitting (such as a different switch when the original pattern could not be found)? Feathers were starting to fly.

Opinions differed widely and analogies were drawn with the 'street rodders' who rebuild and refinish old cars with remarkable creativity and imagination, but not to recreate the vehicles in their original form. Reluctantly, most people agreed that once a person owns a radio, he or she can do what he or she likes with it! Equally, if people wanted nothing to do with chromed radios, they didn't have to buy them.

Bryan Cowan of Sacramento, California made possibly the best point, writing: 'In my other hobby, restoration of automobiles, specific terms distinguish "modified" from "original". "Restoration" means the car is restored to original showroom condition. "Restification" means the car has non-original features added but the integrity of the car remains intact'. Stan Lopes of Concord, California added that 'refinished' implied an altered finish, so when Philip Collins says 'all radios refinished' in his book *Radios – The Golden Age*, refinishing must mean to apply other than the original colour scheme. Confused? So am I. Let's move on rapidly to the next topic.

REPLICA AND REPRODUCTION

Here are another two words that some people use interchangeably and others don't. The Australasian Telephone Collectors Society anguished over this and came up with the following.

Both replicas and reproductions are copies of other things but a replica is an attempt to create an object which duplicates the original in every detail. It may very well be a one-off or produced to order and in very small quantities. A reproduction is most frequently a commercially produced copy, unrestricted by such concerns as medium, scale or colour. Sometimes the word accurate is used in connection with reproduction but often it is not. Whether this concerns you or not is up to you but it should certainly exercise the minds of people who place advertisements for replicas or reproductions.

DOWN TO BUSINESS

So much for the philosophy – let's assume you are ready to roll your sleeves up and start restoring. Stop! If you're going to tackle this in a professional way (is there any other?), it's worth starting a file for the subject. A pocket notebook is fine or even loose papers in a clear plastic wallet. In this you can note down where you put loose components, note parts to be ordered and keep a photocopy of the circuit diagram. Super-brains can entrust all this to memory but ordinary mortals will find written notes extremely handy, especially if a project has to be interrupted and resumed later.

Back to practical matters. You'll need some electrical tools and a basic tool-kit is described in the next chapter. But first we need to set out a work-place for our repair projects. A dedicated work-bench is ideal but not everyone has this luxury. Your repair work may have to share space with other activities, perhaps even the kitchen table, and you will need to protect both your work and the surface below. Some people make a portable work-bench out of wood and this may suit you. Otherwise a good plan is to lay out some carpet to protect both the bench top and the treasure you are working on. Cheap carpet samples or end of roll off-cuts, available at furniture stores and carpet dealers, are ideal. You will also find that if your test instruments are placed on an elevated shelf about 12 in above the bench, and to the rear, they are much easier to use and read.

It's also worth laying in a stock of resealable plastic bags and small boxes; in these you can store all the small parts you remove from restoration projects (feet, knobs, screws and so on). It's amazing how easily these can go astray otherwise – do remember to label each package too! When you have a number of projects on the go simultaneously, it's all too easy to mix these up.

GENERAL CLEANING

Before you can start work seriously, the item under repair must be clean. Ken Owens of the Antique Wireless Association advises that, for cleaning the chassis between valve sockets, transformers etc., you should use a cheap paintbrush with the bristles cut back to about a quarter of an inch. This provides a stiff bristle that will not scratch. When using with solvents, be sure the paintbrush has natural bristles and a wooden handle. For cleaning fine surfaces, such as dials, a cosmetics brush is ideal. These are available in various sizes and shapes, and are extremely soft. Otherwise a soft artist's brush (called a 'mop') will do but this will cost more.

For light degreasing, cigarette lighter fluid (such as Ronsonol or Swan Vestas) is ideal. It will cut light oils and most waxes, but will not harm plastics or paints. Choose one which comes in a bottle with a tip that creates a fine stream when squeezed. *Remember that lighter fluid is flammable, should be used in a well ventilated area, and eye protection should be worn.* For heavy degreasing use automotive brake cleaner in a spray can. This will cut the heaviest grease and wax, but it will also attack many paints and plastics. *Again, have plenty of ventilation and eye protection.*

Some people recommend the use of steel wool but this is risky around a bench being used for radios, televisions, telephones or audio equipment – the tiny strands fly everywhere and may cause short circuits or lead to rusting. They will also play havoc with loudspeakers and magnetized meters. Instead, use plastic scrubbing pads, glass-paper or the foam sanding blocks sold in do-it-yourself shops.

Heavy cleaning will be covered again later but here are some initial thoughts. Straightforward detergent (washing-up liquid if you like) in water is cheap and will remove a lot of dirt. Foam cleanser from an aerosol is a convenient alternative and with both you need to have plenty of towels (paper or cloth) around to mop up quickly – water must not be allowed to settle on wood, fabric or steel surfaces. The foam cleansers are particularly useful for cleaning most knobs and certain painted surfaces; according to professional chemist Barry Ornitz the active ingredients here are glycol ethers which are relatively mild solvents. They do attack some surfaces, however, so be careful and test first. Printed surfaces (such as dial markings on glass or plastic) are especially delicate and can be destroyed by even the gentlest of cleaners.

For ingrained dirt and sticky films such as grease and nicotine, you may need to try stronger stuff such as neat detergent, kitchen or bathroom cleaners but beware, these can leave caustic residues or dissolve transfers and screen-printed lettering. They may also take the shine off glossy surfaces.

Even stronger cleansers and degreasing agents can be found in car spares outlets and do-it-yourself stores. Don't overlook ammonia and methylated spirits, by the way; both of these have their uses as well. Neat ammonia can be caustic, so try adding a little to warm soapy water first. A 50:50 solution is recommended and this can be used warm or tepid.

Drying wet items is important; we don't want to give rust a chance. Paper towels are good for removing the bulk of the dampness, followed by gentle heat (a 60 watt lamp bulb with a reflector shade is recommended).

Whatever you do, if you ever even contemplate repainting any surface after cleaning, avoid completely the use of any waxes or polishes containing silicone oils – Armor All is an example. For its intended purpose it is an excellent product but paint will never adhere properly to any surface treated with it without considerable effort at cleaning. Silicones do give an attractive gleam (for a while) but the old-fashioned natural paste wax polishes such as beeswax and Carnauba polish are far safer and can be removed by conventional solvents if wanted. As Barry says, there is no magical method of cleaning old treasures easily without damage. It is better to go slowly and if you take any shortcuts, be prepared to pay the price!

Metal panels may need harsher treatment, such as rubbing with wire wool or a glass-fibre scratch brush, or else a wire brush in an electric drill. Steel wool is taboo; it fragments too easily leaving tiny strands which will eventually go rusty and may cause unexpected short circuits later. Avoid using a steel wire brush on aluminium too; steel particles embed themselves into the aluminium, leading to nasty corrosion afterwards. This is apparently a well-known hazard in antique aircraft restoration. So use aluminium wool instead, or if that's not available, then stainless steel brushes or wool. And follow that with a thorough cleaning, preferably an Alodine process. Aluminium oxide sandpaper works fine, as do the non-metallic abrasive pads made by 3M under the Scotchbrite brand. You can buy these in supermarkets for cleaning pots and pans and a wider range is stocked by engineering supplies stores and motor factors for industrial cleaning purposes. These are durable and not too expensive.

Paint and varnish can be removed in a number of ways. Ammonia is good for taking varnish and lacquer off brass, although it can lead to stress-corrosion problems. Other paint-stripping chemicals are discussed in detail in Chapter 15, 'Practical restoration techniques – metalwork'.

LUBRICATION

For well-meaning but ill-informed enthusiasts lubrication is a 'good thing'. They apply oil liberally, without realizing that this introduces a whole world of problems. Ian Hopley of the Telecommunications Heritage Group explains.

Oil has the property of acting like a fluid film of millions of tiny ball-bearings and friction is reduced when a film of these 'bearings' lies between two sliding parts. Unfortunately it has three other properties that are less than desirable. The first is viscosity or in other words its thickness. In precision work only the thinnest, lightest oils are used but even these have viscosity and are directly affected by temperature. As the temperature decreases oil gets thicker and causes drag on bearings; similarly as temperature increases, the oil thins down. Ageing is another problem. As oil ages its lighter, more volatile components evaporate so that the remaining residue gets thicker and drag increases. Exposure to the atmosphere causes this residue to oxidize and become acidic, leading to corrosion of the surface involved. The third unwelcome property is a combination of the first two, plus dust and dirt. As the oil becomes thicker with age it is more likely to trap fine dust and dirt, and then, as it breaks down and oxidizes, it becomes a very effective grinding paste which will cause very accelerated wear of parts.

This may not be a major problem for most radios and telephones but it does have profound implications for electromechanical telephone equipment and electric clocks. The upshot is that cleaning and lubrication should be carried out at about three year intervals if accurate operation and minimum wear and damage are required. Only the highest quality oils recommended by manufacturers should be used as these have additives that counter viscosity changes and resist oxidization. Some general purpose household oils can start to impair performance after as little as a year.

For precision lubrication it is worth remembering that Tandy (Radio Shack) shops sell a micro-spout oiler containing oil and Teflon particles. It is not expensive and works very well. In addition Ken Owens has offered the following suggestions on the Internet:

- Instead of dispensing the oil from a can spout, use an oiler such as those used by watch/clock repairmen to transfer the oil to the area being oiled, a drop at a time. An excellent oiler can be made from a sewing needle by grinding off the top end until the eye is opened up at the top. The pointed end can be inserted into a wooden dowel for a handle. Different size needles can be used to control the amount of oil dispensed.
- A flat toothpick is ideal for applying thick lubricants. Where a long reach is needed, such as for band switches, a small bamboo skewer can be used. The end can be made chisel-shaped to help spread the lubricant over flat surfaces.
- Use petroleum jelly (Vaseline) for the contact surface of wire-wound pots. With the winding accessible, wipe a thin film on to the wire with a fingertip.
- For bearings and rotor wipers in variable condensers, use a mixture of light oil and graphite. He recommends a mixture of five parts sewing machine oil to one part of powdered graphite. The graphite will stop the noise from dirty and/or worn rotor contacts. Be sure to shake the oil well, as the graphite will settle out.

CHAPTER 12

First-line electrical repairs and simple conversions

I have already said that this book doesn't intend to repeat what you have in your other books, so it is not a complete servicing guide from the chassis upward. That said, some readers will be coming into our hobby fresh and will welcome some basic information on what's involved in electrical repair. To them, and to nobody else, I offer this chapter. It's not a substitute for a complete servicing guide, nor is it derived from blinding original research (I freely acknowledge that I have picked up most of these ideas from other authors).

Nearly everyone has their own way of doing things, and some people are very methodical, even fastidious, in what they do whilst others take the easy way out and do as much (or rather, as little) as they think they can get away with. This book favours the first approach: radios, televisions, telephones and whatever else were made in a professional fashion for sale on a commercial basis, and that means our aim on restoring them ought to be professional as well, certainly not slapdash. Of course it's easy to make mistakes if you don't have a guide to show you the professional way of doing things, which is why commercial organizations are so obsessed today with 'best practice'. Consultants earn a fortune by lecturing on business methods they have observed, and by comparing and contrasting other people's methods you can usually arrive at one single best method of doing things. That's certainly the idea behind this book; I have read any number of books and articles, also tried out nearly all of the methods described, in order to whittle them down to a 'best of breed' selection. That's the philosophy behind this book and that's also the end of the theory section. Let's get down to business . . .

You have an old radio, telephone or whatever – it really doesn't matter which. What matters is that it – let's call it a set – doesn't work. Why? If you understand the reasons for failure – and can successfully determine the cause – you're half-way towards making the set work again.

Failure is either accidental or deliberate. Deliberate means that someone has stopped it from working, probably intentionally. The set may have been robbed for parts or the subject of a botched repair. It's your job to spot this. If failure was accidental, the cause is generally dirt, moisture or time itself causing ageing of some component part. With telephone devices, this may be something as simple as a broken cord or a 'dry' solder joint, whilst with electronic devices it could be something similar or an internal component. A seasoned repair-man can often sense what is likely to be wrong and go straight to the fault, particularly if he knows the 'stock faults' for that set. But for the rest of us here are some valid hints.

RADIO, TELEVISION AND AUDIO EQUIPMENT

The most frequent causes of failure in valve devices are electrolytic capacitors in the power supply and coupling capacitors between valve stages. The valves themselves are far less frequently at fault (except magic-eye tuning indicators, which in any case are not essential to the set's operation).

BUILT TO LAST

Unlike some of the rubbish turned out today, old radio and TV sets were made to last and I have known many old sets to work as soon as they were switched on, even though they had not been used for years. This is not a wise practice, though. Sets that have remained in constant use for forty years or more can be considered safe but old input filter or smoothing capacitors in particular may not be able to stand the strain of a sudden voltage surge after years of no stress.

If you suspect your set has been subjected to damp, you should leave it in a dry room for a week before switching on. Damp can penetrate transformers and other wound components too; they have a habit of working once after switch-on and then 'dying' forever, presumably because the change of temperature causes expansion and stress, breaking the thin wire corroded by damp.

This is why some restorers start off an 'unknown' set by feeding it around 50 V with a Variac (variable transformer). They then wind up the voltage slowly towards normal supply voltage over a period of a few hours (a substitute method is to run a 60 watt lamp bulb in series with the device for an hour or so; it might even be worth screwing a power outlet and a batten

Figure 12.1 With older television receivers each part is accessible and easily identified, making servicing simpler than the sets of today. Most parts are generic (not specific to a particular set) and are still easily obtainable from the suppliers listed in this book. One warning, however: electrical safety was not the strongest point of these old sets, what with live chassis commonplace and mains-derived EHT on some examples. Read the 'Safety first' chapter if you are not skilled in high-voltage repair work.

lamp-holder to a board for this purpose.) Any faults which then occur are likely to be minor and can be spotted before they do major harm or cause pyrotechnics.

The rationale behind this procedure is that any weak components are less likely to be damaged by stress. The contrary argument is that it's better to discover these weak components now, rather than let someone less technical discover them later. In fact, starting certain electrical devices at a low voltage can be extremely dangerous, as in the following situations.

1. At low heater voltages (and hence cathode temperature), ionized atoms (from residual gas in the imperfect vacuum) striking the cathode can damage it; this does not happen when the cathode is at its proper working temperature. Prolonged low-voltage operation could damage the valves.
2. Rectifier valve filaments need to be in their proper operating range before the operating high-tension (B+) voltages are applied. Rectifiers will not conduct at all at low voltages.
3. Motors can be seriously damaged by low voltage – you should never put a variable transformer in series with most common motors.
4. Relays may operate improperly at low voltages, such as giving inadequate mechanical force to reduce contact resistance.
5. The above four items can interplay, such as a slow-turning motor might not have enough air flow to operate an air interlock which would disable relays from operating, or operate only part of a relay system and so on.

Points 3, 4 and 5 will not apply to simple radio receivers but they could well apply to record decks, projectors, military and studio equipment.

ARE YOU QUALIFIED?

The foregoing assumes you know something about electronics – and valves! – and it's vital that you do, if you intend to tackle the inside of a radio or television set. Remember there are nasty voltages, some of which can persist for a time after switching off. Always wear rubber or plastic-soled shoes when working on high voltages and keep one hand in your pocket. If you have two hands in use, one in the set and the other touching something earthed, a current could flow across your chest and through the heart. End, finito, schluss – one less vintage radio enthusiast.

If you want to drop out at this stage, there's no shame. Take time to develop your confidence, read some books on servicing. Several vintage wireless dealers do repairs (see the 'Buyer's guide' section) or if you want somewhere just down the road, there are a surprising number of local radio and TV dealers who can still tackle 'old' faults! Some of them are rather pleased to have a 'real' set to work on for a change.

STILL WITH US?

I suppose the first concept you need to know is that fundamentally, all electronic devices work in the same way. I'll start off assuming you are thinking of tackling a radio but 99 per cent of this applies equally to televisions and hi-fi amplifiers. After that, I'll mention some points specific to television.

Most books start off saying it really doesn't require a great deal of test equipment or technical knowledge to restore an antique radio to top notch condition. Hah! I suppose these same guys do their own car repairs, with nothing more than a cheap tool-kit from a catalogue store. Chance would be a fine thing.

You do need knowledge but the good thing is that it can be picked up in stages and it will be fun because you are doing something you wanted to. You will gain confidence in stages and at the end of it you will have a fair set of tools and test gear, but because you acquired these second-hand and gradually, they won't have cost you a fortune. Gradually too you will acquire some knowledge of the fundamentals of electronics – how valves work to amplify a signal, the differences between a TRF receiver and a superhet, how diodes rectify, how power supplies work, and so on. As you will read later on, I'm a great believer in self-training and using the resources of the public library (which you have paid for anyway, so why not use it?).

BASIC TEST GEAR

Now as far as test equipment is concerned, you should have three items: a good multimeter (volt-ohm meter or VOM) with a sensitivity of 20 000 (or greater) ohms per volt (an analogue one with a needle dial is better than digital read-out for our purposes), a signal tracer with a demodulator probe, and a schematic diagram of the radio receiver (or whatever it is) you are working on. The multimeter will cost about £40 and can be purchased from Tandy or Maplin Electronics, and the signal tracer can be purchased from Maplin. By the way, the manual that comes with the signal tracer should give basic trouble-shooting procedures. One other item will be helpful at times and this is an RF signal generator ('sig gen') which is also available from Maplin, but you can buy this later after you become more proficient at trouble-shooting. An oscilloscope ('scope') is extremely handy too, almost vital in fact. You can buy second-hand 'old technology' oscilloscopes and signal generators at vintage radio swap-meets; they are not expensive and whilst they may not look as snazzy as the ones in the latest catalogues, they work almost as well. Finally, most experts also mandate using an isolation transformer for safety and it is indeed an excellent idea to supply power to the item you are working on in this way.

TROUBLE-SHOOTING BY INSPECTION

The information which follows applies to mains-powered equipment and is applicable to radios, televisions, audio equipment and more. Low-voltage devices, such as telephones and telegraph instruments, are covered separately later in this chapter.

Your first checks can be made without having to apply voltage. Much can in fact be deduced by careful and methodical visual inspection; this saves time, avoids putting still-functioning components under further strain and has an obvious appeal to those who feel unhappy poking around a live chassis.

Do look at the valves, specifically for discoloration inside the glass. Dark brown stains indicate the valve has had a hard life but it is probably still functional. Clouding of a milky white colour indicates the valve is useless; air has entered, either through a visible crack or by way of a poor metal-to-glass seal. Beware too; sometimes the milky stains are at one end of the glass tube, hidden by the shroud of a rubber top-cap protector.

Electrolytics are sometimes seen to have leaked a corrosive fluid, leaving a powdery deposit, or else the rubber end-seal may be bulging badly. Wax capacitors which have left tell-tale puddles of wax on the chassis below are more 'obvious' signs.

Another vital step is to check the mains lead and all internal wiring for perished insulation. This is especially important where cables pass through the chassis or where connections are made inside the set. Old mains leads may look sound but inside the rubber insulation may be brittle, allowing the live conductor to come into contact with the earth or neutral.

Finally, check the fuses. Not all sets have fuses, but where they do, you will often find the wrong value fuse inserted. If it's open, you at least know there is likely to be a fault.

DON'T SWITCH IT ON YET!

It is not wise to apply volts to any electrical appliance which has not been used for years – you have no idea of the state of the insulation on the internal wiring, whether the mains input filter capacitors are viable or the condition of 101 other components. A gentle start, on the other hand, is kinder and less likely to produce violent fireworks. Remember that a pre-war set could be worth over £2000 – but not if it has been ruined by your intemperate zeal!

Selenium bridge rectifiers (recognizable by the multiple fins) often deteriorate and go open-circuit. Check with a meter and if they are inert (not short-circuited), solder a modern silicon bridge rectifier across the same connections. The new rectifier may produce a higher voltage than the old one (measure the new voltage and compare with details on the original schematic); if so, you may need to provide a series resistor of 5-watt rating to drop the excess volts.

One last point: before you plug in a set do make sure you are applying the correct voltage! Most AC sets are for around 240 V/50 Hz (British), 220 V/50 Hz (European) or 117 V/60 Hz (North American) but don't bank on this. Many older European sets (and some British ones) may be for 110 V AC or they may be for DC mains only. Where sets have adjustable settings for a variety of voltages, do check which has been set.

TIME TO SWITCH ON

Most recipes for trouble-shooting and mending rely on common sense and logical deduction. Applying power to a set of unknown provenance is always a momentous occasion, calling for complete concentration and pressing into service all the engineering faculties – sight, sound, smell and the ability to dodge a flying electrolytic . . .

That's what Mike Izycky says and it's also the point at which he introduces BBC engineering principle No.1, which is:

> *'All circuits run on smoke. If the smoke is allowed to escape, the circuit ceases to function.'*

Often it is uncertain where the smoke actually comes from and this neatly leads us into BBC engineering principle No. 2, which runs:

> *'If at first you cannot see any smoke then switch the apparatus on and let it smoke again on the assumption that all the smoke has not escaped.'*

Joking apart, careful observation can often pin-point the source of the fault. A quick touch may detect a passive component far hotter than it should be or the signs of dripping wax may make it equally obvious. But all this is second-nature to most seasoned technicians, so we'll move on rapidly to some more specific observations.

TIME TO GET SERIOUS

The first step in trouble-shooting a dead (no sound from speaker) radio is to see if the valves light up; of course this applies equally to televisions, amplifiers and any other device using valves. As you can see from an (older) textbook, radios with a power transformer have their filaments wired in parallel and are supplied with AC voltage from a winding on the power transformer. Therefore suppose the radio you are working on is dead, has a power transformer and one of the valves is not lit. You will probably find the unlit valve to have an open filament, which can be verified by removing the valve from the socket and measuring its filament resistance with the lowest range of your ohmmeter. If you don't get a reading, the filament is open circuit ('open') and you have found the trouble.

What if none of the valves lights up? In this case the problem is caused by either a faulty AC mains plug (or fuse), an open-circuit mains cable, a defective fuse in the primary of the power transformer (if one exists), a faulty on/off switch, or a bad winding on the power transformer (although power transformers seldom fail). On the other hand, always be on the lookout for bad solder joints. These may even appear to be sound but one tug usually proves the point. Also a good solder joint looks shiny, a bad (dry) one will look crystalline or dull and grey.

Let's suppose the set you are working on is dead (no sound from the speaker), and has no power transformer. In radios without a power transformer, the filaments are wired in series with the on/off switch, mains cable and plug. As you can see, an open circuit in any one of the above will cause the filaments not to be lit. Again, you can use your ohmmeter, but an open valve filament is probably the cause of failure.

MECHANICAL WORK

Some problems are purely mechanical, such as when you turn the tuning knob, the frequency pointer remains stationary. At this point, you must remove the chassis from the cabinet by removing the screws from the bottom and back of the cabinet. It will also be necessary to pull off the front knobs (which may be fixed with grub screws, concealed with wax). You will probably find that the dial cord which goes from pulley to pulley is broken! Replace that cord with a new dial cord (readily available, don't use string) and work on the radio with the tuning capacitor in the closed position. This will prevent damaging the capacitor and will ensure that the frequency pointer will point to 550 kHz on the dial when the tuning capacitor is closed, which is the way it's supposed to be. But before you replace the chassis and speaker, it would be a good idea to see if the radio then plays.

Other mechanical work is seldom required; occasionally old grease dries out and causes bearings and switches to move stiffly. Relubricate as necessary. 'Sticky' dials and switches usually respond to a quick squirt of WD-40 or a similar penetrating lubricant; if not something has probably broken and you will have to do a full disassembly job (tedious!) followed by washing in white spirit, switch cleaner (watch out for fumes) or petrol. Afterwards lubricate the reassembled item with watch oil. Tandy shops sell a micro-spout oiler containing oil and Teflon particles for a very moderate price.

Here's another tip regarding radio sets – watch out for a bent capacitor plate causing a short in the tuning capacitor and making the radio 'dead'. This fault is remarkably common, yet if ignored can lead to a fruitless waste of time, even an abandoned project.

POWER SUPPLY PROBLEMS

Most of these transformerless AC/DC radios (and televisions) have live chassis and you can easily receive a jolt if you touch the chassis and this can be lethal. The main high-tension power supply (B+ to Americans) can show three faults; either it's missing altogether, is low in value, or has excessive AC ripple content. The causes of missing HT are a defective rectifier valve or else an open filter resistor or choke coil. Valve failure can be verified by substitution with a good valve, and the open filter can be verified with your ohmmeter. A low HT voltage can also be caused by a bad rectifier. Excess AC ripple manifests itself as hum on the audio and is caused by a defective electrolytic capacitor in the filter network. The electrolytic capacitor is the large can on top of the chassis. Replacements are readily available but if you cannot find a can-type replacement, you may install a terminal strip on the underside of the chassis and relocate the wiring from the bad capacitor to the strip, adding new axial-lead tubular capacitors (with identical ratings). For aesthetic reasons you should leave the defective can on the chassis, with its connections removed.

LOST SIGNAL

A handy tool for tracing disappearing signals is the signal tracer. Conceptually it is no more than a demodulator probe followed by a high gain amplifier and loudspeaker but it certainly is handy. However complex a radio, television or amplifier is, it can still be broken down into individual stages, each of which has an input and an output. With this in mind, we can use the tracer to methodically trace the development of the signal through the defective radio.

Take for example a valve radio with no output from the speaker even with the volume control set fully clockwise. The heaters are glowing so there ought to be life and we can use the signal tracer to find the defective circuitry. It helps to set the frequency pointer of the defective radio to a known strong transmitter, such as a local station. We can double-check the pointer accuracy by touching the control grid of the RF amplifier valve with the demodulator probe. We can now hear the local radio station from the signal tracer speaker! Next we touch the anode of the RF amplifier valve, the grid and anode of the converter tube, and the control grid and anode of both IF amplifier valves. At each location we hear the local station, so we shift the probe to the control grid of the first audio amplifier valve. Again we hear the signal, but when we touch the plate we hear nothing. We have found the defective stage! The next thing to do is substitute a known good valve for the audio amplifier valve, but this doesn't help. So we take our trusty multimeter and start taking DC voltage readings. We check the anode and find zero volts DC. The diagram says it should be 65 V, so we turn off the power, unplug the mains lead, and start taking resistance readings of the anode circuitry. Eureka! The load resistor is open circuit; it should read 7500 ohms. With an open load resistor the anode was receiving no voltage, so the valve stopped amplifying. We replace it and the radio works again!

Not all faults are as easy to find but most will respond to methodical tracing. The work does not end here because proper restoration also involves returning the cabinet and fittings to their original condition. It may sound like a lot of effort but the rewards will make it worthwhile and what's more, the restored radio will contain a little bit of you.

REALIGNMENT

This is one activity best left to experts. Don't try to realign tuners unless you have the right equipment and know exactly what you're doing. Unless someone else has already been in and fouled things up, it's highly unlikely that you can make any improvement aligning by ear or by tuning meter. You'll probably make it worse than when you started!

Television sets are more complex than radios, whilst hi-fi amplifiers tend to be simpler but the same techniques apply. Television repair has issues and problems not encountered with radios, the most obvious being a missing picture. Old hands won't need telling this but several people have been fooled. It can easily happen that an old TV refuses to give a picture, yet all the voltages test correct. Perplexing! Yes, but try applying a signal from a test generator to the aerial input, then the set will spring into life. The Bush TV22 is like this, as is the Pye B16T and probably others too.

Another cause of dim or missing pictures is faulty setting of the ion trap on the neck of the picture tube. A quick guide to adjusting these is as follows.

1. Put the ion trap in the correct direction (if it doesn't work after the various steps, turn it in the opposite direction).
2. Put on, correct end forward.
3. Turn on the television and turn the brightness all the way up.
4. Move the ion trap slowly up, rotating left and right till it produces a little raster. Adjust it so that it gets the brightest picture.
5. Turn brightness to a relatively low level. Fine tune it for brightest picture.

Fizzing and spluttering are the signs of another problem which affects television devices only, not radios. Sizzling TVs and monitors are not good news, but these notes may help you solve the problem. The 'sizzle' of leaking EHT is very characteristic (it almost sounds as if a very wet capacitor is spurting noxious fluids or like sausages frying). It's bad news anyway. It is usually also accompanied by a strongly 'electrical' smell of ozone, and whatever you may think about helping to restore the ozone layer, this method is not to be recommended!

'Brushing' is another name given to this effect and it's usually caused by leakage of current from the AC side of the EHT rectifier. The audible hiss will often be accompanied by on-screen interference in the form of a bright jagged vertical bar since, of course, it occurs at line frequency. Upon close scrutiny of a suspected area, brushing can often be perceived visually in a darkened room as a bluish glow. Other common causes are the proximity of loose cables to any line output components, a poorly secured inductor (linearity coil, LOPT etc.) or even internal arcing within the transformer.

First of all you have to determine what is causing the noise – and smell, if any. Remember ozone is poisonous nowadays! It used to be considered good for us, but not any longer. Checking in a darkened room may help: you should be able to see the sparking or whatever. In severe cases of AC arcing, however, there is usually no difficulty in locating the source since the spark is particularly spectacular and destructive!

Generally the problem is corona discharge, either on the CRT, around the LOPT or perhaps on the top cap of the EHT rectifier or stabilizer valve. The area around the connector on the glass of the CRT must be scrupulously clean, so remove the connector and clean with a cream household cleanser such as Jif or Flash. Allow to dry and retest. You may need to seal around the original connector with silicone rubber. The same applies to connections to top-caps of valves; on inspection, they may look decidedly dusty or the rubber safety shroud may have perished. Dismantle, clean, check and restore as necessary.

On the LOPT any spiky soldered joints will give problems. Reflow them and make sure they are nice and round. Carbonization around the EHT rectifier support can also be a problem. Once again, sealing with silicone rubber may be the only solution.

OTHER HIGH-VOLTAGE CONSIDERATIONS

Before working on any set you need to discharge the final anode. Ensure power is disconnected from the receiver. Clip a short piece of cable securely to the chassis metalwork – or to a known earth point in the set. Attach the remaining end securely to the blade of a long screwdriver equipped with well-insulated handle. Slide the blade of the screwdriver under the final anode cap and discharge the connection. It should then be safe to remove the final anode connector and start any necessary work.

Note that it is desirable to periodically discharge the final anode or, alternatively, make a permanent earth connection to it whilst working in its vicinity since the charge can return even if the set has not been subsequently switched-on. Better safe than sorry!

TESTING FOR EHT

Strictly speaking, the only reasonably 'safe' method to check for the presence of EHT is with a probe connected to a meter. This certainly applies to colour receivers where potentials in excess of 25 kV at over 300 mA may be encountered. Where monochrome receivers equipped with valves are concerned, most engineers relied upon the rather spectacular method of drawing an arc. One kilovolt will arc approximately 1.5 mm in free air. Thus 10 kV will arc approximately 1.5 cm. Using a screwdriver with an insulated handle, the blade is placed on the final anode connection and an attempt made to create an arc to an earthed point on the chassis metalwork. An aquadag earthing spring usually makes a convenient earth point; however, it would be unwise to create an arc directly to the aquadag coating in case the spark damages the picture-tube glass!

When searching for EHT on the AC side of the EHT rectifier, it is usually sufficient to hold the screwdriver close to the base of the EHT rectifier when a healthy spark – or at least a blue corona – should be visible.

Touching the top-cap of a line output valve should result in a healthy, though quite small, spark. Again the screwdriver does not need to be in contact with the chassis.

CAUTION

Of course, these procedures apply to sets with EHT derived from the line flyback. Shocks can, as previously described, be nasty although in the majority of cases not lethal. When dealing with colour receivers or those where EHT is obtained by another method (e.g. directly from the mains supply) don't take chances. In fact, if you're nervous about what you're doing the best advice is *don't*!

In the past, service engineers became accustomed to working on live equipment and were usually aware of the hazards; most enthusiasts don't have this background and are more vulnerable. Electricity can be unpredictable, dangerous and sometimes lethal! (Thanks to Malcolm Burrell for these thoughts.)

ODD MAINS

One final point. It's easy to assume that any set you encounter is suitable for today's power supplies, but this may not be the case. The chapter 'Data and more' lists some of the more commonly encountered variations that may present themselves.

In a nutshell the standard voltages can be summarized as 100 V (Japan), 117–120 V (North and South America) and between 220 and 250 V (rest of world). Most of the country of Colombia was on 150 V/60 Hz and an observer noted with considerable interest (and not inconsiderable trepidation) how much 120 V equipment was being plugged into 150 V outlets (including his own stereo, which held up 'just fine' by the way). In most countries DC was replaced gradually by AC, although quite a few towns in Britain did not change until the mid-1950s. DC lasted in some parts of east London well into the 1970s, and in certain 'technical' circles Camberwell was always known as Camberwell DC, recalls Gerry Wells of the Vintage Wireless Museum in London. Not by some strange connection with Washington DC and AC Milan, but because Camberwell remained one of the last boroughs to retain direct current supply mains!

Most countries of the world are supplied with 50 Hz mains, although most of North America and parts of Japan use 60 Hz. Parts of southern

California were on 50 Hz until 1948 and many areas of eastern Canada and the USA and also some Mid-Western locations had 25 Hz mains. Hydroelectric generators were the origin of this low frequency and it is still in use in western New York, mostly in industrial but some commercial loops. Most residential supplies were converted to 60 Hz by the 1950s, although north-west Quebec remained at 25 cycles until 1964 and it is stated that 25 Hz mains are still in use in parts of south-east Asia, such as North Vietnam. Equipment for use in aircraft is generally designed for 400 Hz.

Many radio set manufacturers in the USA made both 25 Hz and 60 Hz equipment from the early 1930s up to the outbreak of war; these included RCA, Stromberg Carlson and Zenith. The Canadian Marconi Company also made many radios for 120 V/25 Hz, and 25 Hz radios and televisions show up regularly at swap-meets in Canada. They will come to no great harm if used on 50 Hz or 60 Hz supplies and, in general, 50 Hz sets can be used on 60 Hz supplies and vice versa. The only problem likely to arise is if the device includes a clock dependent on the mains frequency.

Problems do, however, occur when an appliance is connected to a significantly lower frequency. This is unlikely to happen with most collectors' items (unless a 400 Hz aircraft device is connected to a 50 Hz supply) but it did happen in the past. Gerry Wells of the London Vintage Wireless Museum takes up the story again with a cautionary tale of a puzzled TV dealer soon after the war who had instructions to deliver and install a set in a railwayman's cottage next to Deptford station. When plugged in and switched on, it just went bang! Very strange.

Another set was procured and this one was thoroughly tested before being loaded into the van. And just before turning on the telly, the service hand made sure the house was not supplied by DC mains or anything silly like that. But this set too went bang and had to be repaired. The answer is simple when you know! The houses, like some other parts of the Southern Region of British Railways, were supplied with 25 Hz (cycles in those days!) AC. Gerry recalls that the lighting on London Bridge Station was visibly flickery in those days. In fact London was a remarkable mixture of voltages and frequencies then, not to mention both AC and DC. A small area of Croydon remained, until the late 1960s, on about 205 V AC, too low for normal 240 V appliances. This meant that people buying televisions, refrigerators and washing machines had to buy expensive special versions, which soon became useless when normal 240 V electricity was brought to this block of streets.

All this, of course, is separate from contrary mains, of which a certain John Rea of the Baird Television Company said at a demonstration in 1938: 'The amps are there and so are the volts . . . it's just bad electricity'.

Servicing tips

1. When working on live circuitry, do so with extreme care. A shock probably won't be fatal, but your response from it may cause you to jerk and cause some other problem such as knocking over and damaging the radio you're working on, or upsetting the soldering iron.
2. Keep your soldering iron in a proper stand when in use.
3. A 25- or 50-watt iron or gun is adequate for servicing antique sets. Get one with a small, sharp tip. Maplin Electronics or your local Tandy shop can supply the necessary hand tools.
4. Become proficient at soldering before working on a radio, as poor soldering techniques may introduce more defects in the radio you're working on.
5. Don't change parts on a whim. Do so only after careful investigation.
6. Unnecessary probing and movement of parts may introduce additional defects.
7. Train yourself to see trouble. Always be on the lookout for bad solder joints and overheated parts. Resistors don't smoke without reason; waxed-paper capacitors don't bubble and drip wax without cause.
8. Before turning on the radio, make sure the correct type of valve is in each socket, especially if you suspect the set has been 'got at'. Obviously you will need service data for this.
9. Keep a fire extinguisher in your workshop or working area. You can buy small ones at a modest price in any car accessory shop.
10. When removing a valve from its socket: (a) it may be hot and (b) do so very carefully, as the tube may be weakened due to ageing!

TELEPHONES AND TELEGRAPHS

Telephones and telegraph instruments are, thankfully, a lot easier to trouble-shoot than radios and televisions; this is because their circuits are simpler and they seldom contain any electronic devices. Capacitors and resistors are not subject to the same electrical stress as in higher voltage equipment and seldom need replacement, even after seventy years of use or inactivity. Faults tend to be mechanical in nature and confined to readjustment, bad joints and breaks in wound components such as ringer coils, magneto windings and induction coils. Physical restoration techniques are universal and the methods given later in this book cover telephones and telegraphs equally.

A multimeter is the chief tool required for trouble-shooting telephones; it will rapidly show up any open circuits or poor mechanical contacts. The spring-sets linked to the cradle switch (handset phones) or hook switch (wall telephones) sometimes give trouble but are easily readjusted with a pair of pliers. The permanent magnets in receivers can occasionally give trouble if they lose their strength, in which case turning over the Stalloy metal diaphragm may help. Microphone capsules can also be troublesome; the carbon granules may 'pack' and gentle tapping against a hard surface will help redistribute the granules. The worst microphone fault is a cracked carbon diaphragm on the older type of insets; these are not made any more and an exact replacement inset is the only solution (and then finding such a thing is extremely difficult). Microphone insets can deteriorate and older examples may produce a permanent frying noise. At worst this is annoying; repair is impossible so replacement is the only cure (if you think it is essential).

Some restorers replace microphones and receivers with modern parts, and this is acceptable if the replacements make a good fit. In general it is considered better to repair than to replace. Incidentally, the standardized voltage for powering telephone microphones is 3 V. Accordingly, if you have what is called a local battery telephone, with provision for batteries, this is the voltage to provide. Anything less will result in faint hearing at the other end, whilst a higher voltage will destroy the microphone over time. Local battery telephones include table and wall magneto generator instruments, field telephones and portable linemen's sets. Another kind of local battery telephone is the battery-call set used on railways and some short-distance intercom sets and toy telephones. An alternative generator-calling telephone is the sound-powered sets used aboard ships; these use no batteries for speech (instead they use powerful microphones and very sensitive receivers) and an audio generator for calling, producing a kind of siren sound.

The other essential component of successful telephone (and telegraph) repair is a circuit diagram. Fortunately many telephone instruments have a diagram (known as a 'paster') glued or pasted inside the case or cabinet and this will be most helpful, even if the component symbols differ from the ones we use today. If there is no diagram you will need to look elsewhere, as detailed in the chapter on 'Service data'.

ADDING DIALS

You may be asked to add a dial or tone keypad to a telephone which never had one originally. With some telephones, generally described as 'convertible', this is an easy task. The phone will have a dummy plate covering the fixing place for the dial and is already provided with terminals internally.

All you need do is remove the dummy, fix a suitable dial and make the necessary connections (usually five; you also need to remove an existing strap between two of the terminals). On all other telephones it is considered a sin to add a non-original dial or keypad; it destroys authenticity and the circuitry may well be unsuitable for dial operation anyway.

CONVERSION TO MODERN WIRING

In fact serious collectors avoid any modification to the wiring of telephones, simply because it destroys authenticity. On the other hand, there are many telephones which can be altered without causing harm to work on more modern wiring systems. Two things to consider here – the first is that as far as the United Kingdom is concerned, any such modifications are not BABT approved and can thus be used only on telephones not connected directly or indirectly to public systems. This applies regardless of whether the instrument in question was once a standard Post Office type; once an instrument has been disconnected from the network it is not permitted to be reattached unless it carries the 'green circle' approval mark. The second issue is that only automatic (with dial) and central battery (dial-less) telephones should be connected; local battery telephones (usually fitted with magneto generators) are not suitable for connection.

The two types of British telephone that people most commonly want to convert to the new plug-in (PST) wiring system are the 200 and 300 series of telephones. These have bakelite cases and the standard table instruments are the Telephone 232 (pyramid shape) and the Telephone 332 (squarer shape). Details of the changed connections are given below (note that the telephone's type number is normally found on the base of the phone, underneath). The terminals are numbered with small numerals moulded into the bakelite; you may need bright light and a magnifying glass to see these markings.

- **Tele. 232.** Fit new line cord: white to T1, blue to T2 and red to T3. No further modification is required. Because this telephone has no internal bell, the modification is very simple.
- **Tele. 332 and other 300-type phones.** Ensure terminals 1 and 2, 8 and 9 are strapped together using the original metal links or wire. Remove strap between terminals 10 and 11. Then fit new-style cord: A-wire (white) to T1; B-wire (red) to T9; blue wire to T11.

The ringer impedance remains at 1000 ohms and will tend to 'hog' the ringing current if other, high-impedance instruments are connected in parallel. Ideally you should change the ringer coils for a 4000-ohm unit; some people connect a 3k3 (3300 ohm) resistor in series with the old bell coils.

CHAPTER 13

Practical restoration techniques – electrical components

CHANGING COMPONENTS

There is always a temptation to change all components in an old set and logically this ought to cure a lot of ills. It doesn't. In many cases many of the existing components will perform adequately and replacing them would be a waste of money. Partial substitution makes more sense, although this can also introduce new faults, where higher gain in the replaced stage can cause weaker components in a later stage to perform out of specification.

It's worth noting that opinions vary widely on *how* you replace components, by the way. Here is a widely held viewpoint . . .

'For replacing resistors, capacitors and other components with axial leads, instead of trying to unsolder the original component's connections, simply cut the original component's leads as close to the original component as possible. The leads of the new component can be wound around any handy piece of wire to make two small 'coils' or pigtails, which you can then slide over the ends of the original component's leads and solder.'

To which these ripostes came . . .

'Not in my radio you don't! Solder-wick and lots of patience equals craftsmanship. I don't like my radios to have that "TV repair shop look"!'

'I take my time and restore any repairs to a "like-original-factory" condition, even going to the trouble of "ageing" the new solder connections. I use authentic sleeving and other hardware as required. I try to use superior or improved (but authentic) components when doing repairs so that the repaired component will hopefully outlast the radio its put into. One can be a craftsman and a repairman, too!'

Both systems are equally valid – you pays your money and you take your choice! On the one hand, you may argue that a soldered connection has only two jobs, to be mechanically secure (without stressing the component) and to make a good electrical connection. The 'patch and pigtail' or 'snip and splice' method is excellent for replacing delicate components in old and sometimes fragile equipment. Attempting to unsolder component wires bunched together on a terminal strip is ill-advised; it stresses all else connected to that terminal both mechanically and thermally, something that old components don't like at all. It is easy to underestimate how brittle and fragile an old tag-strip may be. It is also perilously easy to break off a lug or melt and destroy a plastic part in the quest for perfect appearance.

On the other hand, some enthusiasts go so far as to split the cardboard tubes on old waxed-paper capacitors and conceal a modern component inside. The ends of the tube can then be sealed using wax to make it look like the original. Candle wax will do but toilet bowl sealing wax is particularly good since it is soft enough to mould and is almost the same colour as the original.

The last word goes to collector Al Klase, who says that the components in classic electronic gear were generally installed in layers, meaning that the capacitor leads were wrapped around the terminal lugs before the associated hook-up wire. Some enthusiasts may have the patience and ability to take these joints apart and reassemble them to make a reasonable facsimile of the original, but most people will end up with a lot of burned insulation and frayed ends of stranded wire. He adds that there are a lot of joints which cannot be reached unless one strips the whole receiver. He believes the craftsmanship of the original assembly is preserved best by not interfering with it.

VALVES

Taking active components first, let's start with valves or vacuum tubes as they are called Stateside. In several other languages they are called the same word as lamps, such as in French (where 'valve' has the particular meaning of a rectifier valve).

The first thing to establish is whether a valve is really faulty. In his excellent book *Tube Lore*, Ludwell Sibley urges 'Think twice before replacing!'. Despite the fact that books and advertisements used to urge you to bring your valves into a dealer for testing, it's seldom the valves that cause problems. Routine replacements of valves may have made sense in mission-critical commercial operation but in most old sets the valves are quite likely to be usable (audio power valves being the chief exception). As Sibley says, corroded-open inductors, leaky or dried-out capacitors, partly open resistors and similar troubles are far more likely to be present – and it's a rare set that has only one trouble.

Sometimes of course a new valve will perk up a set no end and if you have one to hand, there's no harm in trying it. Its higher gain may of course throw up new faults elsewhere in the circuit too! But, in general, before replacing a tube, consider the following factors.

Wrong valve. Sometimes the valve is not faulty, its position is. Don't trust that each valve in a socket is the correct one. Check it against the schematic. If you don't have a manual (and even if you do) note which socket the tube came from so that you can replace it correctly.

Cleaning. Don't clean valves with water. The marking on the glass will likely come off with one good swipe of a water-dampened cloth.

Missing markings. If you do find a valve with no markings all is not lost. An easy solution is to put it in the refrigerator for a few minutes, then breathe on it and condense the water on the valve to reveal the number. Do not be tempted to use freezing lotion; the temperature could drop so rapidly that it breaks the glass envelope through thermal stress. Another tip is that regular household ammonia will temporarily bring up faded tube markings.

Rattles. Bits of loose cement in the base have no effect on the operation of the valve. Likewise with glass particles inside the valve; as long as the vacuum is good, small pieces of glass are harmless. Loose metal parts like getters are more serious; there is a risk of their falling into other elements and causing a short circuit. Occasionally a large filament-type tube such as a 5U4 or 213 has rectangular white flakes of filament coating loose in the bulb. This is suspicious, and calls for a test of emission.

Visual indications. Newcomers are sometimes concerned about the appearance of their valves, both cold and when switched on. The clearest warning sign and the only one to be concerned about is a milky white coating inside the glass envelope of the valve; this indicates the vacuum has been lost and the tube is useless (normally because the glass is fractured). A silvery patch is normal and is called the getter. Some valves have the inside of the envelope coated with what appears to be a dusting of graphite; to quote the *Radiotron Designer's Handbook*: 'The inside surfaces of glass bulbs are frequently blackened. This has the effect of making them more or less conductive, thereby reducing the tendency to develop static charges, and reducing the tendency towards secondary emission from the bulb'.

Blue glow. Tubes filled with gas (0C3, 074G, 2A4G etc.) or mercury vapour (82, 83 etc.) glow in vivid colours in normal use; lack of glow indicates failure. Some healthy power tubes (6CA7/EL34 etc.) show a small area of blue inside the glass at the top, the result of stray electrons striking the glass. However, blue glow *inside the elements* of a vacuum-type tube indicates a gassy tube which had better be replaced.

One side bad. Duodiode-triodes (75, 6Q7, 12AT6, etc.) often show weak emission on one or both diodes. This diode is only a detector; it has not worked hard, and most radios have the two diodes wired in parallel anyway. Here is a fine case of when not to replace unless a new tube produces improved volume.

Finally, Sibley warns about misinterpreting valve tester results. People naturally like to delegate hard 'go/no-go' decisions to a machine, but a tube tester is simply a guide to the usefulness of a tube, not a hard and fast judge. Moreover, tests have shown that a valve passed as okay by one tester was declared close to the end of its working life by another. Are you sure the valve tester is faultless? As far back as 1934, the RCA warned valve tester users: 'The tube tester cannot be looked upon as a final authority in determining whether or not a tube is satisfactory. An actual operating test in the equipment in which the tube is to be used will give the best possible indication of a tube's worth'.

Hank van Cleef confirms this advice, remarking: 'When I was a teenager, I used to make pin-money fixing "dogs" that various radio shops sent me. About a third of these problems were tube problems, with tubes that tested "just fine" on a tube tester. The place where a tube tester just plain doesn't smoke out the trouble is with RF oscillators. And 6A7, 6A8, 6SA7, 6BE6, 6C4, 955 and 9002 are just a few that I've replaced with a known good tube to solve the problem. There are also a variety of balanced circuits, such as scope vertical amplifiers, where all the tests in the world don't tell you what you find out when you put the tube in the circuit – that it upsets the balance. Another area where tube testers could be counted on to sell the customer a new tube were the larger audio amplifier tubes, particularly 6V6, 6F6 and 6L6. The testers would claim the tubes were low, but in the circuit, they worked fine, right up to full output. And a tube tester will never tell you about one fault in power tubes that a VTVM in-circuit test shows instantaneously: excessive grid gas current'.

Therefore, you need not feel inferior if your workshop doesn't run to a valve tester. As Hank says, 'Tube testers are great – if you have a repair shop and want to sell lots of tubes. But if you want to fix things, they rank below a Q-meter or an impedance bridge on my bench. Both of those instruments are unbeatable for some things, but I never saw a tube tester that would really tell me anything I couldn't find out in a few minutes with the tube in the circuit and scope and Simpson (multimeter) diagnostics'.

Loose base. A valve may be loose from its base. This has no effect on performance, but there is some risk of the bulb twisting around enough for the wires to short-circuit in the base. Many suggestions have been made for refastening a loose brass or bakelite base to a valve's glass envelope. Most involved a celluloid cement or, in recent years, epoxy adhesives and cyanoacrylates ('superglue'). Even white woodworking glue has been seen. All such attempts lead to unsatisfactorily sloppy results and eventually the base again comes loose.

There is a much better method. First, clean away all residue of any previous cement – scraping the glass envelope with a knife is acceptable. If a celluloid type of cement is on the base, it will usually peel off. Epoxy can be cleaned off the base with acetone. With a large darning needle, scrape the cement junction so that any residue of a previous attempt is removed.

Dilute clear fingernail lacquer with acetone: one part lacquer to two parts acetone, Be sure it is thoroughly mixed. Using the small brush in the nail lacquer bottle, carefully apply the thinned mixture at the joint between base and envelope. It will immediately soak into the original cement. Continue application all around the base. Don't be afraid of using too much, although six or seven brush-loads, evenly spaced around the base, are usually sufficient. Stand the tube (valve) vertically and allow the lacquer to set for at least twenty-four hours. It will be found that the base adheres firmly to the envelope.

If any thinned lacquer inadvertently runs on to the base, let it dry. It can be removed later with acetone. Any dried lacquer on the glass envelope can be removed by judicious scraping.

This method has been used for over four decades and never once did it have to be repeated. Electrical qualities of the solidified lacquer seemingly are good enough that even high-voltage rectifiers suffer no impairment. Interelectrode capacitances are not altered.

An alternative, 'vintage' repair was found by Al Klase in an old wireless magazine. Wrap three or four turns of string around the juncture of the glass and the tube base, and tie tightly (this is a good place for a surgeon's knot if you know it). Coat the string liberally with glue and wipe off the excess. I use old-fashioned cotton kite string and Duco cement (UHU and other cellulose cements are similar). Just wrapping with the string is usually enough to eliminate the wiggles and the cement makes it permanent. This, of course, isn't an invisible repair, but it works well and has a genuine old-time appearance.

Valve pins. All valve pins are numbered according to one easily remembered rule. When viewing a valve socket from the underside of the chassis, pins are numbered in the clockwise direction starting from the key or larger area of blank space.

In equipment that has stood idle for a period the valve pins may have become corroded. Remove each valve in turn, squirt switch cleaner into the socket and run the valve pins in and out of the socket a few times. This should disperse any green growth on the pins.

Certain patterns of valve holder are notorious for losing their grip, so if in doubt, go around all of them with a pair of long-nose pliers and pinch *every* contact up. You'll be glad you did.

Recycled valves. Experience alone will tell whether junk-box valves are reliable performers; if in doubt substitute several examples for best performance. In critical applications nothing but the best will do.

Substitution. Many of the valves used in old sets can be replaced with another type (such as a 5Z3 for a type 80) without any rewiring, but a substitution manual must be consulted first. Your public library or antique radio supplier will certainly have them. If you expect to do much servicing, you should purchase your own copy! Bear in mind that many valves had one or more European number and then yet another number in the USA. You may order a valve under one number and receive a box with a totally different (but still correct) number. Then again, military and Post Office valves have their own unique numbering schemes and a good data book really comes into its own here.

In general valves don't just curl up and die (they may glow cherry red first!) but they do lose emission, becoming less efficient in the process. Substituting a new valve may do the trick but it may also introduce new problems if other lowish-gain valves are not replaced as well. If a new valve is not available it may be worth substituting another of the same type from a different location in the circuit; this other valve may have higher gain whereas the first valve may perform adequately in the other, less critical, location.

Removing hot valves. If you want something more sophisticated than a rag for removing hot valves from their sockets, there are proper purpose-designed tools available from Antique Electronic Supply. Sparking plug removers, available from car spares places, will also do the job.

Secret solid-state replacements. For really early valves it may be impossible to find replacements and to remedy this situation an American company has devised what it calls 'vacuum tube repair wafers'. These ingenious gadgets use solid-state devices inside a thin wafer, which is installed simply by sliding it over the valve pins and reinserting the valve in its socket. If the filament is intact but there is little or no emission, it may be left to glow for best visual appearance. On the other hand, if the heater is open circuit, this is not a problem. No change in operating voltages is required and any number of valves in a set may be repaired in this way. If wafers are installed on all valves the HT may usually be reduced without sacrificing volume or sensitivity. Performance will vary between radio models.

The wafers cost around ten dollars each and there are four models, designated 01A, UX199, UV199 and WD11. The first two will also substitute for several other valves, as the supplier's information explains. These wafers are supplied by PTI, 7925 Mabelvale Cutoff, Mabelvale, AR 72103, USA (telephone 00 1 501-568 1995).

TRANSISTORS

Most people harbour a notion that transistors are reliable, stable devices which, if not mistreated, will last for ever. In the main this is so but there are two major exceptions to the rule.

One concerns plastic signal transistors made around 1970. Some of these have become hygroscopic and have absorbed sufficient atmospheric moisture to cause serious problems. The only solution is to replace them by metal-cased devices (which are hermetically sealed). Plastic devices were never used in products made to military specification; for these metal-cased transistor and ceramic-cased integrated circuits were always specified. 'Mil. spec.' devices are more expensive, however, and the plastic transistors were considered good enough for products aimed at consumer and low-end professional markets.

The second 'disease' affecting transistors is the one which afflicts certain early germanium devices. Dead, intermittent or microphonic AF117 germanium transistors are well known amongst those who repair radios of the 1960s period. These are of the same mechanical construction as types AF114, AF115, AF116, OC170 and OC171, which are found in televisions. The cause of death or fault mechanism is rather interesting.

Were you to open it up, you would find that the encapsulation surrounding the transistor contains silicon grease and an air space inside the metal can. From the inside wall of the can in the air space grow microscopic hairs (0.008 mm across) of an unidentified medium which is tough, springy and electrically conductive. After some twenty-five to thirty years these conductive hairs reach the internal lead construction, giving the fault symptoms described. It is not clear if the air space is part of the design intent or is in fact a process error.

Remedies can be effected of course and the simplest one is to replace the faulty device with a stock part of the same type. This cure will be short-lived, however, because the conductive hairs will already be growing within this 'new' item, even if it is unused. The only permanent repair may be replacement with another type with similar electrical characteristics, which goes against the grain for anyone who values originality. A near-original repair is to snip the screen lead, thus isolating the can, but this may cause instability and presumably it is only a matter of time before a second internal lead becomes shorted to the can with the same symptoms.

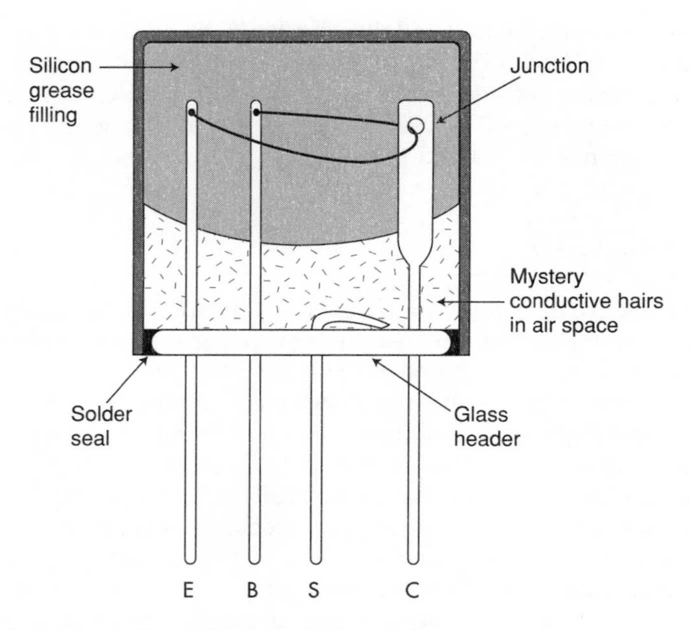

Figure 13.1 Section of AF117- or OC170-type device.

The truly adventurous reader may take action of an experimental nature and there are two courses which have been suggested (but not yet evaluated). One is to blow the short away electrically by discharging a capacitor of 50 microfarads charged to, say, 50 volts connected between the can and the offending lead. The other is to unsolder the can seal, withdraw the offending lead, clean the can interior, refill with silicon grease and reseal.

CAPACITORS

Put five radio restorers in a room and ask them about replacing capacitors; you'll get at least five different opinions on the subject. Let's take them one by one.

The purists will tell you to change as few as possible; this keeps the set as original as possible, even if performance is impaired slightly. 'If it ain't broke, then don't fix it' is their philosophy. The fanatics will tell you it's all right to change capacitors so long as you use new (i.e. not previously used) old stock or else conceal the modern replacement in the shell of the old capacitor (not always easily done). The 'easy life' brigade say yes, change everything in sight in order to minimize any future failures; modern components are extremely reliable.

My own philosophy is a bit of a compromise; so long as the set looks authentic externally, I don't mind what components are lurking inside. I have no problem using modern replacements either; after all if a pre-war set needed replacements in 1960, nobody decries these newer components. What if it underwent further repairs in 1975 or 1985 or 1995 – they are less than thirty years old but does that make them inferior? I don't think so.

This does not mean that you are obliged to change every capacitor in sight; half the skill is in determining which really need changing and which do not. Experience will teach you which capacitors in an old set you need *not* replace but it always makes good sense to change capacitors, since these often go leaky or change value over the years.

Catastrophic component failure is unpleasant and can be dangerous. Small capacitors can act as fire crackers whilst electrolytics can behave like bombs, penetrating the top of wooden cabinets and spraying noxious chemicals over the room (cleaning capacitor muck off the ceiling is a thankless task!). The normal causes are damp (in the case of non-polarized capacitors) and reverse polarity or over-voltage (in the case of electrolytics).

Working voltages. You should never replace a capacitor with one rated below the working voltage of the circuit. It may appear to work (for a while) but will always be a weak point and could lead to expensive damage when it does eventually fail. Using one with a DC working voltage significantly higher than the actual voltage is not a good idea either; don't stick 250 V electrolytics in 5 V circuits since the feeble 5 V potential will not 'exercise' or polarize the capacitor properly.

Value drift. Small capacitors can be notoriously temperature sensitive; when they have warmed up, they change value or break down. Coupling capacitors in valve circuits are prime examples. You can often provoke a capacitor to change value by directing the spout of a freezer aerosol at it, or with mica and ceramic caps, placing the tip of a soldering iron nearby. Any change in performance indicates a faulty component.

Most electrolytic capacitors used in power supply circuits can stay unless they are demonstrably faulty. Signs to check are bulging rubber caps, leaking electrolyte fluid, warm feel to the touch, high current leakage, excessive hum – 50 cycles on AC/DC receivers, 100 cycles on AC transformer types, fuse blowing, instability, distortion, nasty smell, short life of rectifier etc.; they are dangerous in this state.

Replacements, particularly the high-voltage 450 V ones used in televisions, can be quite pricey but, on the other hand, a good set of new capacitors will probably add life to a set. Whilst it's true that even the oldest examples can sometimes be brought back to specification by reforming, this is a technique for experts only. Never compromise on working voltage either; if the circuit says 300 V working, you will not get away with 250 V types in the long run. Electrolytics explode with terrific force when stressed; they can burst through cabinets and redecorate your ceiling with remarkable effect. They are not called smoothing bombs without reason.

Smaller electrolytics (fault symptoms: instability, low gain and distortion) are also used in signal circuits and these tend to dry up and go open circuit. They are cheap to replace and should be changed more or less on sight.

Any capacitors with paper, cardboard or waxed-paper cases will definitely be suspect and should be replaced with plastic film types. Any forty-year-old caps looking like brown or black mint humbugs or caramel sweets (Hunts and Wima brand in particular) are bad news too; they were fine originally but now suffer invariably from moisture entering via microscopic cracks. Testing with a multimeter is deceptive; they will often appear to hold capacity and block direct current. At low voltages they will be fine but under stress of 150 V DC they will leak electrically like anything, acting as a capacitor with a resistor in parallel and possibly causing valves to conduct fully and glow cherry red. Sometimes they heat up physically as they start to break down; they then melt and drip wax on the chassis or on other components.

Typical fault symptoms are:

- Distortion and hum at AF at all levels of volume control means leaky output valve grid feed from first AF, passing positive DC from previous anode, wrecking the negative grid bias and driving the output valve anode current up, causing huge audio distortion and often destroying the output valve, rectifier valve, smoothing caps, mains transformer, loudspeaker field/choke or HT feed resistor in the process if ignored.
- Distortion at RF: distortion only found on strong stations means leaky decoupling capacitor(s) on AVC line, not allowing good AVC swing voltage to variable-mu IF stages.
- Low gain at RF means leaky screen grid decoupling capacitor/s, seriously reducing screen grid voltages to RF stages. Often one capacitor decouples an IF and mixer stage's screen grids.
- Low gain at AF means leaky capacitor feeding anode circuit to AF stage or screen grid if pentode used.

To quote a seasoned restorer: 'In our experience faulty wax/paper caps are the biggest source of faults in valve radios. They will usually still have their capacitance but also will have become resistors passing DC to places there shouldn't be or consuming valuable HT current for themselves – you can almost guarantee that 90 per cent of them will be faulty in your receiver'.

Replacement need not be expensive if you buy the bulk packs of surplus new capacitors that are always on sale at radio swap-meets. Buying capacitors in twos and threes from Maplin, Tandy or Radiospares is not a cheap exercise, though.

In areas such as cathode bias they may be unstressed and fairly innocuous; I tend to leave them, particularly if they are hard to reach, but other restorers will change them as a matter of course. When it comes to decoupling capacitors in AGC and AVC circuits, you will almost certainly need to change them; even the slightest leakage here can lead to RF overload and severe distortion.

The cure is replacement with modern polyester film capacitors (400 V DC working) except in mains filters and output tone correction where 1000 V polypropylene caps are called for. Any capacitor connected directly to AC mains should be X-rated, a marking that signifies it is suitable for this application.

Don't replace ceramic capacitors in RF and IF amplifier stages unless they are physically broken. They rarely deteriorate and their inductance and temperature coefficients are matched to the circuit and substitutes may have different characteristics.

Many people assert that mica capacitors in RF circuits never break down – but they can go open circuit on occasion, preventing you from aligning a stage properly. In some cases they can change value with temperature or with volts applied, causing a set to drift off tune. The same applies to moulded mica capacitors, which sometimes become resistive or change value; the cause is put down to chemical change in the potting material or to stress pressure on the mica as the moulded case contracts.

CAN'T FIND A REPLACEMENT

Finding an exact replacement can be (or appear to be) a problem. It may be that you cannot find a replacement of the same size, pattern or value. Taking these in order . . .

Capacitors have become smaller over the years, particularly electrolytics. This should not concern you unless the leads of the new component no longer reach the solder tags of the old one, in which case you'll have to solder on extensions and sleeve over the joints. With round-can electrolytics, the modern capacitor may be slimmer; all you can do is make a band of insulating tape to increase its girth and make a tight fit in the original clip. Either that or fit a smaller clip.

Certain categories of capacitor are no longer made; waxed-paper, oil-filled and Black Beauty/humbug for example. For all these polyester foil or 'plastic' capacitors will work fine.

The actual values of old capacitors are not always those of today but this is not a problem really; many of the original electrolytic caps quoted their values as minus 20 per cent/plus 100 per cent, leaving you a conveniently wide margin. Again you could argue that competent designers were aware of the way component values altered with age and arranged circuitry accordingly. It may pain us but these old components were not precision devices!

Accordingly a 4 μF capacitor rated at 400 V can easily be replaced by a 4.7 or even an 8 μF one. It is probably a power supply filter capacitor and whilst the 8 μF would make the power supply run at a higher voltage, our radio will not be over-stressed so long as we don't use it day in, day out. You need to exercise greater care over like-for-like replacement with certain power supplies, however. The main electrolytic input filter capacitor in power supplies with a capacitor input filter should not be replaced by a significantly larger value, otherwise it could cause excessive current to be drawn from the rectifiers. In some tuned circuit applications a higher value capacitor could change the time constant or frequency response of the circuit too.

Remember too that you can use capacitors in parallel to make up a value. Two 2 μF capacitors in parallel make a handy 4 μF cap.

Finally, many old paper capacitors appear to be marked as if they are polarized. The marking actually denotes which wire is connected to the outermost layer of foil. For bypass capacitors, this end is to be grounded, minimizing stray pick-up by the capacitor. Modern mylar capacitors make fine replacements – and work well connected either way round.

OLD ELECTROLYTICS – TO REFORM OR NOT TO REFORM

Electrolytic capacitors need to be treated with some care and understanding. Crudely put, an electrolytic capacitor has two plates (made of aluminium foil) separated by a liquid (jelly) called the electrolyte. What actually makes it work is a layer of aluminium oxide (alumina), which grows on the anode foil (or 'plus' side). Because alumina is a good insulator, this gives the capacitor its ability to withstand voltage and have a relatively high insulation resistance. So long as a voltage is applied, this dielectric coating on the insulated plate is maintained.

On the other hand, when a capacitor is not properly polarized (either through storage, disuse or even insufficient applied voltage/wrong voltage rating) the insulating layer deteriorates slowly. After many years, the capacitor will have a low resistance. When first repolarized, it will not block AC (and hence hum) so well and excessive leakage current will be noted. Another cause of failure is operating at high temperatures, which is why switch-mode power supplies in modern appliances such as fax machines and satellite receivers so frequently fail – the electrolyte in the caps is literally boiled away by the surrounding heat (special high-temperature capacitors are now made for replacement in these circumstances).

In fact when a capacitor is reforming and has high leakage it will heat up, just as any other type of resistor would do. If the electrolyte becomes too hot, it will boil, producing gas pressure in the capacitor. The pressure may either cause irreversible damage (leading to failure in the circuit) or the capacitor may rupture or even explode with sufficient force to shatter cabinets.

The oxide film which enables electrolytics to perform properly is normally 'formed' before the foil inside the capacitor is wound, and after the capacitor is assembled, it is generally 're-formed' to repair any minor damage in the manufacturing process. Old electrolytics can in fact be reformed to operate like new – or almost like new – in most cases. Experts say that as long as the electrolyte inside the capacitor is not excessively dried out, a reformed electrolytic capacitor will perform almost as new, although if it continues to show excessive leakage current after reforming, you should throw it away.

But why bother? Reforming is a slow process and people who repair old radios for a living assert that the game is not worth the candle; spending hours on a capacitor which may still perform indifferently is pointless. If a capacitor does reform sufficiently to operate normally there is no guarantee that it will continue to operate for a long time. Old capacitors also present a small but real safety risk; the new product is guaranteed fresh and has better venting to prevent pressure build-up. In fact the only reason for retaining old electrolytics is either for total authenticity's sake or when funds are desperately tight. In this author's opinion, scrimping and cheapskate tactics are incompatible with safety, which is why no circuit for reforming is given here.

RESISTORS

Charred resistors are another obvious candidate for replacement, and any high-value resistors (say 1 mega-ohm or more) should be checked in case they have increased in value or gone open circuit. Some 'dead' resistors show no sign of distress, however, so it is essential to test each component with a meter. Bear in mind that over-stressed resistors can change colour in a most deceptive fashion – red and orange can turn brown, whilst the difference between grey, green and blue can be slight. Certain other resistive components are also worth replacing on sight, without bothering to test them. So in televisions, check the cathode bias resistor of the field output stage, and at the same time replace any cathode decoupling capacitor.

A problem with modern resistors is that, although they are generally more reliable and more closely matched to their nominal value than older examples, they are designed for use in low-voltage circuits. Some may not stand up well to voltages above say 150 V and become completely non-linear with applied voltage over about 200 V. The resistance can rise quite dramatically, around 25–30 per cent, with 1 kV applied. For this reason unused old stock components may be a better bet for critical high-voltage applications, otherwise you could build a string of resistors (on a tag strip) which limits the voltage on each resistor to about 200 V.

When a non-standard value of resistance is called for, just use Ohm's law to calculate the value, and remember that you can 'tailor-in' a non-standard value by using two resistors in parallel. One resistance slightly above the desired value, and a large resistance in parallel to 'pad down' the result, works best.

Why do old resistors change value, by the way? Most of the ones found in old radio and television sets are of the carbon composition type; these absorb water over time and this process apparently causes their resistance to go up. The larger value resistors are more susceptible to this problem. This water absorption problem is still present with modern carbon composition resistors and in high-performance applications metal film resistors are specified.

Identifying component values is seldom a problem. Some resistors have their value printed on them, whilst others use colour codes (see the chapter 'Data and more'). Some companies (such as Atwater Kent) used a different colour for each value of resistor but more commonly a separate colour was assigned to each numerical digit and the total value indicated by a combination of colours. Originally these colours were applied in a 'body-tip-spot' or 'body-end-dot' arrangement, in which the colour of the resistor body represented the first significant digit, one tip or end was marked with the colour of the second digit, and a spot or dot in the centre of the body was the multiplier. Starting in the late 1930s, resistors were marked with colour bands, the first significant digit being the band closest to one end. Silver indicated 10 per cent tolerance whilst gold denoted 5 per cent; these 'noble metal' markings were applied to the 'other' end of body-tip-spot resistors or as a fourth band on band-marked resistors.

A point which sometimes confuses newcomers is that resistors have become physically smaller over the years! More correctly stated, the wattage ratings of resistors in their different package sizes have been revised, the rating for a given package size being increased. When replacing resistors, modern 1-watt metal film resistors generally are about the right physical dimensions for older quarter-, third- and half-watt resistors. It is prudent to derate values by 50 per cent, so a modern 1-watt resistor can conveniently substitute for one with a dissipation of half a watt or less and also look correct.

The actual values of resistors has changed over the years; in older times the resistances available went in 'obvious' steps such as 1000, 1500, 2000, 2500 ohms, and so on, whereas the current range offers 1200, 1800, 2700, 3300, 3900 and so on. This change to so-called 'preferred values' was made with good reason, since the 'new' values increase logarithmically in increments of about 50 per cent and allow the complete possible range of resistances to be achieved with the smallest number of resistor values. Similar preferred values are also used for capacitors.

Substituting modern resistor values for old ones is generally no problem – you merely choose the closest value, so for 4500 ohms you would substitute 4700 and for 250 ohms you would use 220 ohms. Bearing in mind that most circuits were designed for components with 20 per cent tolerance, the apparent mismatch is seldom of consequence. This also applies to replacing resistors whose values have altered; a drift of up to 20 per cent is seldom worth worrying about.

A list of preferred values and colour code systems is given in the chapter 'Data and more'. See also the notes on unfamiliar resistor symbols and notation in the chapter 'Service data'.

POTENTIOMETERS

'Noisy' volume and brightness controls have always been a dilemma. Certain wire-wound types are inherently so, as are other types when used in some DC circuits.

Where skeleton preset controls are used, a replacement is the only reliable cure. Unfortunately, the pattern, type, spindle length, combination and values of user controls on elderly equipment are often impossible to duplicate. Particularly where volume controls are concerned, cleaning is very often effective. Even very modern equipment using various types of potentiometers does not escape the problem. It's therefore probable that correct cleaning where practicable could extend the life of a control almost indefinitely.

THE CAUSE

According to Malcolm Burrell, the carbon track is largely blameless for most problems of noisiness. The only exception was the 'printed' tracks used for presets associated with varicap tuners; this type was virtually unrepairable. The cause of most problems seems to be oxidation of the metal wiper and contact ring which connect to the centre tag of the control. When lubricant is squirted into the housing the entire assembly is saturated and may remain noise-free until the lubricant evaporates. However, dismantling the control, scraping both wiper and contact ring then smearing with silicon grease usually promotes a lasting cure.

DISMANTLING

Disconnect the receiver from the mains supply. When dismantling a control to which a mains switch is coupled, ensure this is in the 'on' position, otherwise difficulty in latching will be encountered upon reassembly.

First unbolt the control from the chassis then gently prise open the tabs holding the slider and connecting tags to the body. This part of the assembly is removed for cleaning. In most cases access to the wiper and contact ring is now facilitated.

In particular, note the method of assembling concentric controls. Proper switch lubricant is recommended. People talk about using light domestic lubricating oil, penetrating oil, WD-40 and even a few drops of car engine oil, but these are insulators by nature. Real switch lubricant is always best.

Scott Robinson has used silver conductive paint for restoring connections on pots; he suggests it could also be used to create taps on untapped pots. The paint is available from RS, Electro-Mail and Maplin in the UK and from GC in the USA; the cost of even a tiny bottle is high but it will fix a large number of pots. A viable alternative is conductive epoxy glue for repairing a potentiometer with a bad connection between one end and the resistive material. The epoxy does not become conductive until it hardens but after hardening and curing works very well. The product contains finely ground metallic silver and has a low resistance; inevitably it is also quite pricey.

SLIDER PRESETS

A feature of some television sets is the panel of slider-type wire-wound presets at the back. These, to quote Mike Izycky, are festering, misbegotten heaps of junk, so clean and regrease with a little Vaseline the metal guides and the surface of the track. Do not be surprised if the tracks are open circuits! The wipers disintegrate too for added value!

RECTIFIERS

More arguments arise over rectifiers than most other subjects – to replace or not? Some restorers feel obliged to replace valve rectifiers with solid-state ones to increase reliability and reduce heat dissipation at the same time … but considered opinion is against this. Original replacements are still readily available; they are not as prone to catastrophic failure as silicon diodes are, neither do they involve a non-authentic modification. Silicon rectifier diodes, on the other hand, need to be of at least 400 V PIV rating and 'stacked' using equalizing resistors and bypass capacitors to form each leg of a full-wave rectifier. Even then, one or more of the diodes in one of the stacks often shorts, eventually overloading the others and causing them to fail in the same manner. What's more, the silicon diode modification causes the old electrolytics to take one devil of a hit whenever you power up, with no chance to accomplish any reforming as the DC voltage comes up on them gradually as would occur with a valve rectifier. The premise is that this creates stress in the set's tubes because the HT is present almost instantly, some time before the set's tubes have a chance to heat up and start drawing current, and that the application of HT to a tube with cold filaments is somehow damaging to the emission capability of the cathode (or filament).

On the other hand, James Binkley points out that the same condition occurs in some valve gear, where the HT rectifier is a directly heated filament type of tube and heats up quite a bit faster than the rest of the tubes which are indirectly heated cathode types. Thus the HT comes up quite rapidly. This rapid rise is aided by the fact that the supply is loaded very lightly (due to the cold filaments) if at all at this time, and thus the rectifier is not called upon to carry very much current and the filter capacitors quickly charge up to the peak voltage. He continues: 'I observed this with an oscilloscope, so the inertia of a meter is not a factor. The HT in my set comes up in two or three seconds to +375 V, sits there for two to three seconds and then gradually drops to the nominal +250 V'. Of course this is why some professional equipment has sequenced power supply relays, applying power to different sections in turn with delays.

Moving on to selenium rectifiers; these, on the other hand, tend to be replaced as standard practice by seasoned restorers who leave the original selenium stacks in place for an authentic look, but disconnected. The reliability of selenium rectifiers is not very high and they are prone to spectacular failure. Modern silicon diodes are more reliable and generate so little heat in comparison, which means they may often be hidden in the circuitry replacing the selenium units while not being noticed.

Barry Ornitz states that if you have the schematic and know the voltages, choosing the proper silicon diode is easy. With no other information, a reasonable rule of thumb is that a single plate handles about 25–35 V. Thus the typical stack for rectifying something off the power line would be five plates (117 V countries) or ten plates (230 V). Another 'eyeball' approximation is that the area of each plate in the stack is about a square inch per 300 milliamps. For most applications where small stacks are used, common 1 amp diodes are fine. Remember that selenium rectifiers had a greater forward voltage drop with current than silicon diodes. Thus when you replace a selenium diode, expect a higher voltage out of your circuit; a series resistor of 15–20 ohms should be added if necessary to drop this voltage.

Incidentally, testing selenium rectifiers can give misleading results because most modern ohmmeters use only 1.5 V, which is insufficient to forward bias most seleniums for accurate indication. Hank van Cleef suggests using a 9 V battery in series with the selenium and voltmeter probes. Using a 20 kilo-ohm/volt meter on a 15 V range, you are looking at 300 K in the loop, which is quite high, but you should see a definite difference between forward and reverse bias. You will also want to take a good look at the filter network capacitors and voltage-dropping resistor. Rectifiers do not fail without reason and excess voltage or excess current is generally the cause of premature failure. If the selenium smells like rotten eggs, you have a bad selenium and the products of a cooked selenium rectifier are poisonous. Do not handle the unit with your bare hands or breathe the fumes.

COIL FORMERS, CANS AND TRANSFORMERS

These can be a source of grief when things seize up. The cores inside coils are frequently sealed with beeswax (used because it does not 'run' as paraffin wax does at high temperatures). The recommended way to free up stuck slugs is to use a very fine-pointed soldering iron (or a paper-clip fixed to the bit) to heat the slug and melt the wax. Never try to force slugs (or use the wrong adjustment tool); once the slug breaks, removing it is almost impossible without destroying the coil totally.

Stuck IF cans and glued coils can be loosened with acetone, sold as carburettor cleaner and fingernail polish remover.

Many mains transformers are covered in pitch. This is done because the laminations have a propensity to vibrate mechanically, whereas the pitch reduces or eliminates the noise caused by 'lamination buzz'. A secondary purpose is to reduce the risk of moisture getting in and causing shorted turns. Finally, because pitch has a much greater thermal conductivity than the air space between the transformer and case, it also improves the overall thermal conductivity between the heat generators and the case.

Beware of employing transformers intended for 60 Hz use on 50 Hz supplies; because of the different impedance they can cause excessive current when used on 50 Hz and end up in smoke. This does not apply to transformers specifically designed for dual-frequency operation.

RELAYS AND SWITCH CONTACTS

It is actually quite rare for contacts to be bent out of alignment, and then a light touch with a pair of snipe-nose pliers should be sufficient to tension the springs back into place. Dirty contacts can be cleaned in two ways, either by passing a piece of hard paper (e.g. writing paper) between the contacts under pressure or by using switch cleaner fluid. Do not, repeat not, be tempted to use emery cloth or other abrasives; they will be too harsh. Professional tool catalogues do show cleaning tools using diamond dust or chamois leather pads on spatulas but these are expensive (unless you have access to these tools at work). In any case the latter can all too easily remove any plating or noble metal contacts and are taboo in many quarters. In telephone practice nothing more than paper is used to clean relay contacts.

CLEANING SWITCHES

Wafer switches and more enclosed switches give problems much more frequently – intermittent electrical connection or noisy contact – but a quick application of good switch cleaner fluid usually works wonders. There are several varieties, the best seeming to be a highly volatile solvent with a bit of petroleum jelly dissolved in it. The difference between good and bad switch cleaner is whether you have to use it again in six months' time or not for years. Cheap products are little more than a fast-evaporating solvent, whereas the good ones create an antioxidant coating so that the benefit lasts. The best-regarded product in the USA is Caig Deoxit D5 and this is now available in the UK from Probus Electronics (see 'Buyer's guide').

There are plenty of fluid switch cleaners and these are preferable to sprays. The latter can leave behind an oily deposit that can soak into the insulating wafers of switches, causing them to lose some of their non-conductive RF insulating properties. Also some 'don't harm plastics' type cleaners can still react with some switch/tuner mechanisms, causing shafts to lock and contact surfaces to stick together. Always be cautious. Almost every 'no residue' cleaner does in fact leave a residue that can upset the small capacitances in RF circuits, causing trouble in some receiver front-ends. Except for potentiometer cleaning the spray cans are better left on the shelf; tiny amounts of lubricating cleaners can be applied to the wiping contacts of switches with a small artist's brush or Q-Tip. This keeps the material off the insulating wafers and on the metal surfaces where it belongs. A superb cleaner is anhydrous 99 per cent isopropyl alcohol, which is also inexpensive.

Sometimes the problem is that the contacts of a wafer switch are not making firm contact with the slider. A bit of a push with a screwdriver to bend them so as to make a firmer contact seems to do the trick in those cases to restore the original contact pressure.

A few switch troubles can be quite perplexing. The Zenith factory used to apply grease to the contacts of push-button switches. This hardens with age, and must be removed with a Q-Tip (cotton bud) moistened with petrol (lighter fuel is recommended) or paint thinner. Then Deoxit switch cleaner can be applied usefully. The other thing that sometimes happens is arcing over to one of the switch mounting posts. This can be cured by carefully scraping the carbonized stuff off – all of it – and recoating with lacquer or some other insulating sealer.

WIRING – INTERNAL

In many cases it will become necessary to renew the wiring in an old radio, TV, amplifier or whatever, and even this operation deserves some care and forethought. For a start, it is more 'sympathetic' to use the correct kind of wire, in other words to match the original kind of wiring, especially if the set is ever likely to be entered in a *concours d'élégance* contest. If the original wiring is PVC or other plastic-covered wire, you'll have no difficulty, because this wire can be had from any electrical or electronics supplier.

Older sets often used rubber wire covered in a varnished fabric and this can sometimes be had from automotive suppliers, particularly the dealers who specialize in parts for restoring classic cars. These enthusiasts have the same goals of authenticity as we do! Alternatively you may see this wire on sale at radio or classic car swap-meets, though it will take some looking for.

The thinner silk or cotton-covered wire is equally hard to find these days; swap-meets may be one source of supply and occasionally it may be worth buying a broken item just for the wire (and other useful sundries) it contains. One enthusiast pulled off a master stroke. He bought, for a song, an old GPO teleprinter, which he immediately scrapped. But it came with a prodigiously long and thick multiway flexible cable which contained many feet of individual pieces of cotton-covered flexible wire. Dipping in fabric dye turned the buff colour of the cotton to the solid reds and blacks that this enthusiast required.

At the time of writing, Antique Electronic Supply (see 'Buyer's guide' chapter) is offering by mail order three kinds of 'vintage' wire, described as follows:

- *Braid-covered power cord*, twin conductor, similar to woven cloth-covered cords in early AC radios.
- *Cloth-covered hook-up wire*, 20-gauge stranded copper wire covered with lacquered cotton braid in eight colours, closely resembles wire used in the late 1920s and 30s.
- *'Push back' cloth-covered wire*, 22 AWG, solid black colour.

Some older devices tended to use solid-section wire, with the so-called spaghetti sleeving slipped on where insulation was required. This spaghetti used to be made of a rayon material and was common in electronics hobby shops. You can still find it (or a close substitute) in some suppliers' catalogues, but if you don't strike gold, you could consider using shoe-laces stiffened with starch.

Really early radios used a square section wire made of tinned copper. In the late 1920s and early 1930s it was also bought eagerly by 00 gauge model railway enthusiasts to use as rail, since in those days more realistic materials were not readily available. Nowadays this square section wire is hard to find – unless you visit a model shop! The better-equipped hobby shops sell a variety of brass profiles and section for model boat and train enthusiasts, and you can use a square profile for radio wiring. The only snag is that it is not tinned or plated but you could arrange for a bulk load to be plated commercially (or at home) if necessary.

Tinning would be best, nickel silver second-best, or alternatively you could, with great effort, draw the section through molten solder. In all cases the brass must be scrupulously clean. Nickel silver does not take easily to brass, however, and you have to flash the brass with copper first. The copper is then plated with nickel silver and a durable finish is obtained. At the time of writing, the real thing (square tinned copper wire) is available by mail order from Antique Electronic Supply so this would be the best source of supply.

WIRING – EXTERNAL

So far we have discussed only the wire you find inside a receiver, instrument or amplifier. Many sets have mains cords, whilst telephones have line cords. Let's examine each in turn.

With mains wiring, safety is a vital factor. It may well be that a pre-war set has a clean-looking mains cable and you feel it would be kindest to leave it just as it is. I sympathize. But if it has rubber insulation inside the flex, can you be sure this has not perished and is liable to cause a short circuit (and fire hazard) at any time? Age is irrelevant, remarkably. I have found pre-war rubber-insulated cables in first-class condition and thirty-year-old cables in a deplorable state.

If in any doubt, your safest course is to replace the cable, not with shiny white PVC stuff but with something more appropriate. Usually this will be a fabric-covered flex, although test equipment and some other sets were originally fitted with black rubber- or PVC-covered flex (which can be replaced by its modern equivalent). Some radios and televisions had gold-colour PVC flex in the 1950s and this is still available. The easiest fabric-covered cable to find is the so-called non-kinking stuff (their description, not mine!). This usually has a herring-bone pattern in the weave and comes in a black and white or black and yellow shade. It is really intended for electric irons and the like and doesn't look too bad on some appliances. To my eyes, it does not look quite right on radios, though (the zig-zag effect is a bit dazzling).

It is still possible to buy more conventional fabric-covered flex, both the round type and the twisted twin pattern. It's not cheap these days but it's the only stuff that looks right and if you are putting a lot of effort into restoring a set, it would be crazy to skimp at this stage. One source of supply is the Stiffkey Lamp Shop (see 'Buyer's guide' chapter) and another is the BHV department store in Paris, opposite the Hôtel de Ville. Just why this store should keep such a superb range of flexibles I don't know but its electrical and ironmongery departments in the basement are a revelation, unlike even any specialist dealer in the UK. They don't handle mail order, though, so you'll have to pay a visit in person.

For people restoring telephones this isn't so helpful; the proper telephone-type fabric-covered cords, whether braided (plain round) or plaited, are really hard to find these days. Some people have tried using switchboard cords (which *are* available but look wrong), whilst others have resorted to using football boot laces as a fabric sleeve to slip over and disguise a modern plastic-covered cord. In the USA the flatter modern type of phone cord (there called modular cord) can be bought by the foot with a brown slip-on fabric sleeving and this looks quite attractive; the idea is that you crimp a modern modular phone plug on at one end and sort out some crimp-on spade tags on the other.

The proper plaited stuff is seen on many reproduction antique telephones but is hard to find in Britain (try Telephone Lines Ltd, Cheltenham – see 'Buyer's guide'); it is, however, freely available in Australia. Anyone attempting to make up their own fabric-covered cords will have to learn how to terminate the ends of the cords. The cheat's way is to crimp spade tags on to the individual conductors inside the cord and slip a length of tight-fitting black rubber sleeving over the end of the cord to stop the cut fabric from fraying. This does not look authentic, however, and candidly the only proper way to do this is the good old traditional (and laborious!) way. The result looks really smart, though.

You will need a darning needle plus reels of mid-brown, red, white and green cotton, and also some fuse wire. It will help if you take an original phone cord as your pattern. You will see that the first job is to cut open and fray the outer braiding of the cord. This will expose the internal wires, usually three but sometimes four or six: for this instructional we'll assume three. Identify which is to be the red wire, and which will be white and green. Fray open the end of one of the wires and you will see internally it is composed of copper wires stranded with cotton or rayon. This is called tinsel wire and it is extremely flexible. The mixture of cotton makes it very difficult to solder or crimp, however.

Expose about a quarter of an inch of the inner conductors (you'll find the ideal length by trial and error) and bind this with fuse wire. Next form a

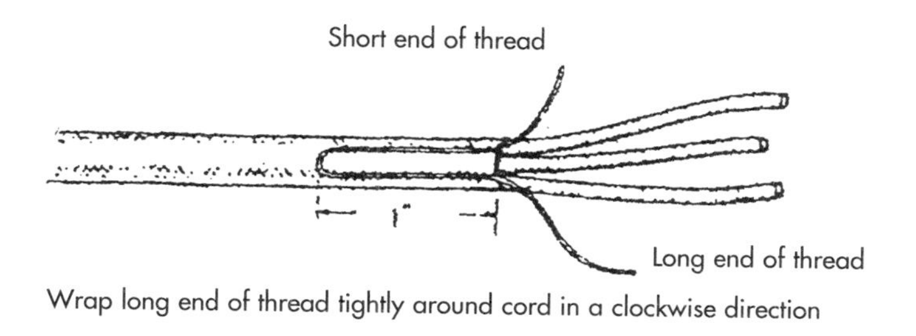

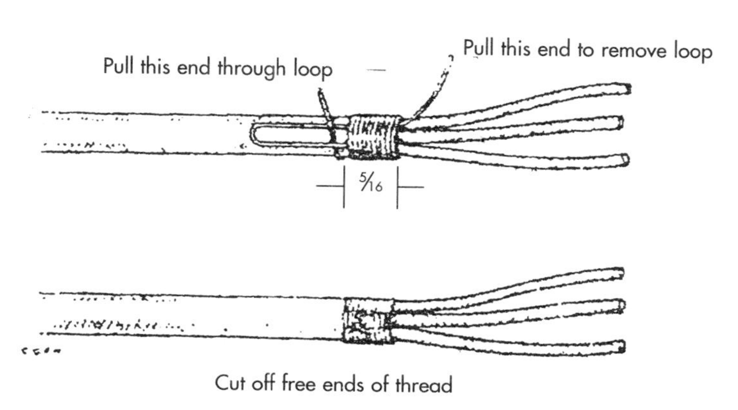

Figure 13.2 Method of whipping cords. Using thread soaked in beeswax you make a loop 1 in long, leaving a short length free. You then wrap the long end of the thread tightly around the cord for 5/16 of an inch, finally pulling the short end tight to remove the loop. This traps the long end. The unwanted free thread can now be cut off. Minor variations of this method also work, all relying on trapping the loose end under the main wrap.

loop from this wire using a nail or knitting needle of the right size with respect to your former. The idea is to go beyond the loop and bring the end parallel to the first strand to make an 'eye' (see diagram; it's easier to draw than describe). With some more fuse wire, bind the eye firmly so that it cannot come apart.

The next job is to bind the end of this cord with red cotton (or green or white, as appropriate) to stop the rest of the fabric from fraying back along the cord; ideally you'll want the proper stuff already soaked in beeswax. This job is called whipping and, using a cunning way of laying down the thread, you can stop it from undoing itself (see diagram). When all three little cords have been given eyes and whipped in this way, you'll need to use the brown thread to whip the outer braiding. Use an original cord as your pattern for copying. The final job is to melt some beeswax neatly into the whipping to seal everything up and make it look totally genuine. Now you know why the makers charge so much for these cords!

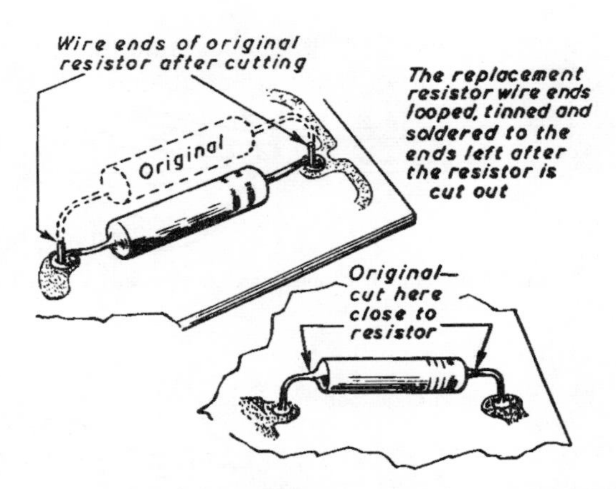

Figure 13.3 When it is inconvenient to unsolder resistors, capacitors and other small components from printed circuits, one solution is to cut out the original part and use its leads as tags for fixing the replacement. Purists consider this cheating, however.

Finally, don't be afraid to try renovating old cords if they look dirty. Flecks of paint can usually be teased off using a sharp knife. Grubby cords can be washed in a biological detergent; one which is particularly recommended is Liquid Organic Cleaner (L.O.C.), sold by Amway agents. A dip in one of these cleaners followed by normal washing in the machine is a tried and tested method. Faded cords can be dyed in cold water dyes (although the inner conductors will be spoiled by this process). Make sure all washed and dyed cords are thoroughly dried in the airing cupboard or over a stove before using again.

FUSES

Fuses are such simple things, they hardly need mentioning – or do they? In fact fuses have a lore all of their own, frequently behaving in a totally unexpected fashion.

Setting aside the ceramic fuses found in British 13-amp mains plugs, three sizes of glass-tube fuse are commonly encountered in electronic equipment. These are: the standard 'radio' fuse (size 0, 1¼ in long), the 'Continental' type (20 mm long) found in new equipment and the minifuse (size 00, ⅝ in long) unique to British apparatus and used to protect meters and other delicate apparatus. Other types are used in special applications; the telecommunications industry has its own 'specials', whilst rewirable plastic-bodied fuses are common in automotive applications.

The minutiae of fuse usage need not concern us here but some ground rules must be observed. Fuses are installed as safety devices and it pays to fuse each piece of apparatus. Some people assume that the fuse in British mains plugs is sufficient security but it is not; its function is to protect the cable, not the apparatus, and it may pass significant overcurrent (up to double the rated amount) before it operates (blows). Think about it – a shorted mains-input filter capacitor can allow a power transformer to cook and burn to a crisp in a few minutes. Even if your particular treasure never had a fuse originally, it's easy to add an in-line plastic fuse holder under the chassis, the fuse being wired in series with the 'hot' conductor of the mains input.

Always replace like with like. If a slow-blow or antisurge type was fitted originally, do the same now. The equipment manufacturer had good reason to specify a fuse type and rating, so you should naturally follow this advice (regardless of what some previous repair-man had in his tool-box). For this reason you should always inspect the fuses in newly acquired 'old' equipment; you may find all manner of horrors – not just fuses of the wrong voltage and type but also 'supply stiffeners', in other words steel nails or dead fuses wrapped in aluminium foil.

Fuses are also rated by the voltage they can pass. The usual figure is 250 V but the 'radio' fuses (10 amps and above) sold at car spares places are obviously intended for much lower voltages. If you need high-amperage fuses for 250 V you must select the ones with ceramic bodies; it's poor practice to use any device at four times its rated voltage.

In general the best practice is to use a slow-blow fuse rated for the correct voltage and for just above the normal current drawn by the device. Failing this, a fast-blow type of perhaps 150 per cent of normal current can be used but this will afford far less protection.

Why do fuses fail continually? Ludwell Sibley tells us an open fuse may hint at the cause of its 'popping'. A simple melting-open of the fuse wire means a slight or moderate overload. A black or silvery deposit inside the glass normally indicates a severe overload or short circuit. However, this is not certain; occasionally a fuse will show the full blackened deposit despite a mild overload. Fuses are not precision devices.

Old fuses can go open circuit or become weaker on account of hidden corrosion under the end-caps (caused by storage in a damp location). The result is an open fuse whose internal wire appears intact. For this reason it's worth testing newly acquired old fuses before putting them in stock; you don't need the mystery of a piece of gear remaining dead after replacing the fuse. Fuses handling heavy start-up loads – motors or transmitters, say – can develop recrystallization or metal fatigue after several thousand starts and simply give up.

Ancient equipment often has the cap or extractor part of a fuse holder missing or broken. Ludwell suggests it is wise to keep an assortment of miscellaneous extractors on hand and always salvage the extractor from a piece of gear that is to be junked. They come in a variety of incompatible styles, so the more caps you save the better the chance of finding the right one.

Finally, he reminds us that fuses can also come in not-so-obvious forms. In an AC/DC radio of 1930 vintage or later, the 'pilot' section of the heater in the 35Z5 or 35W4 rectifier provides fuse action in case the input filter capacitor shorts. Some transformer-type TV sets had the 6 V heater circuit wired through a short link of 25-gauge bare wire to get fuse action. The British 1941 field radio, Wireless Set No. 12, used 36-gauge copper wire as a 5-amp fuse, and included a one-quarter pound roll of spare wire.

LOUDSPEAKERS

We'll confine our discussion here to simple fault-finding and repairs; more complex matters are best left to the professional repairers mentioned in the 'Buyer's guide' chapter.

Minor tears in speaker cones can be repaired by brushing on a coat of varnish or rubber cement, then layering with facial tissue paper and coating each layer of paper with varnish. I usually apply at least four or more layers of tissue. If the speaker cone is brittle, it can be strengthened by painting the entire cone with rubber cement. The cement will be absorbed by the paper yet allow the paper to remain flexible.

Buzzing or rattling noises in speakers can arise when material is trapped between the rim of the cone and the frame or the speaker at the rear of the cone. Another source of rattle and distortion of sound is an off-centre voice coil that is rubbing the pole piece. To check, move the cone gently up and down, using equal pressure on both sides of the voice coil. Any drag can be felt as well as heard (if you hold the speaker close to you ear). The cure is recentring the voice coil, which on some speakers is achieved by loosening a screw holding a 'spider' assembly to the pole piece. On other speakers the spider is mounted under the cone and attached to the speaker frame with screws; these can be loosened to move the spider and adjust the voice coil. If the spider is glued to the speaker frame, this glue can be dissolved using acetone. You then recentre the voice coil and glue in place again.

Other sources of rattling noises (particularly on bass notes) include the cone coming unglued from the edges of the frame, the voice coil coming loose from the cone, the spider becoming detached from the frame, or tears in the cone. The cure is to glue the loose components once more.

CHAPTER 14

Practical restoration techniques – woodwork

Many collectors who feel entirely confident to tackle electrical repairs shy away from cabinet restoration. It's true that different techniques are involved but they are not difficult to learn and you can always develop your skill on cheap 'junker' sets or wrecks. Alternatively you can hand over cosmetic restoration to a specialist but this may turn out expensive and will not give the same satisfaction as if you had done the work yourself. This and the following chapters should give you the methods and the confidence to 'do it yourself'.

This is not to say that you will achieve perfect results immediately and this may not even be the object of your labours. Some enthusiasts do set out to achieve a showroom finish when they restore an object and this is indeed a commendable goal – if you can do it. In a way, though, this is a form of deception because everyone knows the item is not in its original state. A kinder treatment is to make the object look as if it has been looked after carefully, as if a little old lady had dusted it and polished it once a week. In the final analysis, however, it all comes down to personal preference.

WHAT ARE WE TRYING TO ACHIEVE?

Do you go for museum standard restoration? Probably not, because if you could do it already, you would not be reading this book. But if you follow these methods closely and carefully, I suggest you'll get pretty close to museum standard. It all depends how far you want to go but to me it seems daft not to try, especially if the object is valuable. The methods described are not expensive, they just take time and care. But if you can make a super job of restoration, why not go for it?

Bear in mind at all times that some subjects will be easier to repair than others. Radios, telephones and other objects with wooden cases and discrete, generic internal components will not be difficult to restore to something approaching their original finish and performance, whereas a more recent item with a plastic case and solid-state innards may result in compromises. In this respect, our treasures are like old cars; the body building techniques of a pre-1955 car are not hard to learn and its component parts can still be found with no great difficulty. Later vehicles incorporate much more specialist items and techniques and may be far harder to restore to an acceptable standard.

TEACHING GRANDMOTHER TO SUCK EGGS?

Much of what you are about to read is simple common sense and I really did think long and hard before including this chapter. Do you really have to tell someone how to hold a paintbrush, how to varnish wood, how to degrease surfaces? I agree much of this information is straightforward but perhaps it is not blindingly obvious, at least judging by the cack-handed attempts I see at many swap-meets. This, too, is why a nicely restored set stands out at these events, largely because there are so many bodgers who either don't know or don't care about achieving a professional standard. Equally, I often hear the plaint, 'Oh yes, I can do the electrical side of restoration but I'm not a handicrafts expert'.

Well, I give you the choice. If you already know the information contained in these chapters, I give you full permission to skip them and move on to something else. Still with us? I thought so. Well, come along and see if there are any hints and tips you can pick up.

CLEANING AND RENEWING SURFACES

This is where we get our hands dirty but that's fun – so try and enjoy the experience! We will be making a bit of mess, so if you're doing this in the kitchen it might be a good idea to put down some newspaper or, what I find quite useful, the thin flimsy plastic bag things in which your trousers, coat etc. come back from the dry cleaners.

Note too we'll be naming specific household products here. Try to use the products specified because you'll know for certain that you'll get the desired results. By all means experiment but don't assume that similar products for similar purposes contain the same chemicals.

Because there is so much of this material, it has been broken down into separate sections. So on we go . . .

CLEANING WOODEN CABINETS

Wooden cabinets are not all that difficult to restore, especially if you go about it the right way. And going about it the right way means knowing what you are doing. In this chapter we set out some methods that are guaranteed to work. If you have any doubts, though, it would be best to try out these methods first on a part of the cabinet not normally seen to avoid any disasters.

TRICKS OF THE TRADE

Just about everybody has a different method of restoring old woodwork and each system is probably valid. Thorough cleaning and preparation is essential to each system, though.

Terry Burnett has displayed several beautifully restored sets at meetings and this is how he refinishes TV cabinets. Part of the trick, he says, is to remove all traces of old furniture polish and wax before trying to touch up scratches, apply new finishes etc. He says there are a number of polish removers you can find in the shops. Failing this, for small scratches, some aerosol AF spray from the TV workshop is a good solvent. You spray some on, rub away the polish with a rag and then apply scratch-filler polish (Topps or whatever). If you think more drastic treatment is necessary, a gentle application of bathroom cleansing cream such as Flash Lemon Cream or one of the supermarket own-brand lemon creams is good. As Brian Pethers points out, its abrasive powers come about half-way between 1200 grade wet-and-dry paper (used wet) and a paste chrome cleaner such as Solvol Autosol. Used by itself, it leaves a satin matt finish, so you will certainly need to use some polish afterwards.

CLEANLINESS IS NEXT TO GODLINESS

Before you start any kind of cleaning, protect any transfers and be ultra-careful not to rub away or dissolve any badges or lettering. It is better to protect them with matt varnish or wax polish before attempting any heavy clean-up jobs. A recommended product is Micro-Crystalline Wax Cleaner/Polish, sold by Picreator Ltd, 44 Park View Gardens, London NW4 2PN (telephone 0181-202 8972).

OK, down to business. Remove all dust with a dry or damp cloth, and *only* if the wood is filthy dirty get tough and use steel wire wool soaked in T-Cut or a similar car paint restorer. Alternatively one of the water-free hand cleaning preparations is good; it will degrease the wood and leave lanolin behind.

Avoid too much water at all costs as this may stain the wood or cause it to swell and warp. An alternative is foaming cleanser spray (but don't leave this spray on for long). If the wood is very dry it will need rejuvenating with a furniture restorer such as Scott's Liquid Gold. This is a kind of liquid paraffin and wax mixture, which is allowed to soak into parched wood. It is very effective, although professional furniture restorers hate it on account of its high silicone content.

At this stage you should remove any traces of old polish, using the AF spray already mentioned. People used to employ trichloroethylene ('trike') or carbon tetrachloride but this is now considered a hazardous substance.

It is unusual to find radio and television cabinets in really bad condition but wood-case telephones and telegraph instruments tend to live harder lives. Over the years a mixture of polish and dust can build up to an almost black patina and this can be removed to reveal the original finish with this formula offered by John Young of the Signalling Record Society:

- one part linseed oil;
- three parts turpentine (or substitute);
- three parts methylated spirits;
- three parts vinegar.

Clean the affected wood with very fine wire wool dipped in this solution, then let it dry. Afterwards polish the wood with linseed oil or ordinary wax furniture polish.

Occasionally the wood's original finish may have dried out and deteriorated so badly it must be removed chemically with Ronstrip or some other stripping compound. Extreme caution is suggested and for valuable items it might be better to take them to a professional furniture restorer. Note also that many poor surfaces can be improved with amalgamator compound, but we'll come back to this later.

CHOOSING A FINISH

Actually this title is misleading because there is no choice if you are making an authentic restoration. The only choice is a replica of the original finish, which immediately rules out shiny modern polyurethane varnishes! It also rules out French polish in many cases but read on . . .

The first thing to establish is the nature of the original finish. Wood-cased telephones and telegraph instruments, and also radio sets made before the mid-1930s, generally have a shellac varnish finish, also known as French polish. The wooden cabinets of radios and televisions made from the mid-1930s have a cellulose lacquer or varnish. The two are very different in their look, make-up and method of application; the first is made of insect shells dissolved in methylated spirits and is generally applied by hand. The second is more suited to mass production methods and is sprayed on under pressure.

In the main, French polishing was far too expensive and labour-intensive a technique to use on run-of-the-mill items such as radios and TVs. Equally important, it took too long to dry! This is why the sprayed-on cellulose lacquer was used instead and whilst I'll readily agree that French polishing gives a lovely finish and is arguably more attractive than spray lacquer, it will never be authentic on an object which was not French polished in the first place.

To determine whether your treasure has a shellac or a lacquer finish, find a part of the cabinet out of normal sight. Then dip your finger into some denatured alcohol (meths) and rub the finish in a small circle. If the finish starts to get sticky after a short while, the finish is shellac. If not, it's lacquer. Both call for entirely different techniques.

FRENCH POLISH

The distinctive rich orange-brown deep lustre of shellac polish is well known to telephone and telegraph instrument collectors; it gave wooden bell-sets and wall telephones a very attractive finish. True French polishing develops this to a flawless, glassy sheen, good on dining tables and sideboards, which is considered inappropriate by most collectors of telephones but acceptable on some early radio sets.

An expert in shellac polishing is Linley Wilson, although he advises against unnecessary renovation. Once a finish is stripped, he points out, no amount of care can replace 40, 50 or even 100 years of what experts call 'patina'. This includes the fair and expected wear that gives an old piece its character. Just because there are a few rub marks, caused by years of real-world use, is this a reason for refinishing? Serious collectors would not remove that mark for any money – it is a link with history. Other items may be totally unpresentable, however, and here a complete rebuild from the ground up can result in a first class collector's piece.

Wilson's rule 1, is: 'Start with the least damaging option'. A set may look filthy with dirt and grime but a thorough clean with a damp rag followed by beeswax furniture polish and a thorough hand-buffing with a soft cotton cloth can bring about a virtual transformation. If this approach doesn't work, but merely reveals more surface damage, scratches and minor abrasions, do not despair. An old rag moistened with the appropriate shade of

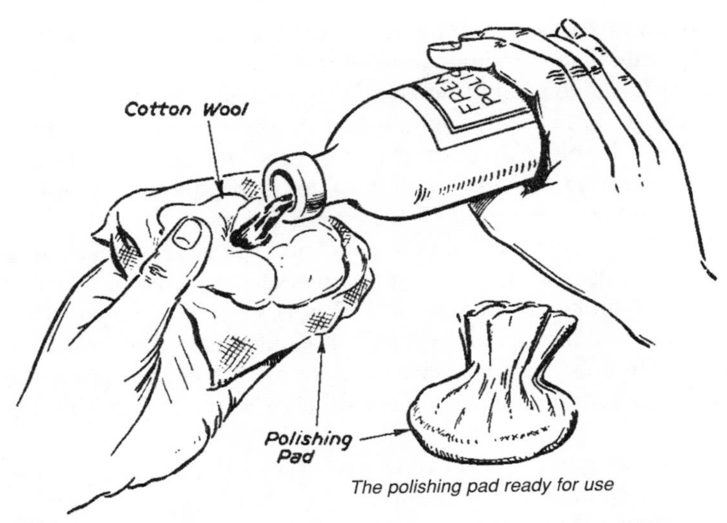

Figure 14.1 French polish and similar liquids should be applied with a pad made of cotton wool and cloth.

ordinary wood stain will cover a huge range of problems. Wipe the whole surface of the affected part and, while still drying, buff hard with a clean, lint-free cotton cloth. Repeat if desired. There are some other approaches described in the section on 'Simpler solutions' later in this chapter.

REPOLISHING

It may nonetheless be necessary to do the whole job from the ground up. A golden rule of any wood finishing operation is that the real quality of the finish depends entirely on what effort went into the surface preparation.

Shellac comes as an orange flake (orange is the variety most suitable for telephones but darker shades may be more appropriate for radios), not unlike naphthalene flakes. It is available from most major paint suppliers and some hardware stores, and because it's a natural product and the 'harvest' varies, it can vary in price. These flakes are dissolved in methylated spirits according to desired strength and colour; about a teacup to 500 ml of meths is satisfactory for most purposes. Simply tip the flakes into the meths in a clean container with a lid, stir a bit, and stand in a warm place overnight. Always keep the mixture covered.

Shellac is best applied with a 'rubber' by hand, according to Wilson, although a brush may be used. To make a rubber suitable for this work, take a piece of lint-free cotton (part of a sheet or pillowcase is ideal) and cut it (don't tear it) to the size of a man's handkerchief. Now place a good

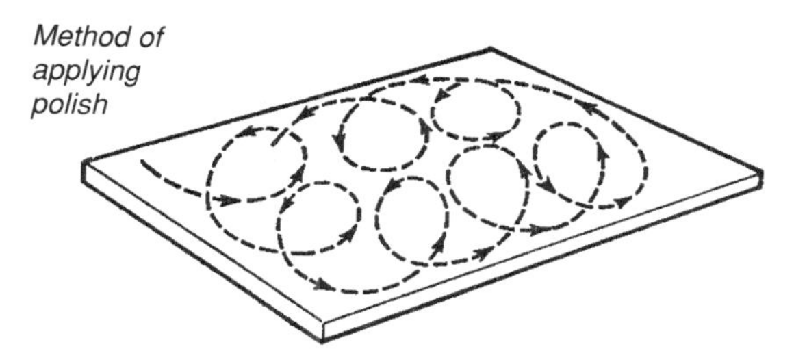

Figure 14.2 When applying polish always work in circular movements and do not stop in the middle of a piece of work.

wad of cotton wool in the middle, fold the ends in, and roll the whole thing into as tight a bundle as possible, keeping all the rolled edges uppermost and with a smooth, wrinkle-free pad below your fingertips. Its end can be pointed and squeezed into appropriate shapes for getting into right angles and odd corners.

Before applying polish, he advises you do a final scrub down of the surfaces with fine steel wool (after sanding). It is recommended to unroll the pad of steel wool and wrap it around a small wooden block, which prevents uneven dips and hollows appearing. Dust off with a clean cloth or brush.

Temperature and humidity are important. Too much dampness in the air will cause white 'blooming' in the polish (harmless but annoying – it will generally disappear with the next coat in a warm environment), so a heater is recommended in cold weather. Too hot an environment will cause rapid evaporation of the meths solvent, increasing the tendency for the rubber to drag up and scuff previous coats.

Wilson's advice is that you begin on raw timber by working the rubber up and down and round-about (professionals work in figure eights all over the surface). Beware of runs over the edges – collect them with the rubber as you go. On raw timber runs may cause a semi-permanent mark. To aid the process, pour a little raw linseed oil into a saucer or lid, and start by dipping the rubber into it, then into the polish. Occasionally repeat this process throughout the polishing process. The oil will not affect timber or polish, but will lubricate the rubber and prevent scuffing. When the polish dries, the oil will lie on the surface in a matt film – it may then be wiped off or left to lubricate the next application.

Only practice is needed now – keep applying coats – with one and a half hours between coats in a warm, dry atmosphere until you are satisfied. While the polish is wet you may go back and forth almost as often as you please, but beware of runs and drags caused by inexpert overworking. Three or four rubs are probably enough at any one time, always with the grain after the initial few coats. Steel wool rubbing, or wet and dry (very fine) with water is needed occasionally to level off humps and to get rid of runs. Water will not harm the surface if it is wiped dry immediately afterwards.

When the desired depth of finish is acquired (anywhere between three and twenty-three coats), rub off with steel wool to get a matt finish, which may be buffed to any level of brilliance with a soft cloth. There are many ways to colour polish – one is wood stain, tipped straight into the polish until the desired shade is obtained.

Experts say the best polish is produced over a longer period of application, i.e. leave the piece to stand for a few days every few coats. Even an apparently bad application may look better since the polish tends to sink down and level off if left to stand. Conversely, polish in open grain and cracks may sink lower, necessitating further applications. Finally, small dents and nicks may be filled by buying some 'button' shellac (comes in hard lumps), and using a blade or chisel heated in a flame to spread it like butter into the defect. Once it has set it looks like toffee, but when sanded and polished over a few times you'll be hard pressed to pick it out (don't get the blade too hot; shellac burns black quite easily).

CELLULOSE LACQUER

Cellulose lacquer gives a superb gloss but it is also rather a brittle finish and can be prone to scratching. All the same, it can frequently be renovated without too much effort, simply by softening the original surface and allowing it to settle and then harden again, leaving a surface looking like new. This technique, known as re-amalgamating, is a marvellous cure for old cabinets where the finish has blushed, crazed or cracked.

'Simply' is perhaps the wrong word and the mistake made by some beginners is using the lacquer thinner as it comes out of the can. The problem is that it evaporates before the finish is liquefied.

To re-amalgamate a lacquer finish you need to first clean the set. A recommended technique is to use naphtha (or mineral spirits) and a bunch of paper towels, pouring some on and rubbing until the towels stop picking up dirt. Then you need to brush on a mixture of lacquer thinner and lacquer retarder (three parts thinner to one part retarder). Use about a half-inch soft brush (from an art supply store) and stoke the thinner mixture gently with the grain on a horizontal surface. The thinner will dissolve the old finish, and as it liquefies it will flow out again. Allow it to dry, and repeat as necessary. Do not scrub, don't use steel wool. All you want to do is to turn the finish back into a liquid, not move it around. (There is also a ready-made product called Mohawk Amalgamator, sold by Antique Electronic Supply.)

After this, spray a new coat or two of lacquer on top and rub out the finish with number 0000 steel wool (make sure it is fresh, clean and will not leave dirt or dust), followed by pumice and then rottenstone. A possible alternative to brushing is to spray Lacquer Blush Eliminator (Star Products No. 1491) on to the surface; this is the same 3:1 mixture as above so it should do the trick.

In the USA you can get Lacquer Retarder as well as the Blush Eliminator spray and lacquers from Star Products (Liberon) and some of these are distributed in Europe – see the 'Buyer's guide' chapter or check their Web site:

http://www.mcn.org/MenComNet/Business/Retail/Lib/liberon.html

for product lists and distributors in your area. The Web site also has several technical articles that might help with your refinishing. Restoration Materials in Bury has these lacquers as well as toners (dyes) for preparing the wood, and materials are also readily available by mail order from the USA.

The same process as described above can be used for shellac finishes, but the solvent then is simply denatured alcohol instead of the lacquer thinner/retarder mixture. It evaporates slowly enough by itself to do the job without any additives.

If you are connected to the Internet there's a valuable article that details a method of reproducing lacquer finishes. It sounds complicated, but isn't really hard to do. It's posted at:

http://www.antiqueradios.com/lacquers.html

DEEPER REFINISHING

If your cabinet is heavily scored or peeling, your only solution is to strip and refinish the piece. This involves removing the existing surface treatment. An electric sander is useful, although as the work gets more delicate you may decide to use lacquer thinner and some no. 0000 steel wool. Apply the thinner with the steel wool with the grain, and wipe it up with clean paper towels.

Before you start stripping, however, you need to need to choose the right shade of lacquer; one way to do this is to get some stain colour chips from your paint store and note which colour chip comes closest to matching the colour of your piece. This will give you something to compare later to make sure you get the colour right, and without a reference it can be difficult to remember what the exact shade of the piece should be.

After this we need to renew the base surface and then build up a new tinted lacquer surface. In commercial catalogues you will find two types of product, toning lacquer and ultra-flo toner – in both cases the satin finish is recommended as being closest to the kind of finish we want to replicate.

Toning lacquer, also known as tinting toner stain or shading lacquer, is a combination of lacquer, sealer and pigmented stain. It can be used to return a cabinet which has been stripped to its original finish. Toning lacquer is relatively transparent, so the more coats you apply, the more opaque is the result. Colours available include Van Dyke brown (a dark shade with a hiding effect), fruitwood (light caramel), mahogany, walnut, cherry and blonde (for 1950's sets). Burnt umber toner colour closely matches the original finish on most of the walnut-veneered sets, whilst ultra-flo toner uses a dye instead of a pigment and is more transparent than the normal toning lacquer.

Before we start this some preparation comes first. A wood filler paste should be applied to fill the grain and the Americans can buy this in both walnut and natural shades, whatever natural is. Alternatively you can mix with it some stain to match the colour of the finish. You can buy these filler pastes in Britain also, as well as sanding sealer which is used prior to the final finish for filling and levelling the surface. You can use several coats to eliminate grain lines in wood. Afterwards wipe away the excess, following the instructions on the can, and when this has all dried, buff the surface with no. 0000 steel wool. Wipe off the sanding dust with a hard cloth and follow up with a tack rag (sold at decorating stores).

Now spray on clear cellulose lacquer as a base coat and to seal the filler paste; this is vital.

When completely dry, mask off the veneered sections if the cabinet had a two-tone finish. Rub down smooth and remove dust.

It's time now to apply the lacquer. First spray several light coats of the toner; the more coats you add, the darker the colour. Within reason, you can add as many coats as necessary until you achieve the desired shade; drying time is fortunately quite short.

The last stage is to apply several top coats of clear lacquer finish to these reworked sections when you apply finish to the rest of the cabinet; this is essential. After these final coats are dry, true experts sand with 400 and 600 grit and then use automotive polishing compound, several coats of carnauba car wax, and a final coat of wax furniture polish. It sounds like a lot of work, but the results prove it's worth it.

SIMPLER SOLUTIONS

What if you just need to patch up a small area of cabinet and cannot get hold of cellulose lacquer? A sympathetic or 'friendly' substance is button polish, which you can buy in old-fashioned hardware shops. I believe this is shellac and colouring dissolved in methylated spirits. Whatever it is, it works well. You rub this on with a cotton wool pad and get a very authentic effect. When it is dry you can judge if another coat is required.

Minor scratches can be hidden with Topps Scratch Remover cream (sold by the bottle in three shades: light, medium and dark). This is a very useful product and hides a multitude of sins if given a good wax polish when dry.

PEELING VENEER

Traditionally veneers were fixed using animal glues, which would soften under damp conditions. What comes off will go back, although it is usually better to replace tatty or incomplete veneers with new, although this upsets some collectors. To assist removal, use a thin, sharp knife-blade or spatula (if this will do) or else pour on boiling water (take care!) and then let the cabinet dry out thoroughly and remove all traces of the old glue before attaching the new veneer. There are specialist dealers in veneers who will be pleased to help you select a new veneer if you bring a sample of the old stuff.

I am assured by experts that applying veneer is not a tricky exercise, it just needs a bit of care and patience. I believe them too. However, I am not a woodworker (I got the school's lowest ever mark, 8 per cent, in woodwork), so you must not look to me for practical advice. Instead, read a library book on the subject or else take the piece to a professional cabinet-maker. The cost may be less than you expected.

Sometimes the veneer is not damaged but has merely separated from the rest of the wood. You could probably get a hypodermic needle in the space and inject white PVA glue but this may not be necessary. Furniture books suggest placing a warm iron (just hot enough that you can't touch it or about 140° F) on a towel and heating up the old glue. This usually works if the object was made with hide glue. If that doesn't work you can inject white glue (or yellow or brown) glue into the space and clamp it using small flat blocks of wood and wax paper to spread the pressure evenly. Alternatively use liquid hide glue in the same way.

With hide glues you need not worry about glue oozing out since it is always soluble in warm water (with a little vinegar if you want it to go faster). The same is true with white glue as long as it hasn't been set too long. Just use a cloth that is damp with warm water to wipe it off. It may take a little while but it won't damage the finish. Just be careful not to soak the wood or you'll find it separating there also. Some people use contact cement to fix new veneer but you don't have a second chance with this stuff; once it makes contact, it stays.

Some more insights on veneers. Some veneers are thin with a paper backing, so do not sand these aggressively. Other veneers have wood backing and are more rigid. Take advice from your supplier of veneers and ask plenty of questions. Before you apply new veneer, sample stain the original and new veneers for colour matching. If you nick or damage the foundation for the veneer, fill and sand the surface to make it flat.

MINOR WOODWORK REPAIRS

The proprietary wood fillers and plastic wood products found in most do-it-yourself shops are ideal for our purpose; some can even be had ready tinted to match the colour of the timber to be mended. Small surface chips and holes can be filled with a knife and rubbed flat with glass-paper after curing. If a larger area needs to be built up, you can do this by placing successive thin layers of plastic wood, using templates or formers of wax paper and modelling clay (Plasticine) – thicker layers take much longer to cure. Alternatively, small pieces of wood can be glued or pinned into place, shaped as necessary. Remember to finish off by sanding along the direction of the grain.

WOODWORM

Small holes are a warning sign that must not be ignored. It's quite possible that the active insects have long since flown the nest but only a fool would not wish to make sure. Having woodworm in the house is the last thing you want, so any item showing signs of infestation must be treated with Rentokil woodworm fluid before bringing anywhere near. Once treated, the holes can be filled with semi-hard wax, plastic wood or some other preparation compound if minor. Otherwise the panel will need to be re-veneered.

REMOVING ODOURS

It's an unfortunate reality that some sets exude a musty smell or reek of nicotine. This, however, is better than the odour of mice or other domestic animals which have mistaken the set for a toilet. In these cases the inside of the cabinet must be washed down but don't get the outside wet as the finish will probably suffer. If plain water doesn't do it, try water with a little sodium bicarbonate (baking soda) dissolved in it. Rinse with plain water afterwards. Apparently the bicarbonate turns organic acids (which have the odour) to organic sodium salts (which don't). Do not use this method on the metal chassis as it should be possible to clean it in other ways (hot water and a small brush), and the bicarbonate is likely to leave a mildly corrosive residue. It is always advisable to use a generous last rinse with distilled water to prevent future corrosion from the cleaning agents you used.

An alternative to bicarbonate of soda is chlorine-based domestic bleach (a solution of sodium hypochlorite) which is also effective for removing smells.

When the problem is merely a musty, damp odour other techniques can be tried. Here are two. Place the offending item in a robust plastic bag and throw in a pound of cheap ground coffee, then tie the bag up and leave it for a couple of days. The coffee absorbs the smell of the mildew (it is alleged). A technique for smaller items is to wrap them in paper, then bury them in a covered box of cat litter for three or four weeks. Cat litter is good at absorbing smells as cat owners know. If these two ideas do not work, then dowsing in disinfectant or household air freshener may help.

THE FINAL POLISH

If the original woodwork has a smooth polish finish, ordinary wax or liquid furniture polish will be found to be adequate but use a real wax polish, not something containing silicones. The finish is quite different and the look of wax polish is far preferable. In some cases, a coat of brown boot (shoe) polish before the final wax helps as this gives a richer, deeper appearance. A final wipe with a soft duster should finish the job, which should never be rushed. If you are in doubt, a few trials on the underside of the woodwork will not go amiss.

PAINT SPOTS

Many radios and telephones are spoiled by paint spots on the top of the cabinet, caused when painters were too idle to lay down a groundsheet before starting decorating. Sideways pressure with a wooden spatula (or lolly stick) will sometimes remove them, as will the skilful use of a scalpel. Others report success in removing paint splatters using a furniture polish cleaner such as Murphy's Oil Soap. The cleaner dissolves the layers of polish and, so long as the paint lies on the surface of the polish rather than the finish of the wood, the paint can be removed.

STUBBORN SCREWS

One of the most frustrating problems is a screw which refuses to turn. Often the temptation is to drill the head off or cut around the wood surrounding the screw head and 'bodge' it out with pliers or a mole grip. This might well work, but it's hardly professional.

Tudor Gwilliam Rees has the solution. Instead of attempting to force with the screwdriver (you want to keep the screw slot in good condition), you should clean the flat surface of the head with emery cloth until it shines. With a good size soldering iron, tin the screw head with solder, then bring the screw up to the iron's temperature (the solder helps conduct the heat efficiently). Remove the soldering iron and quickly attempt to undo the screw once more; this time it will turn easily. Before replacing the screw, treat any rust on the thread with a wire brush and smear petroleum jelly (Vaseline) on the surface, which will preclude future rust and sticking.

CHAPTER 15

Practical restoration techniques – metalwork

In this chapter we tackle all kinds of metal surfaces – metal cases, cabinets and chassis. The techniques are also valid for smaller individual piece parts made of metal too.

PAINTED METAL SURFACES

PAINTED PANELWORK

Some of the equipment you will want to restore – for instance test gear, broadcast studio equipment and metal-case telephones such as the upright 'candlestick' variety – have a painted metal finish. Generally this is some kind of stove enamel or japanned finish. If it is really badly scratched or rusty there is probably no alternative to stripping down and repainting but the snag then is that the new finish will look too, well, new and you may not be able to match either the shade or the finish (cellulose paint sprayed with an aerosol can doesn't have the thick, hard look of stove enamelling).

TYPES OF PAINT

Before we get down to any kind of refinishing, it is worthwhile just reminding ourselves of the kinds of paint on the market and how to choose the right paint when refinishing.

Enamel paints are the sort you find in tinlets in hobby shops and in larger tins at cycle shops. You can find some aerosol cans of enamel paints as well. They are available in gloss and matt finishes, and you use white spirit to thin them and clean brushes afterwards. They are cheap, easy to find and handy to use but take a while to dry. Another source of 'synthetic enamel' paints is supply houses specializing in commercial vehicle parts. Here you can buy trade paints in British Standards (BS) colours, useful because a lot of industrial radio/TV equipment, test gear and telephone equipment was painted in BS shades which never altered from year to year with fashions. Typical BS colours are mid-greys, signal red and other plain, utilitarian colours.

Cellulose paints are generally found in car accessory stores and can be had in aerosol cans, tins and in touch-up canisters with a small brush inside (these last ones are ideal for touching up scratches). Cellulose paint dries quickly and, being thinner, is less likely to 'sag' or run on vertical surfaces. Of course, it is best to turn surfaces horizontal before painting. Cellulose paints applied on top of enamel ones may well dissolve the old finish or bring it up in bubbles – be warned! Use only cellulose thinners to dilute these paints and clean brushes.

Note that the 'obvious' shades are often not the ones you want! This applies particularly to whites and silvers. The standard snow-white and aluminium shades will look far too harsh on most restoration jobs, mainly because these really pure pigments were not available to paint manufacturers in times gone by. By way of preference, look for BMC 'Old English' white and Volkswagen antique or pearl silver shades; the car names are only examples but you will need to explore for these off-white and greyish silver shades in the paint ranges specific to a particular car manufacturer.

Hammer finish paints (Hammerite, Holts etc.) and their smooth equivalents (Smoothrite) are, unlike others, applied in one thick coat which has a kind of self-levelling effect. They need no primer generally and, although they 'go off' quickly (like cellulose paints), they can take two weeks to harden. It is not a good idea to recoat these paints in less than two weeks, otherwise you can bring up the first coat in a ruffled mess. These paints also attack enamel paint layers underneath. They are very prone to 'sagging' so don't try to paint vertical and horizontal surfaces in one go – it will look a mess. Far better to lay the surface to be painted horizontal and paint this alone. Then come back the next day and move the job round so that another surface is now horizontal.

Note that hammer paints stick like the proverbial to a blanket and normal brush cleaner does not work. Use the (expensive) Hammerite thinners or else a paint stripper such as Nitromors. The solvent is very volatile and will cause severe headaches if used without good ventilation. Special thinners are sold, although commercial xylene may work as well; cellulose thinners are definitely not suitable.

These paints are easy to find in DIY and car accessory shops and come in both brushing and aerosol spray varieties; each gives an entirely different finish. Spraying gives a fine, consistent, slightly rippled pattern, where as the brushed-on paint gives a thicker, glossier finish with a more pronounced hammer pattern. Experiment on scrap-metal first as it takes a while to get the knack of applying this paint. It will not take to even slightly greasy surfaces (small oases or pimples of bare metal will appear, very annoying!) and it helps if you roughen the surface first with scouring powder or an emery cloth.

OTHER TEXTURED FINISHES

Hammer, wrinkle and crackle finishes are often confused but each has an entirely different look.

Hammer finish paints, already mentioned, give an undulating effect not unlike orange peel. Sometimes you will see veins of a darker colour running through the paint, a bit like marble.

Wrinkle finish paints give exactly that; a surface with many small wrinkles. The term 'crystalline' describes a similar finish with extremely regular, small wrinkles, whilst 'rivel' is something also very similar.

Crackle finish is not the same thing, although many people use the name to describe wrinkle finish. Proper crackle finish is very decorative, with 'islands' of a dark colour (e.g. black) surrounded by 'sea' of a paler colour (e.g. gold). The overall effect is a bit like crocodile skin or, if you prefer, like the craters of the moon; it is very attractive in a dated sort of way. The effect is created in a three-stage process and is not easy to replicate. The base colour is applied first and allowed to dry and harden. After this, the overlay colour is applied and baked at a very steady temperature in an oven. During this process the crackle finish appears. Finally, when all is hard and dry, a coat of clear lacquer is applied for protection.

All these finishes were devised to add a more attractive look to plain metal surfaces cheaply and also to disguise any surface imperfections. Hammer and wrinkle finish paints are available in the shops but the crackle finish seems to be a trade secret, and it was probably baked on. Wrinkle finish paint is available only in a few colours but you can always spray a black wrinkle, then recoat with a plain cellulose spray paint of the desired colour. Black and brown are the most popular colours on radio and TV equipment, and you can find the black wrinkle paint in most car accessory shops. Brown is harder to find but you can get it from the Eastwood Company (UK) or from Antique Electronic Supply (USA) by mail order and there are other suppliers (see the 'Buyer's guide' chapter). If you need, say, a gold finish for some 1950's hi-fi probably the best idea is to paint it in black wrinkle, let it harden off for a few days and then spray gold over the top. Cellulose on top of cellulose sometimes causes a reaction, so an enamel paint aerosol would be best for the second or final colour and these are available in DIY shops, and also in some model and hobby shops and artist's supply places.

APPLYING WRINKLE PAINT

Experience shows that four relatively heavy coats give about the right amount of wrinkle. It helps to spray one coat on in one direction and the second at right angles. Spray the centre of each succeeding pass so that it will be right on top of the edge of the previous pass; this gives a 50 per cent overlap for each pass; very important for uniformity. Wait about five minutes and then repeat the process; the important thing is to paint only one horizontal surface at a time (the paint will run or accumulate at one edge if the surface is not horizontal).

After coating the surface as above, a heat gun can be used to cure the paint. It will wrinkle a lot sooner with heat drying, which is almost essential in the winter. On a hot summer day the paint will wrinkle fairly quickly by itself. It is important not to stop the heating process too early as the wrinkling continues for a while after the initial onset. Wrinkling takes about three minutes to start and probably ten minutes to complete. Then wait about half an hour and go on to do the next surface. Be certain that the surfaces are very clean and oil-free before starting, otherwise strange figures will appear in the finish.

POWDER COATING

This is a finish found on relatively modern equipment and is recognizable by its extreme durability and minute dimples in the surface, rather like 'orange peel'. Another powder finish looks rather like small grains of sand or rice in the paint. Both of these are commercial epoxy powder finishes and not available as a brush or spray paint. The process puts an epoxy powder coating on electrostatically charged metal; the powder thus clings to the metal. The part is then baked (at about 350°F) for ten or fifteen minutes. The epoxy melts and flows over the surface. Scratch resistance is excellent, and adhesion (impact) is good and can be made excellent if the metal is pretreated (phosphate). This is an industrial process and cannot be reproduced at home. However, most large towns have a powder coating contractor who will do one-off jobs for cash-in-hand.

DOWN TO WORK

So much for paint technology; back to the work in hand. How far do you go? Generally you have three options for your treasure:

- leave it as it is and clean as well as possible;
- strip it and repaint by brushing or spraying;
- take it to a professional and have it repainted in the original fashion.

OPTION 1: THOROUGH CLEANING

You may well consider that thorough cleaning is preferable to repainting, especially if the panel has lettering which would be difficult to replicate. A gentle touch is always required, especially if there are screen-printed legends, lettering or transfers you wish to retain. Try a test on an area that isn't normally seen, such as the underside.

The simplest form of cleaning is a good wash down with a thick solution of washing-up liquid (detergent), followed by a quick drying off with paper towels – you don't want damp getting into where it shouldn't go. Alternative cleaning solvents are cigarette lighter fuel and liquid dishwasher soap. I prefer the industrial ones sold by Maplin Electronics and the trade supply houses (RS, CPC etc.). Tougher grime calls for the kind of household product sold for cleaning ovens or really dirty kitchen floors (e.g. Mr Muscle); these products are very powerful so read the instructions and proceed gingerly.

Afterwards you may find a quick wash over with cellulose thinners will help; if lightly done with a paper towel it can help soften the paint and redistribute it into very shallow cracks. Leave to harden again afterwards and protect the soft skin with a coat of wax polish or Armor-All (see below). This technique needs a bit of experience and you should practice first on something of low value before attacking some rarity.

For really dirty surfaces you are best using T-Cut car paint restorer or the Flash Lemon Cream used for cleaning sinks. The latter is quite abrasive so go gently and even T-Cut will remove the top layer of paint as well as the dirt. Wash off and leave to dry. With care you can do a good job of blending the glossy and less-glossy areas, leaving a flat and uniform finish. It will look much cleaner and brighter than before but, because it is unprotected, it will need sealing to prevent rapid abrasion and absorbing of greasy marks. Spraying on some kind of varnish would solve the problem but this will look too glossy. Some people use a silicone car polish such as Armor-All (liquid) or Back-to-Black (aerosol) but this will make any subsequent painting very hard; far better is traditional wax furniture polish. This will give the right protection and dry to a dull sheen, which is just right for what we want.

There may be a few scratches that need repainting, even though the rest doesn't. If the original finish was spray painted and the scratches are only superficial, you can try amalgamating the surface. By this I mean flowing cellulose thinners generously across the work surface and loosening the paint with a hard cloth (not a soft one or bits of lint will be left behind and they will stick to the paint). Try a small area first but, if you're lucky, this will leave a smooth (but soft) surface with the scratches far less visible. Leave to harden off for at least two days, then apply the polish as described above.

If the scratches are too deep, your secret weapon is a small touch-up tin of grey cellulose thinners, the sort sold in car accessory stores. Buy the tin that looks like a cigar case and has a small brush inside the can. A deft touch with this quick-drying paint will cover up the scratches and, after it has dried thoroughly (try to leave it a day), you can use a small artist's paintbrush to touch in the spot with the correct colour of paint. Most shades dry darker than how they look when wet, by the way.

OPTION 2: REPAINT

Only you can judge if you can paint over the existing surface or whether stripping down to bare metal is necessary. Often I find I can get away with filling in scratches with grey cellulose primer (use a small touch-up can from a car accessory shop, the sort with a tiny brush built into the cap). Then, when this is dry and hard (leave twenty-four hours even though the paint is touch-dry much sooner), rub down with very fine wet-and-dry paper.

Sometimes, however, the surface is so bad that nothing less than stripping will do. For paint stripping I always use a chemical stripper such as Nitromors. Not a lot to be said about this, except to remind you to use rubber gloves (the caustic stuff stings like you-know-what if it gets on your skin), use a really stiff brush and hot water to remove the stuff afterwards and make sure you get every bit off. The chemical actually works by releasing gas, I understand, and seems to work best if left for five minutes or so. With really tough paint, you may need to make several applications, stripping off a layer at a time. A tiny screwdriver may be the best way of removing stubborn patches of paint that the Nitromors is powerless to free. Bear in mind that several of these chemical strippers are *extremely* harmful if inhaled; read the warning remarks in the 'Safety first' chapter).

For repainting I'll consider both hand painting and spraying, but first how do you match the paint? You certainly need an eye for colour but there are many sources of paint. It helps if you can bring a sample of the original to the shop where you'll be buying the paint, so where do you start? With the car accessory store: they have a huge range of cellulose paints in many shades plus a range of smooth and hammer finish paints in aerosol and brushing form – don't forget brushing cellulose, it goes on cleanly and dries quickly.

MATCHING A SHADE

In some (perhaps most) cases, you will not be able to find an exact match among ready-made paints, so you need to have some paint made to order. There are four ways of going about this, three of which will not be cheap but will give ultimate satisfaction.

The first way is the cheapest. If you have an eye for colour, you can attempt your own matching. Buy a selection of tinlets of hobby paint (the enamels made for painting models) or indeed cellulose paints (don't mix enamels with cellulose though!) and you can try to make a match. Bear in mind that some mixtures dry darker or lighter than they appear as a liquid, so you may need to experiment a bit. Making exactly the same mixture later may also prove difficult, so perhaps it's better to leave this job to the experts after all.

House paint is our second option. Large DIY stores such as Sainsbury's Homebase have machines which can mix a large number of shades of emulsion and gloss brushing paint (we need the gloss of course). They also supply shade cards which you can take away and check against the items requiring repainting. This can be fruitful, although the range of colours is not huge and this kind of paint dries rather faster than optimally.

Option three requires visiting a car accessories shop that offers custom paints. Many of these shops now have a service where they will make up spray aerosols and cans of brushing cellulose to any shade in their album of older car colours. These include the RAL shades (not the BS ones though) and the shade cards have a hole in the centre of each colour panel so you can place the card over your paint sample and hopefully establish the correct match. The price is not cheap but not horrendous either, and the only snag is that the stuff they put in the tin for you doesn't always look exactly like what it was supposed to do.

Finally there are the trade paint stores who sell enamel and cellulose paint in gloss, semi-matt and matt finishes. These people are experts and can match anything! Their main customers are vehicle refinishers and they can be found under the classification 'motor factors' in your *Yellow Pages* directory. If you're fussy about getting an authentic match (and I am), this is the only paint that will do. Their minimum order is usually half a litre, which will cost you between £8 and £12, or a one-litre tin. This is expensive but should last you a lifetime.

If you can establish the official name of the original paint colour, you're home and dry of course, otherwise a specialist will quickly find it for you. If you think about it, every shade of paint must have a name, and equipment manufacturers need to be able to find repeat supplies of the same shade over what may be extended periods. And whereas some paint shades change from season to season with fashion, equipment manufacturers need paints that can be supplied without variation over a lengthy period. In fact many of these shades have not changed since before the last war (e.g. the mid-grey used on Post Office and services equipment) and you will be able to get an exact match.

This is why manufacturers generally choose from standard colours with names and numbers agreed by all paint manufacturers; in Britain these are known as BS (British Standards) shades and other countries have similar arrangements (e.g. RAL colours in Germany). Thus the greys and other plain colours found on test gear and 'technical' equipment almost certainly come from this range, and when you find for instance that the mauve-grey shade used by Pye on its broadcast equipment is known as BS22 B25 Tornado, you're home and dry. Don't waste time looking for these 'trade' paints in a decorator's showroom, however; the motor factor is the place to visit.

THE BASICS

Remember all the basics of painting. Proper surface preparation is mandatory. No rust, no old paint, no oil! A good undercoat (grey primer, red oxide primer) is essential and if there is any rust or other corrosion on the metal surface, remove this with emery cloth. If necessary 'bond' it in by giving a first coat of smooth Hammerite because this sticks like the proverbial you-know-what to the blanket. Then rub this down a bit and build up your top coats on this.

RUBBING DOWN

Notice how I casually mentioned rubbing down. In fact rubbing down is an extremely painstaking operation or should be if you want a decent finish. The foam-plastic blocks covered with grit are useful but they are also 'conformant' – in other words, they adapt to the surface you are sanding. That may be an advantage if it's a curved or lumpy surface, but it's certainly not when you are trying to smooth out hard lumps of paint in an otherwise flat surface. Then the only solution is a good old-fashioned wooden sanding block and emery cloth. Emery cloth works better than glass-paper (and lasts longer too, even though it's more expensive). 'Wet-and-dry' used wet can achieve extremely good results. But sanding is a boring job and when you just want to take the gloss off a paint surface so that new paint can key in well, there is a convenient alternative. You'll need to visit a trade motor factor's depot and buy Farécla G-Matt compound. This white paste comes in a plastic tube; in fact it's not unlike toothpaste. You rub it on with a damp cloth (an old face-flannel is ideal) and the harder you rub, the more of the old paint surface you take off. It's easy, it works like lightning and afterwards you just wash it off and wait for the surface to dry. Then you can brush or spray on the new paint in the sure knowledge it will look superb.

In theory those tough foam blocks covered in abrasive ought to be just as good for sanding, although they tend to clog very fast. You can wash them out and then use scouring powder for rough sanding or sink-cleaning cream (e.g. Lemon Flash) for fine sanding. Removing the dust and debris is a bore, so a clever alternative is to wash down using a nailbrush and Lemon Flash Cream (or similar cream bathroom cleanser) which not only gives a superb dull finish but also removes any grease from the surface.

If you are cleaning up a lot of panels an alternative to hand sanding is a decent sand-blaster, from a trade motor supplies dealer. Better hobby shops sell a 'hobby abrasive gun', which uses a very fine aluminium oxide powder (you can also use pumice powder); it is similar to a jeweller's sand-blaster. People have used the latter to recover some front panels considered hopelessly lost; with a light touch you can even sand-blast and clean up roller coils and variable caps. Try to keep the aluminium oxide away from ceramics though. Pumice powder is better for that – less risk of flashover. (Be sure to get all that stuff out of bearings and the like or the part will be destroyed over time as the bearings, rollers and contacts are ground away.)

KEYING IN

Whatever you do, you want a smooth surface into which the new paint will key – and no dust at all! You may not be able to see the dust that's left behind but it will sure enough grin back at you under the new paint surface. So accept the fact that a duster just isn't good enough for this; copy the professionals and use a tack cloth. A tack cloth is open-weave cloth impregnated with a barely tacky mineral oil and it mops up dust and grit like a magnet. Tack cloths are disposable, cheap and can be found at your local trade motor factor's depot (see *Yellow Pages*).

Painting is more a mind-set thing than anything else, and like any other skill-based project, if you attack the problem in a professional and methodical manner, you are assured of success. But your existing skills at painting window frames (or model aircraft) may not help. Let's analyse the problem; what are we trying to do? In fact we're trying to recreate a factory-applied finish using different techniques and at home. We have two options, to spray or to brush. Spraying is certainly feasible, using aerosol cans, but brushing is a lot cheaper and simpler.

So how do we go about brushing? Well the remit is to achieve an even finish with no brush marks, probably not too glossy. No problem, but forget about achieving this with one coat. Yes, you can use thick gooey paint and the sort of artist's paintbrush sold for children's painting sets, and yes, you will achieve coverage with one coat (it will probably look awful too). You can do better than this, however. The trick is to let the paint down with thinners (cellulose, turps substitute or whatever is appropriate) and using a very soft 'mop-head' brush, gently flood this thin paint loosely over the surface with virtually no pressure at all. Brush marks are impossible this way and because the paint is diluted with thinners, it will dry quite fast. Don't confuse touch-dry with hardened, though; paint needs several hours to set chemically and if you apply a second coat too soon, it will merely lift the first coat. So have some patience and leave the job overnight, then give it another coat the next day!

Three-dimensional items (metal cabinets and so on) are best painted one flat surface at a time. You can try to cover the flat top and the vertical sides in one go but there is always the risk of sags and runs marring the finished job. This applies particularly to paints you must apply thickly, such as hammer finishes.

In all cases several thin coats are preferable to one thick one. Thickly spayed coats never seem to dry and may take on an 'orange peel' effect, whilst hand painting may leave sags and runs which will always grin at you afterwards.

In general, spraying gives a better finish than brush painting, especially when a professional spray-gun is used (a so-called touch-up gun is ideal); brushing needs a soft, fine-haired brush and very careful 'laying-off' to avoid brush strokes remaining visible. Even then, it may be impossible to achieve an even layer of paint. Spraying has its disadvantages as well – to achieve a good finish several light coats are called for, which takes time. There is also a risk of overspray if you don't mask off the surroundings.

Most people end up using aerosol cans for spraying and with care these can give good results. The 'care' involves assessing air temperature, humidity and turbulence, and also achieving an even, steady pressure from the can. Ideally you should spray outdoors on a dry, still day. If the atmosphere is damp, the paint will not dry properly and any wind may draw dust or debris on to the work. The low-cost pistol-grip device known as a can-gun makes the aerosol can easier to handle and has a trigger to operate the spray button. It costs less than a can of paint and is highly recommended to make the job easier.

Professional spray equipment is much better but it does not come cheap, however; reckon on laying out several hundred pounds (or dollars) on the gun, compressor and filter traps. You will also need a proper breathing mask together with the proper filter cartridges for the type of solvents in the paint you're using. Permanent brain, liver, kidney, nervous system and visual damage is no fun, nor are soft-tissue cancers.

With practice you will learn that the finish you achieve depends on the way you use the gun; the finish on a lot of professional radio equipment was achieved with a lot of air pressure, a fast-drying reducer and holding the gun about three feet off the workpiece so the paint was about half-dry when it hit. There is also a bewildering array of spray paints, including acrylic lacquer (easy to work and ideal for a low-gloss industrial finish) and two-part polyurethane epoxy paints ('two-pack'), not recommended for beginners. All told, spraying is probably best left to a professional; the cost is not high.

One exception to the rule of many thin coats: simple items such as tuning scale needles can be repainted in one go. For this application and for filling in small scratches, the 'touch-up' pens sold at car accessory stores are ideal. You have a small container of runny cellulose paint with a paint brush fixed inside the cap; this is ideal for simple touch-up jobs and this paint is available in the useful colours of red, black, white and silver. Incidentally, if they have a 'broken' or 'old English' shade of off-white, choose this by preference. Today's ultra-white pigments look quite wrong on old equipment, whereas the slightly yellowed look of these off-white shades looks far better.

When you have finished your paint job it will look great – except for one thing. It will almost certainly look too glossy, as if it was applied only yesterday. No problem – cut it back! For this we use a very light application of paint restorer such as T-Cut or Color-Bak (you can also use Flash Lemon cream), just enough to skim the shiny surface off the paint. Do this once the paint is really hard, not while it is still setting. Then rub on some wax polish (beeswax or carnauba, not silicone) and buff it up with a soft cloth. This will look great and because it's a natural wax, it can be removed later if you need to repaint it. Silicones, on the other hand, will resist all attempts to repaint. On other occasions you may find the finish too flat, for instance if the only good matching paint is, say, a matt modeller's paint. You can correct this by spraying with a satin finish varnish or by applying wax polish over the original matt finish.

There are occasions when spraying is impossible, yet brushing gives too 'heavy' a finish. This could be the case on an item with an original wrinkle finish which will be hidden by a brushed-on coat. Here's another secret! Using cellulose paints, make up in a jam jar a very weak solution of, say, ten parts cellulose thinners and one part paint. This will look very weak and watery but when you flow it on to the surface to be painted, it will cover it easily and dry off within five minutes or so. No brush marks will be visible, no surface detail will be obscured and you'll find that you can then flow on a second coat. By the time this has dried, the coverage will be remarkably opaque and you may not need another coat. Building up multiple coats of very thin paint in this way is an old craftsmen's trick and it's very effective. Of course, you can paint only one surface at a time and it must be held level to avoid sags and runs. It's fiddly but gives a really good job in situations where other methods are impossible.

Brushing by hand often leads to disappointing results, with all too visible brush marks or pieces of brush hair in the paint finish afterwards. First of all, do not skimp on brushes. Use the best you can afford and then spend a little more. Think like a professional; time is money for him and he cannot afford to waste his time. Cheap brushes drop hairs; good ones do not. The best brushes for our purposes come from a commercial vehicle factor, not the DIY shop – you want a coach-painter's brush.

Brush marks are all too common when you use quick-drying paints . . . and most house paints and enamels sold for model makers are intentionally quick-drying. Most people can't wait for paint to dry. But we can, because we want our paint to flow out flat, without any brush marks. For this you want a slow-drying synthetic enamel, the sort that coach-painters use to brush-paint buses, vans and trucks. This you will find at your nearest motor factors. Incidentally, there's no point in using a hair dryer to accelerate the drying time of this kind of paint; it may work with cellulose finishes but enamels just soften with heat!

Many paints will 'run' or 'sag' if you apply them on vertical surfaces. You can avoid this by rotating the object to be painted so that you paint only horizontal surfaces.

For small parts, dipping is an effective alternative; after thorough degreasing tie a piece of wire around the object to be painted and dip it into the tin of paint (or a bowl into which paint has been poured). Then retrieve the object and let it dry. Whichever method you use, make sure that dust will not settle on the drying paint.

A word on masking. Spray paint has an uncanny ability to reach unintended places, which is why masking tape and newspaper were invented. For covering intricate work a product called Maskol can be very helpful. Made by Humbrol Ltd in Hull, it is sold in hobby shops to modellers who assemble and paint plastic kits. It is a mauve-coloured liquid which is painted over the surface to be masked; the liquid evaporates to leave a thin film of latex, which can be peeled off afterwards, removing paint.

SECRETS REVEALED

Now, a good cheat's trick! What do you do if, just as you are applying a nice fresh coat of paint, the brush begins to drag and you realize the new paint is lifting the old surface? This, by the way, usually happens when you are putting cellulose, Hammerite or some other solvent-based paint on top of oil paints or acrylics. It can also happen if you didn't leave the last coat to harden off properly.

Well, you have two choices. One is to find a rag and plenty of solvent, and wipe off all the congealing mess. Then you must strip everything down to the bare surface again. What a bore! Or you can take the cheat's way out. Carry on painting, so as to cover all the old finish. It will look pretty dreadful of course. Now the master stroke. Let it dry a few minutes, clean your brush in thinners and . . . pour a small amount of thinners on to the freshly painted surface and watch it dissolve the top layer of paint. It will smooth out any lumps but it will still look pretty awful. Now the *pièce de résistance*! Pat the by now rather runny surface gently with the flat side of your paint brush and you can create a loose stippled effect which will smooth itself out and end up looking rather like hammer finish paint. It won't look perfect but it will look acceptable.

Another cheat's tip. What if a piece of metal is a bit rough? Easy – spray it with the plasticized stone chip spray that people use for coating the underside of their cars (you get it in an aerosol at any car care store). This will dry out into a 'bobbly' or slightly knobbly surface, and when it is really hard (allow two days), you can spray it with, say, satin matt black cellulose and it will look a million dollars.

One last tip on paint. Silver paint always looks awful – until you spray it with clear lacquer. Then it looks like a professional silver paint finish. Do not overlook this. In fact all metallic paints need a coat of clear lacquer to protect them.

OPTION 3: PROFESSIONAL REPAINTING

Here we are talking about stove enamelling, epoxy powder etc. – this could be expensive so get a few quotations (try *Yellow Pages*). You will probably find the charge is less if you do all the preparation and strip off the original finish.

ENAMELLING AND ANODIZING

ENAMEL PANELS

You don't normally encounter these, but if you acquire an old enamel 'Marconi Dealer' or 'You may telephone from here' sign from the outside of an old shop, or one of those many 'Danger high voltage' signs that were recovered from the BBC's Moorside Edge transmitter, you may wonder how to remove those orange stains which indicate rust below the surface. A solution of dilute oxalic acid is said to work a treat. General powdery dirt can be removed with oven cleaner, T-Cut (my recommendation) or even scouring powder, although the last-mentioned distresses the shiny enamel surface a bit.

Taking oxalic acid in greater detail, you can order it at most chemists if they don't have it in stock. Be prepared to say why you want it because they will probably be mighty suspicious if a 'normal' person requests it. If they offer a solid version, you'll want about 500 grams, which will cost about £4.40 at 1994 prices. Mix this quantity with five litres of water to make a 10 per cent solution. Alternatively they may offer to make up the solution for you. Be advised that oxalic acid is nasty stuff, so wear rubber gloves, goggles and a mask (the fumes can give you a sore throat) and don't do it anywhere near children or animals. Dip the items in the solution and expect results after about ten minutes. Remove the panels as soon as they look clean as some colours (blue for example) may be bleached if immersed too long. Heavily stained items may need a soak of two hours, though.

Whether you attempt to fill in scratches and holes on the surface of old enamel signs is a tricky decision. With practice, you can fill in depressions with epoxy glue (Araldite) smoothed off and then painted with gloss enamel paint. You can always remove the desperate glossiness of the new paint with T-Cut, or even rub in a little dirt from the garden to 'age' the restoration.

ENAMEL AND PAINT-FILLED ENGRAVED BADGES ON OLD SETS AND OTHER EQUIPMENT

These need careful attention if the badge is damaged. Small cracks can be painted over but may not look very good. Any attempt to alter the shape of a badge after enamelling will result in cracking of the glass; this can also be caused by over-tightening of the retaining nuts. There are professional repairers who will strip out the old glass or paint, replate the metal and provide new glass or paint (see 'Professional services' chapter). There is also a product called cold enamel which is stocked by specialist watch and clock repair supply shops; it is available in white only and is a traditional remedy for filling small holes in enamel dial faces.

ANODIZED ALUMINIUM BADGES

These are very prone to scratching and fading. Sometimes a deft touch of paint will do the trick, otherwise it may make more sense to have an exact reproduction made photographically (see 'Professional services' chapter).

BRIGHTWORK AND PLATING

CLEANING BRIGHTWORK

To clean brass, chrome and nickel use an impregnated wadding such as Duraglit, or if it is really dirty use a cloth and either a liquid cleaner like Brasso or a paste such as Solvol Autosol. All these will clean the metal so that it needs only a polish with a soft duster to look perfect. If the metal is corroded do not use emery cloth (the scratches will be all too visible) but instead use a very fine grade of wire wool or a glass-fibre scratch brush (from large ironmongers) to remove verdigris and other dirt without ruining the original finish on the unaffected areas.

Be very gentle to begin with. Some bright plating on top of brass can turn out to be very loose. It looks corroded and you feel obliged to use chrome polish or a scratch brush. Too late you realize you have removed all the plating. Go easy!

Some decorative trim is made of aluminium. The detergent used in dishwashing machines (Sun, Finish) is very good for cleaning aluminium. A soak in hot water and a dash of the detergent will work wonders on aluminium decorative trim, if you can remove it.

PROTECTING BRIGHT METAL

Authenticity is important. In general, iron and steel parts are best preserved with silicone wax applied as a liquid with a small brush. Brass can be protected with acrylic lacquer (Krylon or similar), preferably sprayed on. This is clear but in older times a spirit lacquer was used which had a rich golden tint; this is what you see on old telegraph instruments. It was applied with camel-hair brushes to workpieces which had already been heated in a flame and thus dried before any airborne dust could settle. A close replica, applied cold, can be found in shops specializing in clock repairs, under the name green bronze varnish (it dries brown, not green).

PLATING METAL

Sometimes it is necessary to renew a plated finish. Years of cleaning may have removed plating or corrosion may have set in, helping the plating to peel off by itself. Worse still, someone may have 'brassed up' metal parts by removing the nickel or chrome plating; in this way they think they have made something look more 'antique'. In fact very few items ever employed bare brass finishes, simply because brass tarnishes rapidly. No radio, telephone or similar item ever had fittings in a brass finish; they were always either plated or else chemically blackened. Even the telegraph instruments of the nineteenth century had a thick layer of spirit lacquer applied over brass components and the only time these were exposed and polished was on the block telegraph instruments in railway signal cabins where the signalmen (with time on their hands between trains) kept everything spotlessly clean.

Do-it-yourself electroplating is not desperately difficult, but unless you intend to devote time to building up skill, it is far better to leave the task to experts. Professional platers can be found in the *Yellow Pages* directory and in the 'Professional services directory' chapter of this book. If you are putting the job out to a trade house, do explain exactly how you want the work to be done; many will produce a finish that is too bright and shiny. A dull nickel finish can always be polished brighter, whereas a bright nickel finish can jar the eye. Also be sure to copy like with like; if the original was bright nickel plated, then chrome plating will look terribly wrong (and vice versa). If you are not sure and cannot tell the difference, rest assured that the people at the plating company will be able to (as will all the people who see your wrongly replated ex-treasure!).

You can also buy at silversmiths' shops a rub-on solution of real silver which claims to apply a permanent silver plate to base metals. This works after a fashion, but the layer is very thin and easily rubbed off with polishing; it also only works when the metal for plating is scrupulously clean (try degreasing and dipping in mild acid first). You may find this liquid satisfactory but I have not been impressed. In the same line of country is 'Tinnit' electroless plating solution, sold by electronic component suppliers to give a silvery coating for home-etched copper-printed circuit boards.

FADED GOLD TRIM

Many 1950s radios and other devices had gold trim, which often fades or becomes scratched, and collectors may wish to repair this. If the only problem is scratched or missing lacquer (with the gold finish untarnished), you can use metal polish to clean it up or use varnish remover to take off all the lacquer and attempt to revarnish it. You will need to use considerable care removing the trim; the varnish is best sprayed with an aerosol can as brush painting may look rough. In some cases it would appear that the gold finish was a tinted lacquer applied over chrome plating.

If the plating is defective, one solution is to remove all the old gold covering; it will come off quite easily, either with a small wooden spatula (lolly stick) or a wire brush. Soaking in a 50:50 ammonia and water solution is a last resort for really stubborn cases. Underneath the gold you will find what looks like solid nickel, which is quite attractive, almost as good as chrome, and you may opt to leave it like that. If you want the full original look, you can try the gold or brass plating sets sold by Antique Electronic Supply or better, take the job to a professional plater. Someone who has used the AES kits reports pretty good results, with a slightly mottled and old-looking effect, which is what he was aiming for. The gold plating kit is, understandably, more expensive.

Alternatives to plating are either a gold-toned clear lacquer or brushing on 'antique gold' paint from the local hardware or art store (also sold by furniture restorers). Sometimes the 'gold' or 'chrome' trim is a thin layer of metal vacuum-deposited on a plastic moulding and for this only a specialist can help. See the specialist magazines for automobile restoration; *Hemmings Motor News* (a USA title) is said to have advertisements from plating companies that specialize in replating plastic dashboard parts.

PLAIN PANELS

Sometimes all that's needed is a plain aluminium or stainless steel panel, either refurbishing an existing one or creating a replacement. How to get rid of scratches and other 'distress' features?

The answer is the metal finishing system developed and marketed by 3M for paint removal and metal preparation. It's a type of Scotchbrite and there are different grades or degrees of roughness. Dennis Fox recommends the drum sander type, mounted so as to use the whole flat side across the panel. You can obtain it from motor factors (see *Yellow Pages*) and other trade counters which serve the car body refinishing trade. Dennis says you'll need the mounted mandrel as well as the cylinders that cover it and suggests that you practice on something other than what you want to finish so as to develop your 'touch'. Remember that, unless you anodize or clear-coat varnish your panel, it will still oxidize and show fingerprints.

Barry Ornitz finds that a Scotchbrite kitchen scouring pad does a good job, especially if used in an electric sander. An orbiting sander, he says, will produce a random orientation to the finish while a finishing sander will make the scratches run in one direction. Sandpaper or emery cloth will work too, but these kitchen scouring pads do such a good job on even the hardest metals. After you have properly 'scratched' the surface, either coat it with clear Krylon (or an equivalent acrylic varnish) spray immediately or you may etch it to produce a nice satin finish. With soft aluminium (one that does not contain much copper) a bath of dilute sodium hydroxide (lye) in warm water for about thirty minutes is used. If much copper is present in the aluminium alloy, black spots may appear. These can often be removed with dilute acetic acid (vinegar). Rinse the panel well after etching and allow to dry. Again a quick spray of clear Krylon will help preserve the finish. Wear gloves and safety glasses when working with the caustic solution. Avoid getting fingerprints on the panel before or after etching.

FIGURED PANELS

A popular finish on plain metal panels, particularly brass and aluminium ones on telegraph equipment, test gear and early radio sets up to the crystal set era, is the repeated 'peacock's eye' pattern also known as 'satin whorl', 'engine turning', 'damasiening' or 'jewelling'. The overlapping circles can be as large as an inch in diameter, although the small ones are more attractive to most people. It used to be a common treatment on custom-made parts of high quality and was a silent testimony to the high craftsmanship required to achieve the finish. This method of decoration is also called 'figuring' in the UK, and was commonplace on the brass front panels of electrical instruments (particularly Morse telegraph equipment) in the nineteenth century.

The actual methods of achieving the finish are many and various. In telegraph workshops the figuring was done with a piece of glass-paper or emery cloth glued to a wooden dowel, mounted in the chuck of a pillar drill, then lacquered afterwards for protection. Usually the circles produced were repeated so as to just touch or overlap a little but there was also an alternative 'snail's trail' which required even more skill. The machinist started in one corner and went in random directions, in one continuous trail, until he had filled the total surface of the brass.

In mechanical workshops it was called engine turning and was common practice back in the days when people had more time, and craftsmanship was regarded as something to be pursued. Typically a part would start out as a rough casting that was machined to its final dimensions in a lathe or milling machine. It was common practice to give exposed flat surfaces a precision look by finishing with an end mill (the cutting bit) in a vertical milling machine. The milling head is brought down to the final dimension. Then it is lifted and the work is moved over a fixed amount and the next vertical cut is made. It is easy to get an even pattern on a milling machine because the table 'X' and 'Y' controls are calibrated. (It's probably really easy on a modern numerically controlled machine!)

Another way of making the patterns is by bringing a rotary wire-brush into contact with the material to be finished in a series of overlapping circles. The process has been done with pencil erasers or corks and valve grinding compound or with a piece of dowel rod, plastic or hard rubber. It can be done on a drill press but it's hard to make a regular pattern without a guide fixture; a milling machine is better and makes it easy to obtain some very attractive patterns. You can make a fixture out of plywood and with a little practice get a beautiful finish on a drill press – just don't try it freehand with an electric drill or you will get a sort of random pattern that looks home-made.

Another variant runs as follows. Chuck-up a wooden dowel of a diameter of the desired swirl in a drill press. Spin the end of it on some 150 grit sandpaper at about 900 rpm until the end is squared off. Then apply jewellers' rouge (or toothpaste) to the end of the dowel and bring it lightly against the work. Try it on some scrap stock first so you get the feel of how much pressure to use and how long to spin it for the desired effect. It also will help you to practice doing the swirls with the right amount of overlap and in straight lines across the work.

PREPARING ALUMINIUM FOR PAINTING

Aluminium is easy to work but a swine to paint successfully; it is notoriously difficult to make paint stick. The 'formula for success' is:

1. Clean the piece thoroughly.
2. Acid etch the surface. A quick dip in hot caustic soda (lye) will also work. You want to get rid of any shine on the surface. Using lye (sodium hydroxide) to etch aluminium as a preparation for painting works quite well, but the reaction produces aluminium hydroxides and hydrogen (very flammable) so you need to do this outdoors with plenty of ventilation and no ignition sources around. Generally five to ten minutes is enough. Wear eye protection and rubber gloves.
3. Prime with zinc chromate primer.
4. Finish with a coat of colour. Any of the automotive paints will work.

The caustic soda actually serves two purposes here. First, it thoroughly cleans the surface of the aluminium of any oil or grease. Secondly, it etches the surface allowing a good bond with the primer. Just be sure to completely dry the aluminium before painting it and try not to touch the surface if possible, as it will readily pick up oils from your skin. Baking the panel after etching is a good idea; this not only dries out any residual moisture, but can chemically seal the surface too. Let it cool before priming, and also remember to let the zinc chromate primer dry thoroughly before applying the top coat(s).

In situations where it is impossible to chemically clean and etch the surface, degreasing with solvent is vital. Scouring the surface well with a Scotchbrite pad to roughen it will help the paint adhere better. A zinc chromate primer is still advised.

METAL CHASSIS

CLEANING ALUMINIUM CASES AND CHASSIS

The normal suggestion is to use lye (NaOH) but this is a highly caustic and hence hazardous substance and the 'victim' metal needs neutralizing afterwards. Much simpler, suggests their correspondent, is a lye-based oven cleaner and, because this is a paste, it does not run off into places where it should not go.

Another, less aggressive, method is to use stainless steel kitchen scouring pads (not the cheaper wire-wool type, which break up) and *undiluted* washing-up liquid or liquid soap as a lubricant. Then wash with clean water and dry with a cloth.

The aluminium cans surrounding coils are generally less badly soiled and the detergent used in dishwashing machines (Sun, Finish) is very good for cleaning aluminium. A soak in hot water and a dash of the detergent will work wonders on these coil cans.

CLEANING STEEL CHASSIS

Chassis rust offends some people, although it's worth considering that it doesn't spread and is not contagious. More than anything it's a matter of opinion and aesthetics. If you decide that some attention is essential, you have two choices – wet or dry.

The wet method is to use a phosphoric acid-based liquid such as Jenolite or the various rust jelly products sold in car accessory shops. Most of them tell you to wash off the 'slurry' after five minutes or so, otherwise damage can occur. Some of these products claim to actually remove the rust chemically, although most of them convert it to an inert, passive substance.

If a wet treatment is inconvenient, you can use abrasives, either wet-and-dry paper, Scotchbrite pads, a glass-fibre scratch brush or even some kind of wire brush in a power tool. Steel wool is another option but if you are using this, be very careful to blow away the swarf, otherwise you could have potentially dangerous short circuits later. Rubber gloves are advised to avoid possible dermatitis. Some workers like to give the chassis a quick rub with a cloth or paper towel soaked in 3-in-1 or similar lubricating oil.

Alternatively, the chassis can be brushed with clear varnish (any kind) afterwards; both methods will delay the onset of rust.

Note that some steel chassis have a dull grey finish which may be loose cadmium sulphate resulting from previous cadmium plating. Be careful not to breathe in the dust created by cleaning this; it is said to be toxic. The dust can even pass though skin and certainly through open cuts, so you would best wear rubber gloves.

TO PAINT OR NOT TO PAINT

It all depends. If you are fussy, you might care to spray paint the chassis with clear varnish or silver cellulose paint from a car spray aerosol (or if it's an old valve amplifier of the 1950s, a bronze or gold metallic shade). Geoffrey Dixon-Nuttall points out that a good match for Pye MM panels is Ford Roman bronze, whilst early Ekco chassis can be repainted quite well with Ford silver fox (this is a pearl silver-grey colour; never use aluminium paint because it looks too flashy). Ford Rio brown is a useful dark chocolate colour for speaker frames and the like. If you visit your local car spares store, you will soon familiarize yourself with all the truly useful shades in aerosol cans of cellulose paint.

Alternatively, you can brush paint the chassis with smooth or hammer finish paint, which covers up rust and other corrosion rather well.

FILLING HOLES

The occasion often arises when you need to fill holes in panelwork drilled in error or made by a previous owner. A simple solution is the ordinary hole plugs sold by trade electronics suppliers. They look relatively inconspicuous, especially when painted the exact same colour (and at the same time) as the panel/cabinet. On black wrinkle cabinets they virtually disappear altogether. For small holes the old 'put a bolt in it, whether you need one or not' trick works well. You just find a screw or bolt whose head and finish more-or-less matches other screws/bolts already there; another makes little difference. Remember to put a lock-washer and nut on the other side. If it is impossible to gain access to the inner side, you can use bolt cutters or a saw to shorten the screw, leaving just enough to enter the hole. This can then be fixed with epoxy glue.

For a more professional and 'invisible' cover-up, this remedy suggested by Ben Hall sounds good. The first task is to deburr and countersink the holes, with the countersink on the 'good' (front) side. This is a simple task on aluminium and should work as easily with steel. You then take a stiff piece of aluminium sheet, and lightly grease it with shortening (from the kitchen). This is done for two reasons: as a release agent so the epoxy you'll be using will not stick to it and second, so it will form a close seal against the panel. You then clamp it to the good side of the panel and apply slow-cure epoxy glue from the rear (the five-minute variety stiffens too fast and does not allow the air bubbles to escape). You need to make sure the epoxy fills all niches in and around the hole and make sure it reaches into the countersunk area, which helps keep the epoxy plug in from the front. Let this set for a long time, preferably a few days.

After this you can remove the aluminium panel and, using some fine wet-and-dry paper on a sanding block, you can level out any slight imperfections in the epoxy to the extent that, after a coat of paint, you cannot tell from the front where the epoxy-filled holes are.

Some hardware and car-repair stores sell a product called epoxy putty (it has other names too) which rolls out flat from a tube. It is intended for plumbing repairs but could also have possibilities for repairing metal panels. Mike Hanz, however, reports he has never had much luck with large-section epoxy fillers and says it is somewhat disconcerting to put your thumb through an aluminium tape-backed epoxy patch, which he has done more than once. He states that the only lasting repairs that have succeeded through the rough and tumble of everyday use have been solid aluminium filler disks made on the lathe. They have a shape like a top hat, with a thin flange on the back to provide adhesion strength. For 1/16 in aluminium, the total thickness only needs to be about 3/32 in. This has an added advantage of providing an almost seamless fill when you paint the patch with black wrinkle. The wrinkle paint seems to have a slightly different texture over epoxy than over aluminium, making the patch visible in certain lights if you use the large epoxy patch approach.

CHAPTER 16

Practical restoration techniques – plastic cabinets and cloth coverings

PLASTIC CABINETS

Generally these are made of bakelite, moulded celluloid or ABS (a tough styrene), and this applies whether we are talking of radios, televisions, telephones or any other communication or entertainment device. Some highly decorative pre-war American radio sets were made from a moulded resin known as catalin, whilst some older, coloured radios and telephones were made of beetle or similar formaldehyde plastics. Finally, some utility items (such as German field-telephone cases from the Second World War) are found in a very low-grade plastic made by soaking rags in resin and moulding these under pressure. This is not the place for a detailed treatise on plastics but some useful information will be found in the chapter on 'Materials and finishes' at the back of this book.

Bakelite and ABS tend to be dimensionally stable, whereas the cellulose acetate plastics are not and can both distort and deliquesce (dribble away under the influence of atmospheric moisture). There's not much you can do about advanced cellulose acetate decay except look for another set with a better cabinet.

Bakelite can be cleaned and repaired as described below, whilst cellulose acetate and ABS should be treated as for dial glasses made of the same materials. You can use harsh paint strippers and solvents on bakelite but not on ABS and other styrenes; cellulose paints also attack styrene plastics – you have been warned!

DISCOLORATION AND DEGRADATION

It needs to be noted that certain plastic materials degrade with time, a process which generally cannot be reversed. Certain self-coloured plastics (ABS in particular) discolour as well. These problems are discussed in the chapter on 'Conservation techniques'.

CLEANING FINE DETAILS

The modern biological detergents are very useful for cleaning fine details, such as the grooves and mouldings in some plastic knobs. A soak in a biological clothes stain remover such as Biotex for instance will be found to shift almost everything. Assistance with an old toothbrush may also help; never throw these away!

BAKELITE – KEEPING THE GLOSS

Everyone agrees, there are few finer sights than a gleaming article made of bakelite. Bakelite, named after Baekeland, its Belgian inventor, has been described as one of the most gorgeously tactile materials ever made; it is hard and glossy yet capable of being moulded into fine detail. When new it carries a lustrous sheen which is difficult to better, even in black, which is otherwise a pretty lifeless colour. In fact there are few man-made objects more functionally attractive than a gleaming black bakelite telephone, although it doesn't really matter whether it's a radio, telephone or a TV receiver . . . there's just something immensely appealing in the fine details and hard lustre of that plastic.

It's clear that bakelite was made in several different grades; that used around 1950 for interior components of motor cars (dashboards and steering wheels) seems to age and discolour very little. On the other hand, if you compare a Bush TV12 television with the later TV22 model, it is all too clear that the older set had the better finish. The bakelite of some telephones was even worse and soon came to look more like pumice stone. Quite possibly the better variants contained more resin and less filler material. No doubt it was cost-cutting that reduced the resin content.

What a shame then that the lustre vanishes all too fast. So many artefacts made of bakelite seem to lose their gloss and end up brown and porous. It doesn't seem to have anything to do with the age of the item, so why does this happen and what can be done about it?

First, let's look into the technology. Bakelite is a thermoset plastic, that is a plastic which starts molten as a liquid but once solidified, does not revert to its liquid state when heated. In crude terms it consists of a resin (which has the glossy appearance) plus a bulk filler material, usually wood flour. The shiny surface you see and admire is the top layer of resin but this is often very thin. Once rubbed away through atmospheric action, over-enthusiastic polishing or by scorching in the sun's rays, it is lost and nothing will bring it back. You are then left with a pitted mixture of resin and wood flour (or asbestos powder), and wood flour, being very fine sawdust, is not a particularly glossy material. It is this wood flour that looks brown and porous once exposed.

Cleaning should precede any physical restoration of plastic mouldings. Degrease first, if necessary, with methylated spirits and a hard cloth. The restoration techniques you use will depend on whether the top layer of resin is intact; let's assume first that it is. Bakelite and other plastics which are only lightly soiled should be cleaned with Paste Polishing No. 5. This is a waxy polish compound, with a mild abrasive, and extremely effective. It will remove any film of 'gunge' and leave a smooth, glossy surface. Everyone who sees this cannot believe how good it is, far better than the metal polish that people used to use to clean bakelite! PP No. 5 is the stuff the GPO formulated for refurbishing old bakelite phones to new condition, by the way, and you can buy it either in bulk from the manufacturer or in small quantities as Baykobrite from The Radiophile.

The paste should be rubbed on with a hard cloth, then removed and buffed up with a soft cloth. Alternatively you can use Solvol Autosol or similar paste (remove with white spirit) to rub down to the clean plastic underneath, but this always leaves a white powdery deposit which is difficult to remove from cracks and mouldings.

If the surface is already rough and porous the technique described will not help. The remedy then is to cut away the discoloured layer with a hard brush and grinding paste ('soap') if you have a buffing machine. Otherwise use an abrasive polish such as T-Cut or Color-Bak (car paint restorers) or Bluebell (metal polish). Then let this dry, wipe off the residue and inspect the bare material exposed. If it has retained its pigment, polishing with a liquid, real-wax polish will suffice. Buff it up to a high gloss and be prepared to repeat this exercise every twelve months or so.

If the bare material is discoloured, you will need to add new pigment. For black there is an intense black stain in car accessory shops called Back to Black – use it but try not to get it on your clothes. Otherwise you can use shoe polish (work it well into the pores of the bakelite) and buff it off several hours later. Several applications may be required. Then finish off with liquid wax polish.

There are some people who prefer to avoid the hard work and paint low-gloss tung oil varnish on badly dulled bakelite (this usually takes about three coats and gives a very fine finish, not as shiny as lacquer). You need to use a lint-free cloth to apply the varnish and for details and crevices, use a tape-recorder head cleaning swab – these have foam-rubber heads instead of spun cotton). This method, which aims to restore the finish by filling the microscopic pits with a clear substance, is valid if the bakelite has not lost its colour. If it has started to go brown, on the other hand, sterner methods are necessary.

Small cracks and chips in the bakelite can be filled with soft furniture restoration wax, boot polish (my favourite) or car body fillers (cellulose paste or fibre-glass resin types – you can buy these already tinted black, intended for filling cracks in black plastic car bumpers). Clean breaks can be joined with superglue (cyanoacrylate) or Araldite (epoxy resin), though care is needed to prevent smears of glue showing.

These techniques work. I know they do and they work very well. There may be other products and methods, so let us know if you can help.

REPAINTING BAKELITE

Finally, some bakelite products were painted another colour and we may need to restore these paint finishes. Bakelite itself is generally either brown or black, simply because the moulding powders for bakelite had to be dark colours to mask the wood flour and asbestos dust filler material used. What many people call coloured bakelite isn't bakelite at all; these coloured plastics were often too expensive to use for utilitarian products. Paint was often used for bright colours, this being a lot cheaper than choosing self-coloured plastics of these colours (even if not quite so attractive).

Certain brown bakelite radio cabinets were spray painted to provide a contrasting colour (e.g. primrose panels on brown bakelite); equally some bakelite telephones, such as the STC 'Antwerp' phone, were given paint finishes over black bakelite to provide additional colourways.

For repainting bakelite an enamel finish adheres best, preferably an eggshell paint which is not too glossy. A hardware shop will have some spray colours which may match, otherwise you will need to go to a specialist paint supplier (see previous chapter).

Vintage TV and Radio Supply (USA) has two products, Barrier Coat and coloured lacquer enamel. The former is a basic primer for rigid plastics and bakelite; it ensures good adhesion and provides a barrier between the plastic and finish coatings. The coloured lacquer enamel is fast drying and has a superior hiding power. It comes in just two colours, gloss black and antique ivory. In practice, most cellulose or enamel spray paints could substitute for this product.

ADVANCED BAKELITE REPAIRS

It is possible to repair broken bakelite in a way that is very hard to detect. The first thing is to make solid joints and, if you are lucky, you can piece the bits together and glue them in place with superglue (cyanoacrylate). Doing this slowly and carefully is the main secret; it may take quite a few 'dry runs' to get the three-dimensional jigsaw to go together properly. Plasticine will help keep the pieces together while the glue cures. For pieces under mechanical stress, superglue will probably not work; in this case you will need to use a modeller's drill to make holes for inserting short pieces of wire to pin the joint in several places. A traditional slow-setting epoxy glue such as Araldite is used to make the joints and this can be built up behind the joint out of sight to give added strength. When nearly set, surplus Araldite can be trimmed away with a sharp blade.

An alternative method which also has a lot to commend it is the use of glass-fibre cloth and resin (bonding paste), as used for patching car bodies. The materials are used on the inside of the case to be mended and the trick is to use a file or grinding tool to thoroughly roughen the surface. Using a putty knife or wooden spatula, the roughened area is then covered with a thin layer of resin (about 1/16 in thick). A piece of glass-fibre cloth is then pressed on, smoothed out and left to cure. Once fully set, another thin layer of the bonding paste is spread over the cloth for extra strength. This can be sanded for a smoother finish if desired when cured. Yet another possibility is Loktite BlackMAX adhesive, which is a cyanoacrylate that has rubber dust loaded into it. This helps the impact resistance and also gives a black colour.

Some hair-line cracks will remain and these can be filled with the self-curing black resin body-filler paste available at auto accessory shops. Once cured, you can use very fine wet-and-dry paper (as used for car body work) to smooth the joints until they can no longer be seen or felt. Finally polish with Paste Polishing No. 5 (Baykobrite). For brown bakelite you could mix artist's powder paint with clear epoxy glue but be sure to mix in the powder extremely thoroughly.

REPAIRING OTHER PLASTICS

A lot depends on where the break occurs, how visible the damage is and how much room there is to create a repair. One proven method is to use glass-fibre cloth and epoxy resin to add strength behind cracks. Special packs can be had at car parts stockists and the repair is made on the inside of the cabinet where it's mostly invisible.

With Perspex, ABS and styrene cabinets, joints can be remade by applying the appropriate solvent with a thin artist's brush. A good place to buy bottles of solvent and the brushes is good hobby shops – the sort which sell plastic construction kits and plastic sheet and profiles – whilst a specialist plastic sheet distributor would be another good choice. Because these are thermoplastics, not thermosets, you can also dissolve shavings of the plastic in the solvent and make this into a kind of sloppy putty to fill cracks and gaps. The solvent will evaporate and the filling can be polished with very fine wet-and-dry paper and then polished with plastic polish.

Minor cracks can be prevented from spreading by 'stop-drilling'. This involves finding the end of the crack with a magnifying glass and drilling a hole (maybe 5/32 in diameter) just beyond the end of the crack. This will relieve any stress that may be present and the crack will not travel further. It's worth loosening chassis mounting bolts a little; this will lessen the risk of cracks due to expansion when things get hot. Hair-line cracks can be repaired by running superglue into the crack (from the inside of the cabinet). Be sure to flex the cabinet a little to ensure the superglue does find its way into the crack.

SEIZED BAKELITE PARTS

A recurring question is how do you unscrew the earpiece cover on the handset of an old bakelite phone? With great care is the facile answer. One method is to drip some penetrating oil into the minute gap around the cover (the oil won't harm the bakelite), leave for twenty-four hours and then get a stronger grip on the cap, either by wrapping several large rubber bands around it or, better still, by using one of those thin, flexible neoprene mats specially made for getting a grip on obstinate objects. There are also special tools sold in kitchen shops for unscrewing lids on jam jars and at car accessory shops for removing oil filters; the tools work in similar ways and may be of use. Failing this, heat can help; you place the offending object in a bowl of scalding-hot water (and take great care!).

On bakelite handsets (of the British pattern) it's not always clear how the mouthpiece is released. Actually this is very easy; there's a small hole in the rim of the mouthpiece, behind which is a metal detent. Forcing a thin metal point (such as a thumbtack or drawing pin) into this hole will release the mouthpiece.

PUTTING A GLOSS ON OTHER PLASTICS

Minor scratches and haziness can be polished out of most plastics with suitable abrasives. The Plastic Polish made by Greygate Chemical Company has these in a liquid suspension smelling strongly of camphor and I have found this superb on plastics of the Perspex, Diakon and ABS variety. Others swear by a repair kit called Micro Mesh, available from Antique Electronic Supply. The latter consists of several grades of abrasive paper and cloth that are used wet to gradually remove all of the surface blemishes. Both of these are industrial products, used commercially in the airline industry to clean and polish aircraft windows. Both are messy and take some time, but they will take care of deep scratching and leave you with a surface looking like new.

Experimenters who would like to do a bit of kitchen chemistry can try equal amounts of toothpaste and baking soda, mixed together into a paste, and then applied in a thin coat with a piece of soft cloth (such as T-shirt material). Results are said to be slow but rewarding.

BAKELITE, EBONITE AND VULCANITE PANELS

These materials were all used as panel material at one time and are confused by some people. Technically, they are entirely different in that bakelite is a plastic material and, on panels at least, has a hard, glossy black appearance. It has a disinfectant smell if you drill it and so it should, because it contains a phenolic resin.

Bakelite and ebonite were both used for panels on early radios and pieces of test gear; strictly speaking the bakelite panels were frequently made of bakelized paper, that is paper material soaked in black bakelite resin.

By contrast, ebonite and vulcanite are forms of very hard rubber (and smell like rubber when you drill them); sulphur was added to achieve the hardness. Ebonite in particular tends to dry out and fade to a pale brown shade, although its started off as ebony black (some people say the change is due to the sulphur 'sweating' out of the compound). Vulcanite on the other hand is red, sometimes with black swirls in it, and tends to keep its looks better.

As well as the front panels of radios and other electrical equipment, some knobs are made of ebonite, as also are the handles on lever key switches and even the ear-caps and handset stems of telephones from the pre-bakelite era. Ebonite was also used for the safety coating on the bell receivers so common on early telephones, and the cases of propelling pencils was another common application. Vulcanite was often used as a basis for terminal blocks but so was red fibre; a kind of vulcanite with swirled patterns (looking like old rubber bathroom tiles) was also used for radio panels.

Fading is very common but it is probably a mistake to try and recapture the original colour of these materials, even if you know the shade you want. If you must try and restore these materials, don't try painting the surface but instead brush on ink (Indian ink for ebonite) or rub on liquid wax polish. Silicone polishes (such as Armor-All or Back to Black) are not recommended as they are so difficult to remove later.

LEATHER CLOTH AND REXINE

Some wooden cabinets were covered in artificial leather cloth or rexine, and some steel cabinets had a textured-vinyl plastic surface applied which looks and feels similar to rexine. The problem is that dirt collects in the texture and the surface can fade or dry out. To remove the dust and dirt you can use a nail-brush or toothbrush together with a foam cleanser or plenty of soapy water. Several attempts may be needed for stubborn, greasy dirt.

Mould is another problem which sometimes affects cloth-covered cases. Washing with a bleach-based cleaner is a good idea. Some enthusiasts talk of 'nuking' the mould by putting the case in a microwave oven for a minute but this seems a shade reckless to me.

At this stage you may decide the colour is so faded that a spray-paint session is necessary. Specialist car accessory shops will sell you special vinyl paint but, at a pinch, cellulose or enamel paints will do as well.

To cure the dried-out look I recommend applying baby oil with a paper towel. Several applications may be required to get the desired degree of glossiness. You could also use liquid silicone car polish (Armor-All or Back to Black) but this will prove very difficult to 'undo' if you wish to change the finish later.

GRILLE CLOTH FOR LOUDSPEAKERS

A number of manufacturers used expanded metal to cover grilles and, however unattractive this may look, at least it does not deteriorate as much as cloth does. Over time cloth inevitably becomes soiled; it often fades and may suffer from holes or tears.

If the cloth is damaged, it will need replacement. Some enthusiasts set great store by suit-lining material, a sort of shiny rayon cloth in gold or brown colours. This can be had new from a dressmaker's shop or taken from old clothes – charity and thrift shops are a good source. Generally, however, proper grille cloth usually looks and works better. A quick glance at the 'Buyer's guide' chapter will reveal that there are several suppliers, some of whom even make exact replicas of original patterns on which some collectors insist.

If you feel the original cloth should be preserved, it may be possible to take it out and wash it in a gentle solution of soap flakes. Other collectors prefer to use the services of a dry cleaner's shop. For stains try dilute hydrogen peroxide on a hidden corner. If it does not bleach the cloth, it removes some organic stains. (It also removes them if it *does* bleach the cloth, of course!) Gerard Tel says: 'On two occasions I have used Poudre Éclat, an Amway product, with very satisfying results. Poudre Éclat is a household bleaching agent, strong against stains but safe for cloth. It does not remove rust marks'.

Incidentally, grille cloth can be stretched on an embroidery hoop for gluing to a board or cardboard. By using the hoop, the cloth can be stretched tight and the weave straightened up prior to gluing.

CHAPTER 17

Practical restoration techniques – backs, knobs and dials

BACK PANELS

Twice while writing this book I have answered the phone only to be asked where the caller can find a back for a Bush TV22 television. The answer of course is about 12 inches behind the front. It's not so funny when the back is missing and goodness knows where (or why) all these back have disappeared. I certainly have no idea!

The good news is that replica backs are now being made for popular sets, such as the Bush DAC90 radio. Supplies are variable but if you go to the vintage radio swap-meets, you may see them there. Alternatively, you could make your own. Nigel Pollicott (0181-840 1075) is now supplying tough insulation board for making replica back panels for these (and other) sets. He says it can be drilled, cut (use a lino knife), punched (use a leather punch, obtainable at craft supplies shops) and stained (wood stain from DIY shops) to produce very presentable replacement back panels. Sounds good and he'll willingly send you a sample piece. The material is 2.3 mm thick and, at the time of writing, comes in sheets 800 × 1200 mm for around £10 each.

Probably the best way to cut round holes and small round-ended slots is with leatherworking punches; the slot types are called bag punches (because they cut holes for the straps of bags). Bobbi Barmore reports that with a set of decent punches, a soft-headed mallet and a proper cutting surface (soft wood, like pine, seems to be about right, or dense rubber over wood), you should be able to do nice work. A mat cutter or other sharp knife will do for long straight cuts, with punches for rounded corners.

The only thing missing on your DIY back is the lettering. On 'the real thing' this was often stamped on, using silver and red paint. Unless your lettering skills are first-rate, you are better off omitting this.

In some cases you may be able to restore an existing back and this excellent 'recipe' appeared on the Internet from an anonymous correspondent going by the initials PFJW. He writes that, if the back exists at all in any form, it can be salvaged with some work and a bit of preparation.

Preparation: You need to make a 'back press' consisting of two 18 in squares of 3/4 in C plywood, laminated with 1/4 in masonite. Each panel should be a nominal 1in thick. Drill for eight 1/4 in bolts, three across each side. (If you want to be clever here, countersink the laminated side for a 1 in compression spring, enough for the spring to become fully compressed when the panels just touch . . . this need only be done at the four corners.) Get some good carpenter's wood glue, the yellow stuff, and cut it three parts glue to one part distilled (not tap) water.

The drill: In a shallow pan just larger than the back to be repaired (or larger if you have the glue for it), pour enough of the glue mixture to almost but not quite cover the back. Put the back in the mixture, printed side up. Using a very soft brush, brush the glue across the printed side, while letting it soak into the unprinted side . . . for no more than a minute. Now is the time to pour all the broken and flaking bits you carefully saved into their proper places – the shaped masonite grommets in the broken-off corners, the missing bits in place, whatever. On the bottom plate, as described above, place one full thickness of cling-film food wrap and one full thickness of waxed paper (in that order), the back, printed side down, silk-span paper (available at hobby shops in many colours) a bit bigger than the back, waxed paper, plastic-wrap, the four springs and the top plate. Using wing-nuts, tighten down the bolts evenly until not quite dying-strain tight. Let this sit for three days, or seventy-two hours, whichever comes first.

The result: What should result (and has often for the writer of this recipe) is a passable back which only requires the punching out of the silk-span with either a knitting needle or a craft knife. The silk-span acts as a bridge to hold the repair bits in place and, at a pinch, can be laminated on the print side as well if the almost clear rice-paper type is used. The springs help spread the plates apart as inevitably some glue leaks through. The wax paper, if a decent brand, will not stick to the work, but if this is a worry, use two layers of plastic wrap. This may leave wrinkle marks, however.

RENOVATING PLASTIC KNOBS

This is easily done if you follow these steps.

1. Clean the knobs with strong detergent, a toothbrush and running water. Let them dry.
2. Apply a coat of black (or brown, as appropriate) shoe polish (the old paste type) and buff.
3. Repeat step two with some acrylic car polish such as Armor-All.
4. Give final polish with a soft cloth.
5. White lines may be replaced, added or fixed with white acrylic paint. It may be worthwhile rubbing in some household dust to take away the brilliance of the paint once dry.

GLASS AND PERSPEX

DIAL GLASSES

Tuning scales are generally made of glass, celluloid or Perspex and each of these materials presents its own problems.

Glass is the easiest to clean – if you do it correctly. Of course, if the glass is cracked you do indeed have a problem, although there are some cyanoacrylate 'super glues' which are very successful in rejoining glass. Look in car accessory stores too for special resins for filling cracks in glass windscreens; they might conceivably be of use. The usual problem is dust on the underside of the tuning scale and the temptation is to take out the glass and sluice it down with soapy water or foam cleanser. Don't! Instead be very careful! These scales are often screen-printed or use some kind of water slide transfer and any dampness or even touching can cause havoc. So please be careful – I have seen a cherished radio ruined by soaking the dial glass so don't destroy another one.

Often a flick with a soft brush or duster is the kindest treatment. If the dirt is really tough, then Tony Statham's trick is to obtain an 'everlasting eraser' (Staedtler is one brand) from a stationer or drawing office supply house. This consists of a plastic holder (the size of a fibre-tipped pen) and a long, snake-like white plastic eraser material. Once in the holder, the eraser is extremely controllable and you can safely rub close up to where the paint starts.

There is a special liquid acrylic preservative, called CYG Lite, which is recommended for fixing the designs printed on radio dials and any other piece of glass with designs printed on the back side of it. Apart from radios, this includes some antique clocks and, of course, the back glasses on pinball machines. If the ink is starting to flake off, CYG will prevent further deterioration. If it hasn't gone bad yet, CYG will guarantee it never will. Just don't get it colder than 10°F. One quart will do a couple of dozen radio glasses; just be sure to get the glass truly level before proceeding and use the supplied eye-dropper.

REPAIRING BREAKS IN GLASS

Scott Robinson reports a good way to repair breaks in flat glass panels. First, go to your local model railway or aircraft hobby supplies shop and get some miniature brass channel section stock that fits loosely over the edge of the glass; bring along the glass to ensure a close fit. Then, using clear epoxy, glue the glass pieces together and lay them on wax paper on a flat surface while the epoxy sets up. Longer set-up times usually give stronger resin. Then epoxy the brass channel pieces to the edges of the glass along its full length, bridging the crack(s). Incidentally, the purpose of the brass is to reinforce the glass so that it can be handled normally without breaking again at the crack.

Another method involves using an automotive windscreen glass repair kit available from some auto-part stores and windscreen specialists. The kit consists of a syringe-like tube of clear resin and a suction cup. The cup fits on the tube and is then pressed to the glass, making a good seal. You then simply inject the resin into the crack. For long cracks you might have to use more than one injection point. The resin sets up in about four hours and hardens in twenty-four hours or less, depending on temperature. The crack virtually disappears. You should be able to use a similar procedure if the crack is relatively clean and you can clamp it together gently.

CELLULOID

Celluloid is another matter. Sadly it often distorts and goes yellow and brittle with age . . . and there is little you can do to correct this. Best to treat it as a natural function and live with it. Alternatively, there are people who sell acrylic plastic replacements for the dials in popular American sets (Antique Electronic Supply; see 'Buyer's guide' chapter) and there is also an enthusiast who will make custom replacements for you (L.W. Terrell; see 'Buyer's guide').

This is what one enthusiast reported on a warped dial strip which was very brittle and yellowed. After removing it very gently to avoid cracking he heated up some water in a pan very slowly with the dial strip in it until the dial strip became pliable again. This also seemed to 'clean' up the yellowing almost to its original white colour. The water should not be hot, but only warm enough to cause the strip to bend easily. The next step is to take the dial strip out and rapidly lay it flat underneath a cutting board with some weight on top. This keeps the strip flat until it has cooled and avoids unintended warping. After it had cooled, the strip was almost like new and could be wrapped around the dial strip holder and secured with no problem.

The same applies to coloured filters that have become distorted due to the heat of dial bulbs. Assuming there is no silk-screening to be harmed by heat and liquids, you put the warped plastic into water as hot as you can stand. After a while, you can start bending the item gently to the shape you want. Then you can take it out of the water and allow it to cool back to room temperature. Another proposal is to heat the object with a hair-dryer set on low, avoiding any tendency to overheating. If you can somehow set the object in a drill press (with the motor off), you can turn it periodically as you gently shape it with your fingers; this will quickly show up any wobble that needs to be adjusted (obviously this works only for circular objects).

Two collectors report success in straightening bent plastic dials by putting them into an oven at about 130–160°F, sandwiched between two pieces of glass with a weight on the upper piece of glass. The oven must have an accurate thermostat; baking time is a few hours. Afterwards leave the 'sandwich' to cool with the weight still on top. Different plastics soften at different temperatures of course and the heat indicated should be about right for older cellulose ester plastics (celluloid and the like). For polystyrene and plexiglass something around 212°F would be more appropriate. All experiments are carried out at your own risk!

STYRENE DIALS

Cleaning clear plastic dials requires great care; you only get one chance. My favourite remedy is the official plastic polish used for cleaning aircraft windscreens and made by Greygate Chemical Company (see 'Buyer's guide' chapter). This gives a very good finish when applied with a bit of elbow grease. Antique Electronic Supply (Tempe, Arizona) has an alternative approach using a micromesh cushioned abrasive cloth which, it says, is always effective in removing scratches and haziness from plastic cabinets and clear dial windows.

SUBSTITUTE GLASS SOURCES

George Gonzalez writes that where the dial 'glass' is broken, cracked, yellowed or just plain missing, it makes sense to check the usual sources. Antique Electronic Supply has several sizes of replacement dial covers. Your local watch repair shop may have a source of circular glass replacement dial covers for clocks. Try these places first unless you like to experiment, but if that fails, you can try making one.

First, you need a mould shape. If you have the original cover, even badly cracked, it can be used as a mould. Cracked covers can be patched with plastic tape. If you don't have the original, you need to find something similar, for instance the base of a sugar bowl or a household canister. Supplies necessary include a small bag of plaster of Paris from the hardware store and some thick cloth gloves. Visit a plastic supply house and obtain from their scrap bin some off-cuts of 0.050 in Lexan sheet at almost no cost. Lexan is polycarbonate plastic and there are, of course, other names for it. The thinner plastics are easier to work with but some older radios used (and need) much thicker material. These methods work for up to 1/8 in thick plastic. You will also need plastic polishes which you can obtain from the plastic supply house or from Greygate Chemical Company (see 'Buyer's guide' chapter).

Gonzalez says you'll also want to stop at the local junk shop and find an electric skillet, one with a good, calibrated temperature control and a close-fitting top. You don't really want to use your good household one. It's also a good idea to have a spray can of silicone spray or a similar slippery mould-release agent.

Work starts by making up a batch of plaster of Paris. You spray the inside of your mould with silicone spray, then pour the plaster of Paris into the mould, up to the very top if it's the original dial glass, or up to the 'right' level in the substitute mould. The right level is the level at which you want the dial to flatten out around the edges. You may have to do some eyeball guessing to find this level.

Wait several hours for the plaster to harden, then remove the mould from the form. If the top is bowed around the edges due to surface tension, grind it flat by rubbing it on some flat concrete surface. If the original cover had cracks, you'll have to take some glass-paper to the mould and erase the crack lines. The mould must end up perfectly smooth, with no bumps, bubbles, cracks or dirt specks. Any flaws in the mould will likely be transferred to the final product.

Now find a place outside to work where you can plug in the skillet. Turn on the skillet to 250°F. Bake the plaster mould for two hours. This drives off the excess moisture which would otherwise condense under the plastic.

Then turn the skillet up to 375°F. This seems to be the best temperature for moulding Lexan. If you got the plastic at a picture frame supply shop, or another place where you can't be sure of its exact composition, you should experiment to find the best temperature. Put a small strip of the plastic in the skillet and keep raising the temperature by 25°F at a time, until you find a temperature where the plastic will sag slowly when you lift up one corner. If it sticks, melts or forms small bubbles in the plastic, the temperature is quite a bit too high. A very slow sag is optimum.

When you've found a 'good' temperature, you need to spray the plaster mould with more silicone. Cut the plastic with sharp scissors to the approximate shape, leaving about half an inch around the edges. If the cut edges form millions of tiny cracks, you have a piece of styrene, which is quite troublesome to mould. Now place the plaster mould in the skillet. Put the cover on and let it warm up for at least half an hour. Periodically lift the lid to let out any moisture.

After thirty minutes, spray the mould one last time with silicone. Place the plastic sheet on top of the mould. Put the cover back on and wait. Every ten minutes, take off the lid and inspect what's happening. If you are lucky, the sheet will be sagging around the mould. It may take quite a while for this to complete.

If you are lucky, after forty-five minutes to an hour you'll find that the plastic has sagged into place, or close enough. If you think it is ready, turn off the skillet, uncover it and let it cool off with the plastic in place. When cool, you should be able to separate the plastic from the mould. With a soft cloth, wipe off any white plaster residue. If it looks reasonable, try it for fit. You'll probably have to trim the edges a bit for best fit and appearance. Some plastics will get brittle after all this heating, so trim it with care. A light polishing with plastic polish will remove any scratches.

If this does not work perfectly the first time, try it again, at a higher or lower temperature, perhaps with more or less help from the tongs, scissors or fingers. You can afford to experiment a bit. Failing this, you might approach the local art college and see if one of their staff or students would take on this task as a paid project.

CORRECTING YELLOWING

A technique for cleaning yellowed dial drums advises the use of 'dilute (0.1%) sodium hydroxide solution' to safely clean and slightly whiten yellowed dial drums. Caution is advised, however; many collectors actually expect old radio dials to look yellowed and will be suspicious of anything which looks too fresh. As one collector said, old sets lose the romantic 'flavour' of age if you remove all the visual clues.

REMANUFACTURING

If all else fails you may have to resort to remanufacturing missing parts. How keen are you to make a really good job of that restoration? Of course there are some people who say including newly made parts in a restoration job is unethical (that's not the opinion of most museums) or you may prefer to wait for the correct part to turn up; but otherwise remanufacturing is the only answer.

The enthusiasts who restore old cars and commercial vehicles are past masters at this art and indeed, many of the owners' clubs for each marque do a healthy business supplying replica in-demand parts. But even where these are not available, they can be recreated. It helps, of course, if you have at least one of the missing parts to use as a pattern, otherwise dimensions (taken from the original or from photographs) may help. You may decide to make just the number of parts needed for your own purposes or you may make a short production run (and recoup some of your costs).

A glance at the Eastwood Company's catalogue shows catalogue no. 6611, *The Complete Handbook of Sand Casting*. This book tells you how to set up a foundry in your own workshop and pour metal castings with sand moulds. It shows you the tools needed, types of sand required, techniques of pattern and mould making, and so on. This assumes you are working in iron, brass or lead. If this idea doesn't appeal, there are several model manufacturers who make die-cast kits for model trains, road vehicles and model soldiers from a low melting-point metal similar to pewter. If this latter material would have sufficient dimensional stability for your application, then it is easy to make replicas by pouring the alloy into a mould made from a special type of silicone rubber. These model makers will generally add your part to the mould they are making from their own patterns, and will turn out the castings you require. The cost will probably be quite modest.

The same kind of mould can also be used to make castings in resin (which is ideal for making replica volume control knobs and similar items). Look in a good craft supplies store for supplies.

The Eastwood Company has another book (no. 6608) *How to Cast Small Metal and Rubber Parts* and in fact they supply various grades of liquid rubber, primer and flex additives for this purpose. You will also see in our Trade supplies section the names of a couple of American antique radio specialists who remanufacture parts.

Many metal components can also be fabricated by turning and milling from solid brass, steel and the like, and for this you need a precision engineer. Ask your local technical college if they have any people who take on small, private jobs or apply to a local engineering company. I have found these firms are not too expensive if you make it clear you are not in a blinding hurry for the job and are prepared to pay cash. Suddenly nothing is impossible!

Another potential field for home manufactures is casting small components, such as instrument knobs and other small mouldings, in resin. Supplies and handbooks are available at your local craft or hobby shop, although achieving perfect results may require the acquisition of some skill. It may be worth asking the shopkeeper if any of his regular customers might take on bespoken tasks. A similar caution applies to lost-wax casting in brass and centrifugal casting in pewter and white metal; the former technique is used by manufacturing jewellers and dentists, whilst the latter is used by firms producing soldiers and fine-scale models. With a little persuasion, any of these organizations might be prepared to take on the occasional custom job along with their regular work.

MISCELLANEOUS

REMOVING KNOBS

Control knobs should never be levered off, be they stuck grub-screw types or push-on clip types; there's a risk of damaging the bakelite knob or the veneer on the cabinet. A safe technique is to use a duster or cloth wrapped around and behind the knob. Pull the cloth – you would not believe the force you can exert this way – and you will find it works almost always and without damage.

Some knobs refuse to be parted from their shafts for renovation. Applied force is not the answer – patience is. The recommended method is to place a few drops of proper penetrating oil in each screw hole in the knob . . . and let this soak in overnight. As if by magic, the fluid works its way around the set screw and breaks it free when you apply the wrench the next day. Experts say this works every time.

DIALS AND DRIVES

The stringing of complicated dial cords can be a real pain, and you swear that you don't have enough hands. Ken Owens reports that the use of a tool that will retain its grip on the cord can be a great help. It can act as a weight to hold the cord in place when draped over a pulley or hung over the edge of the bench. An angler's fly-tying hackle plier is the perfect tool for this. Also, locking tweezers, a surgeon's haemostat with the serrations taped over or an alligator clip filled with solder can be used. A crochet hook is a handy tool for stringing dial cords up over pulleys and around drums etc. They come in several sizes; the larger sizes seem to work best.

Always use a quick drying glue, such as Duco or Radio Cement (USA) or UHU (Europe), on the knots in the cord. This will keep them from coming untied later on. Replacement drive rubbers can often be made by cutting short sections of hose. The hose can be shaped by mounting on a bolt, chucking-up in a drill, and cutting down with a coarse file or sandpaper backed by a piece of wood. For slipping metal-to-metal friction drives, clean the drive surfaces with a degreaser. The spray degreasers can be sprayed into a can to provide a small amount of liquid for soaking of submerged metal parts.

SLIPPING DIAL CORDS

Experts say you need to tackle this problem from two angles – lessening the drag and increasing the pull. You can lessen drag by making sure the bearings of the tuning capacitor roll smoothly and also making sure all pulleys over which the cord passes run freely. A very small drop of oil on each pulley shaft often helps. Be sure that the oil does not get on the dial cord though. The tuning capacitor sometimes gets very gummed up with old grease and oil. A thorough degreasing and re-oiling is the solution for this problem. Extra drag may come from having too much tension on the dial cord and decreasing the tension can sometimes remove more drag than pull, which is of net benefit.

To increase the pull, the following deserve attention:

1. Clean the cord. A solvent such as alcohol will remove a long-standing build-up of greasy dust coating the dial cord.
2. Roughen the knob shaft. This is a last resort but roughening with coarse glass-paper will often add sufficient grip to prevent slipping.
3. Add stickiness. The special kind of rosin used on violin bows (from a music store) is helpful. Put some on the dial cord to increase its grip on the knob shaft. Candle-wax or beeswax makes a good substitute.
4. Add tightness. Add an extra turn of the cord around the shaft.

ENGRAVED PANEL LETTERING AND DIALS

Many instrument panels were lettered by engraving the letter forms and then filling in with paint of a contrasting colour. Cleaning often loosens this old paint, so what do you do?

Antique Electronic Supply sells paint pens (sticks of nearly solidified paint, made by LA-CO Markal) especially designed for repainting lettering engraved into panels. They are easy to use; you rub the paint into the lettering, then wipe off the excess with a hard cloth. Any residual smear can be removed with a paper towel made slightly damp with paint thinners or some other common solvent, but take care you don't remove too much paint. These Lacquer Stiks are made in white, black, gold and other colours.

There are other techniques, such as overfilling the engraving with paint, allowing it to go tacky and then using a rubber squeegee and sponge or paper towels to remove the surplus, but these require far more skill. If you do intend to try these, use water-based acrylic paint – the solvent for removing this is water and you will not risk ruining the main paint finish. Any haze of white paint remaining around the lettering afterwards can be removed by putting a squirt of window-cleaning fluid on a nearly dry cellulose sponge and gently rubbing the haze away.

ENGRAVED IDENTIFICATION LABELS

These are sometimes damaged or missing but can be replicated – at a cost. Plastic labels are generally made in one of three ways: (i) by engraving plain material and filling with paint of a contrasting colour; (ii) by engraving through the top layer of a 'sandwich' of laminate to reveal a different colour below; or (iii) by heat stamping, imprinting a legend with foil or paint. Metal labels are sometimes produced by method (i) above or, more frequently, by a process of photo-engraving (chemical etching), printing on to anodized aluminium or even die-casting.

Several of these processes are considered old-fashioned today but there are still firms which can replicate any or all of them. Engraving on to plastic is generally done by computer-controlled machines today but some firms still have the old Taylor–Hobson equipment which used a pantograph device and separate metal templates for each letter. The best thing is to show a sample of the work you want copied and see if the firm has the right equipment. Your only potential problem is that the celluloid used in former years for labels on telegraph instruments and early radios is no longer easy to find; you may have to sacrifice some material for this. Modern engraved plastic labels are made on a durable laminate called Traffolyte.

For printed and photo-engraved metal labels there are now a number of companies specializing in meeting the needs of classic car enthusiasts, who also have a need for accurate replicas. An accurate photograph of the original to be copied is essential, otherwise these firms will have to produce new artwork. This is generally done by computer and many computer operators do not have access to the traditional typefaces and letter-forms used in the past. With care, however, superb replicas can be created.

See Chapter 24, 'Professional services directory' and your local *Yellow Pages* directory for details of companies who may be able to help you.

LEATHER STRAPS

Many older portable sets have leather handles which have seen better days. Whilst some shoe repairers may be able to help you with a brand-new one, a better idea is to visit the local charity shops and boot fairs, where you should be able to find a period replacement on an old suitcase at modest cost. With luck you will be able to find a good match for the colour and these suede straps will be a little worn, making them look like they belong on the radio.

FLOCKING

Record turntables used to have a flock surface and renewing this finish is problematic. One correspondent writes that, although he uses the right kind of flock, his turntables always come out looking like a thick layer of felt rather than a bristly, light, flocking job. The way he was told to do it was to paint the platter with a slow drying epoxy paint of a similar colour as the flock then, while still wet, toss the flock at the turntable. Norman Lehfeldt has used a normal duster with some success for flocking and another idea would be to use either baize cloth or the self-adhesive flocked plastic sold in furniture and hardware shops for putting below ornaments. In each case, the result does not reproduce the original effect, but is possibly better than the worn, scratched condition in which most old turntables are found. Norman believes one of the problems is that the flocking he has been able to find is composed of shorter fibres than the original and surmises that the original flocking process may have been electrostatic.

DEALING WITH ASBESTOS

Many old radios (and other electrical items) contain pieces of asbestos sheet, which is used as a heat barrier. Spraying or brushing this old asbestos with plenty of clear lacquer is a good way to make the set safe to use. Asbestos is a health threat only if inhaled or ingested and it is often far safer to simply do something to stabilize and encapsulate it rather than to remove it (which creates all sorts of airborne fibres and necessitates elaborate precautions). The lacquer may reduce the value of the asbestos pads as fireproofing but at least the fibres stay put.

BEATING CORROSION

Battery terminals can suffer from corrosion. The solution is the household product for removing limescale around bathroom taps, to be found at most supermarkets. Just soak a Q-Tip or old toothbrush in it and scrub for a few seconds. You may find that the chrome plating has been eaten off the contact springs and a little solder will make them look better.

REMOVING MARKER INK AND LABELS

It's remarkable how many savages use spirit markers (Magic Markers and the like) to price items at swap-meets. Afterwards the ink bonds with the paint or plastic and you cannot remove the 'ghost' image. Americans have a product found in hardware stores called Goof-Off. Another good product is called D-ink, sold for removing ink from typewriters. Other products to try are Mr Muscle, acetone (finger-nail polish remover), lighter fuel and various household cleaning fluids. Toluene is also effective but it is a hazardous substance. Some of the products mentioned will attack plastic finishes so take care.

Sticky labels are best removed with lighter fuel but understand that this

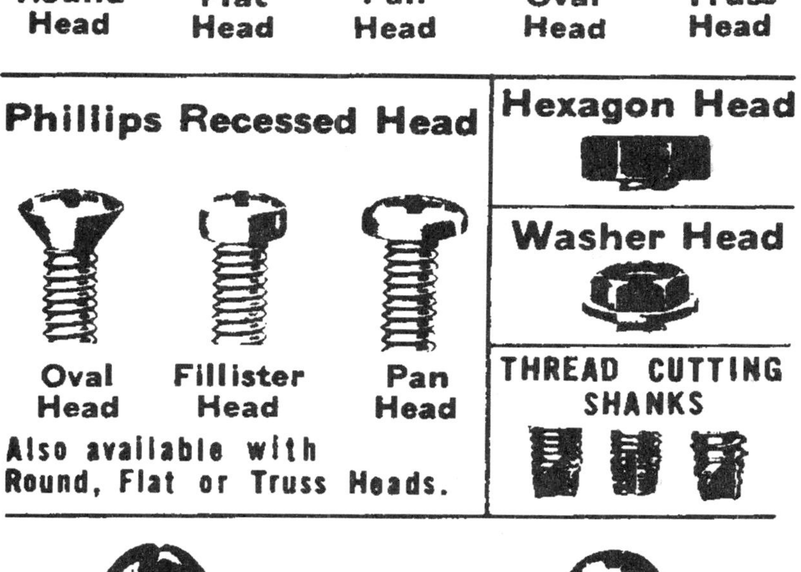

Figure 17.1 Common screw types. Oval head is also known as raised countersunk. Flat head is also known as countersunk.

will take several minutes to dissolve the adhesive. The citrus oil based Mykal Sticky Stuff Remover is also good.

SCREWS AND FASTENERS

It goes without saying that modern cross-head screws look wrong on older equipment, as do screws with violently shiny plating. It is worth keeping a tin full of old screws in good condition retrieved from 'donor' items which have been dismantled. Today's metric screws generally cannot be used in place of BA and UNC screws, and the day may yet come when these older screws are unavailable. It's worth hoarding them now!

CHAPTER 18

Mechanical matters

I couldn't think how else to describe this miscellany, which covers matters as diverse as telephone dials, drive belts and the pulleys, gears and other parts concerned with tuning and wave changing in radios.

TELEPHONE DIALS

These devices have a wonderful ability to run sluggishly even after cleaning, lubrication *and* replacing the spring, which is most frustrating. The cause is probably a combination of the solid paste of dust and oil which accumulates over the years plus mechanical misadjustment. The spring problem is a real bogey; the metal is very brittle and so particular to the pattern of dial in question that cannibalization is sometimes the only solution.

A proven course of action is as follows. Disassemble the whole thing (remove finger wheel and number plate), then wash in petrol or other solvent to remove dried oil and grease. Check that the coil spring has not, in fact, broken inside its cage (observe great care when you release it, as it will fly across the room and could do someone real harm). Lubricate the bearings of the reassembled dial with watch oil; do not allow any oil in the governor. Tandy shops sell a microspout oiler containing oil and Teflon particles for a modest sum. Check all adjusting screws on bearings and the speed governor, ensuring that the mechanism runs freely but without excessive play. The dial should now run at ten impulses per second; in the absence of a dial speed tester you can dial zero and, whilst the finger wheel returns to its home position, count 'one thousand, one hundred and one', preferably not out loud unless you wish to confirm you are mad.

The various telephone administrations and manufacturers have produced far more detailed instructions but these will suffice for a careful person.

Disassembling dials and changing telephone dial labels (dial cards to our colonial cousins) foxes some people, simply because many dials were designed to make access to their internals as difficult as possible!

The easiest dial to 'get at' is the BPO type 10, which is found on most 200-, 300- and earlier 700-type telephones. The spring circlip retaining the celluloid cover for the dial label is easily removed with a small screwdriver, and once the label has been removed, the centre screw holding the finger wheel to the rest of the dial mechanism is clear for all to see. The no. 21 dial which replaced it has a moulded plastic finger wheel and a label cover which is easily damaged if levered out. The trick here to remove the clear plastic cover is either to use a special rubber sucker (the official method) or to insert a small screwdriver blade at the 3 o'clock position to the right of the finger wheel. Levering upwards will cause the cover to pop out, allowing removal of the dial label and exposing the keeper screw. You can, of course, flex the finger wheel and ease in a finger nail to remove the cover.

Foreign dials work differently. Older American dials have a pressed tinplate frame ring for retaining the celluloid dial card cover, and to remove this you need a special tool, inserted in a small slot in the left-hand side of the ring. The moulded clear plastic ring of the no. 7 dial (used on 500 series telephones) has a small hole between the finger holes for 9 and 0. A straightened paper-clip inserted here allows the finger wheel to be lifted away from the rest of the dial mechanism. Some other American dials have a simple sprung wire circlip holding the dial label (and protective celluloid cover) in place; this is easily removed with the help of a watchmaker's screwdriver. Occasionally the celluloid discs for protecting the dial label are missing. Genuine replacements are hard to find, so a good source of stiff clear plastic is the material used to stiffen men's shirts (normally discarded when you wear the shirt for the first time but now you know better!).

On other dials you will just have to use your ingenuity as to how to disassemble them, but in all cases the first task is to remove the finger wheel. On many Continental European dials this is held in place by a centrally located hexagonal nut, which is easily removed. Care should be taken, however, to avoid scratching the anodized aluminium or plastic plate immediately below it.

DRIVE BELTS AND WHEELS

It's as inevitable as it is unfortunate that rubber deteriorates with age. It perishes, it stretches, it loses its elasticity and it even loses its grip, meaning that drive belts and pinch wheels can effectively bring otherwise well-preserved relics to a full stop. NOS (new old stock) spares, even if available, may provide only minor respite since they are likely to have suffered in the same way.

For slipping wheel surfaces you can try cleaning loose material with ethyl alcohol or you may use emery sticks to roughen the surface. GC Electronics in the USA sells a special fluid for restoring these surfaces (stocked by Antique Electronic Supply) and it seems to have some success.

Slack but otherwise intact drive belts can be cured temporarily by dousing them in scalding hot water (in a washing-up bowl) but the effect does not last long. The days when wholesalers kept comprehensive stocks of pattern part spares are long gone, although some of the trade houses (such as CPC in Britain) still carry a good selection, many in fact intended for video recorders. Other sources are heavy-duty rubber bands and various automotive products, and also the neoprene 'rubber' O-rings sold in some hardware stores. Where the belt is too large, good results are had by cutting out the surplus with a sharp scalpel or razor-blade and then sealing the freshly cut joint with cyanoacrylate adhesive ('superglue'); a properly made joint usually ends up stronger than the rubber itself. Trim the O-ring carefully to length or a tiny bit less so there will be some tension when assembled. Apply a small drop of fresh superglue on one end, and then hold the two ends together carefully until the superglue sets (a minute or so). Long-nose pliers can help here.

CLEANING MECHANICAL ASSEMBLIES

Moving parts, such as telephone switch hooks and tape transport controls, will work far better if they are kept clean. A quick squirt with WD-40 may appear to do the trick but thorough cleaning with methylated spirits and a felt pad is far better, followed by careful lubrication with light oil (the lubricant containing minute Teflon particles sold in Tandy/Radio Shack stores is particularly effective). WD-40 is said to leave a non-conductive residue, which obviously does not help current-carrying contacts.

RESTRINGING DIALS

Many old radios used elaborate arrangements of drums, cords and pulleys for indicating tuning and, frankly, they can be a right pig to get right. For determining the correct routeing of the dial cord you will probably need a service manual or the appropriate 'trader' sheet. Old dial cords usually stretch or break, so this will need to be replaced. You can buy dial cord from Maplin Electronics or even at your local Tandy shop (where it comes in a packet with some handy springs for tensioning the cord). Don't use any old string, it will stretch. Remember to remove any fluff from the dial window, whilst defunct dial lamp bulbs should be replaced (again try Tandy or Maplin).

FAULTY CASTINGS

Some radios and TVs used die-cast alloy castings as capacitor frames, dial drives, dial drums, bushings and other mechanical parts. The metal, known as mazak, zamak, muck metal or pot metal, is made of zinc, magnesium, tin and copper and has a grey appearance. It can take fine detail (it was used to make Dinky Toys and many parts in model trains in the 1930s, 40s and 50s) but its long-term dimensional stability is not so good, turning dry and brittle with age or high levels of humidity. Worse still, it can distort badly and when forced back to its proper shape, break into many pieces. The folklore is that workmen used to throw the silver paper from their cigarette packets into the ingots to 'add grist to the mill' and this of course added impurities or destroyed the careful balance of ingredients. The nett result is the mazak disease collectors have come to recognize (known technically as intergranular corrosion); it can take thirty to fifty years to appear.

Never attempt to straighten a mazak casting by bending or pounding it back to shape; it will almost certainly break. Repairing affected parts is seldom possible. Broken parts can sometimes be joined using cyanoacrylate glue ('superglue') or epoxy resin (Araldite and equivalents), although the repaired items may not be able to stand much mechanical stress. Soldering is possible with special fluxes or paste solders (Solder-It paste in the USA) but is not recommended.

In practical terms the solution generally is to find a replacement, perhaps from a junk set. Failing that, a kindly collector may lend you an undamaged part and you could make a new casting using a plaster or silicon rubber mould and casting metal (the sort used to make 'pewter' soldiers and white metal model railway and car kits). This is not an easy process either and it might be better to take advice from a specialist casting company. An alternative is to grind a completely new part out of 6061 aluminium using a computer-controlled machining tool, which will neither be easy nor cheap, although no harder than creating a new metal mould. The worst part is that there is no guarantee that other surviving mazak parts in the device are still dimensionally correct, and in some situations this may be crucial. Sorry it's not very good news!

DRILLING OUT SET SCREWS

Set screws (also known as grub screws) are normally hardened steel and a regular drill bit will likely not cut through one. The titanium nitride coated bits will do a little better but not much. To really do the job, get cobalt bits. These are the solid-gold coloured bits with a point that is ground at a slightly different angle; the titanium nitride bits are only coated down the length of the flutes. Cobalt bits usually cost less than the titanium nitride coated bits too. Another tip is to use a left-handed drill bit. You can find these in many hardware stores. As drilling proceeds and the vibrations and rotary impact hopefully loosen the screw, it will tend to back out, not get tighter.

CUTTING SCREWS TO LENGTH

An old but still very valid tip concerns machine screws which need to be cut to a specific length. The trick is to first run a nut on to it. After cutting the screw, make the cut end slightly pointed with a metal file. Now remove the nut, which will square up the thread and ensure that the screw will now enter its appropriate hole with ease.

CHAPTER 19

Finishing off a set

Even people who appreciate little of the effort we put into restoring our treasures know when something looks 'spot on'. Generally things look right when they are right, so why not take that extra pain to make your restoration perfect?

WIRES AND PLUGS

These are the little touches which make things look right, like using an appropriate fabric-covered mains cord and a period plug on the end of it. To my mind there is nothing more jarring than a nice old radio (or TV) fitted with a gleaming white PVC mains lead and a shiny white plug.

Fabric-covered wire looks better generally, although some sets in the fifties were sold with a black or gold coloured PVC-covered lead. This flex is still available; the gold stuff looks pretty tasteless to me but it would be entirely appropriate on some living room TVs and radios.

A brown bakelite plug finishes off the set. Now I'm not suggesting you go back to round-pin wiring but a brown 13-amp plug looks more harmonious than a bright shiny white one. They are still made, although only to special order, but you can find plenty of them at car boot sales, in charity shops and the like. If I see them I buy them up by the dozen; the cost is usually ten pence each and people give me pitying looks which I assume means they think I cannot afford new plugs. If only they knew why I was really buying these plugs they would probably give me even more pitying glances . . . you know, the 'it's treatable but not curable' look. Still, you did want to restore your old sets to museum standard, didn't you? Do check the plugs for safety and reject any which are cracked or have missing cord restraint fittings. Keep things safe out there.

KNOBS AND TRANSFERS

Knobs and transfers probably belong in this chapter as well; they are certainly details which can make or mar the appearance of a set. Nothing looks worse than a missing knob or a wrong replacement and we British are not as well catered for as the American market in this respect. Companies like Antique Electronic Supply (Tempe, Arizona) and Vintage TV and Radio Supply (Cleveland, Ohio) stock a wide range of reproduction wooden and plastic knobs to fit out virtually any American radio or TV set. Over here we are not so lucky, so finding that elusive job probably means a lengthy search round the swap-meets or a trip to a thoughtful dealer such as Rupert's Vintage Wireless (Ealing) where you can root through a large box of knobs. Alternatively, you could try a small ad in one of the vintage radio magazines. It will be worth it in the end of course.

APPLYING TRANSFERS

Nowadays we are spoilt for ways of applying legends to equipment. If we need to make up individual words, there is Dymo tape (the embossed plastic tape), or there is the new P-Touch tape, where the printing is done beneath the top clear laminating layer (magic!).

Alternatively, you can use commercial decals or transfers, which may be screen-printed on self-adhesive vinyl or applied by rubbing on (dry-transfer lettering or Letraset). Another type is the waterslide transfer and yet another is the varnish-fixing transfer. If you look in, say, the Antique Electronic Supply (USA) catalogue, you may find both kinds of transfer . . . and this is where some people come unstuck! Some helpful hints follow in a moment.

Many old sets used transfers for identifying controls and for applying their trade mark. If your set still has its original markings, try to preserve these at all costs. You can sometimes pick up old sets of water-slide transfers or radio legends at swap-meets or you can get new sheets of reproduction lettering at modest prices from Antique Electronic Supply. They have transfers (they call them decals over there) for many panel legends and trade marks, and I believe it is they who make very convincing HMV transfers (see the Hard-to-find parts section in this book). You can also have transfers made to order and indeed you could then start a sideline selling the surplus stocks. Some photocopy shops can even put waterslide transfer material in their colour photocopier (instead of A4 paper) and this would be another way of getting tailor-made transfers. Avoid using modern rub-down letters if at all possible; they are useless for restoring old sets because the letter shapes are all wrong. I have seen several sets restored with modern Letraset typefaces and they look plain terrible.

When applying transfers, ascertain first whether they are rub-down (dry), waterslide or varnish-fixing. Each requires a different technique.

RUB-DOWN

Rub-down transfers are difficult to apply well and beginners often end up with cracks in the design because they started rubbing down from one corner. Better to apply soft sweeping strokes from the centre to gently tack the transfer down and then apply progressively more pressure overall. A soft pencil or the proper plastic Letraset burnishing tool (from drawing office shops and art material dealers) is the key to successful application. Afterwards place the blue protective paper over the applied transfer and burnish it to make the transfer conform with the surface and reach into any minute cracks.

WATERSLIDE

Waterslide transfers are fairly easy to apply. We probably all played with them as children or applied them to plastic model aircraft when younger. Make sure no air bubbles are trapped under the clear carrier and that the transfer is really straight. It is easy to align when wet, and impossible later once set. Make sure also that the transfer conforms to the surface on to which it is being applied, i.e. pat it down well with a wet finger. Any air trapped underneath will look grey once the transfer has dried out. And do let transfers dry thoroughly, then protect with clear varnish.

VARNISH-FIXING

Varnish-fix transfers are the hardest to apply. Here you need to coat the surface to receive the transfer with clear varnish (such as Humbrol from a hobby shop) and when the varnish is very tacky and almost dry, you apply the transfer together with its tissue paper carrier sheet. Allow to dry thoroughly and then soak off the tissue paper with water and allow to dry.

That's the theory, anyway. *Radio Constructor* magazine used to supply panel legends as transfers (later they were reissued as sticky vinyl decals) and a lot of readers could not get them to work properly. Here is a detailed explanation of how you should use them, courtesy of an ancient issue of *Radio Constructor*.

'One or two readers have had difficulty in applying panel signs satisfactorily when using varnish. On investigation it was found that in each case the fault was the same. Some notes are given here in the hope that they will enable other readers to obtain the same high degree of satisfaction as we ourselves have enjoyed.

The trouble was simply impatience – these readers were too eager to see the final result. Now, take for example one of the potentiometer panels. This consists of a thin film of paint held in position by a water-soluble gum to a sheet of tissue paper. If a coating of varnish is applied to the paint film, and then the transfer is mounted on a sheet of metal, it is obvious that no air can penetrate to the varnish and that it will take a very long time to harden; in fact, a thickish coat of varnish would need five or six days to dry!

Point number one, then, is to apply the varnish thinly. Number two is to let the varnish become very tacky, in fact almost on the point of hardening, before applying the transfer to the metal surface. Point number three is to let the transfer stand long enough, preferably overnight, to get really hardened off before attempting to soak off the tissue paper covering.

Remember that all types of transfer need a protective coat of varnish; a satin finish is probably best. Take your time – don't rush this seemingly easy task!'

OLD PROGRAMMES FOR YOUR SETS

I feel that old sets should play old programmes, so why not buy some tapes to relive the good old days (assuming they really were that good!)? Disagree if you will, but I personally think that vintage loudspeakers and screens should produce vintage programmes. You'll find full details on entering the spirit of the age in Chapter 21. Another pleasing touch is to scatter a couple of period magazines 'casually' next to your set – copies of the *Radio Times*, *TV Times*, and the pre-war publications *Radio Pictorial* and *World Radio* are not hard to find.

PAPERWORK

I thought I would slip this in here, mainly because I couldn't think where else it would fit in the book. Most enthusiasts end up collecting a number of old books, illustrated sales leaflets, magazines, programme papers, service manuals and the like. Hopefully these are all in perfect condition but, if not,

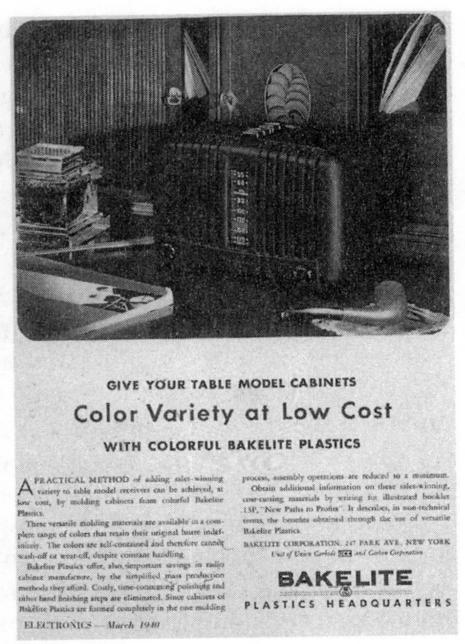

Figure 19.1 Period paperwork has its devotees as well. Above the advertisement from the Bakelite company extols the virtue of moulded plastic cabinets for table radios, whilst on the left, the poster from a telephone kiosk reminds new owners of television sets that they can obtain a rebate on their radio licence when they trade up to a combined television and radio licence.

here are a couple of tips that may help you: they are not 'secret' by any means but they do work and I have never seen them in print anywhere else.

First of all, conserving old paperwork. You may well come across an old invoice, sales leaflet or whatever – it really doesn't matter what – which is grubby or covered in pencil marks. The paper is old, brittle and yellowing, so how do you go about removing these marks?

Well, don't use any old pencil eraser! You'll probably end up raising the surface of the paper or, worse, tearing it. The hot tip is a Rowney Kneadable Putty Rubber. Don't waste time with anything else. You'll have to go to an artist's materials shop for this, as it's a speciality item. But it is marvellous. The rubber is made of a very soft kind of material and it is marvellous for removing ingrained dust on paper. It's not so effective on pencil marks but that's fine; when you have used the putty rubber to remove the general dust and grime, then use a plastic eraser (the white sort that generally comes from Germany) to get rid of the scribbling.

Sometimes you will need to make a reproduction of some old label, perhaps a tuning scale, a warning notice or the dial centre label in a telephone. If you haven't got the original, a reproduction is probably the next best thing. What you don't want, however, is a photocopy that is gleaming white; it just doesn't look the part. In fact 99 per cent of all the reproduction labels I have seen look totally unconvincing, even when people have tried to age the paper artificially, using a variety of methods from baking in an oven to staining them with weak tea to achieve a slightly yellowed period look. Needless to say, none of these methods is really perfect and certainly none of them would fool an expert. But my sneaky method does!

So what is the object of the exercise? To create the authentic look of old paper. And what looks most like old paper? Why, old paper of course! So you toddle along to your local charity shop where they sell old books (the ones that nobody at all wants) for pennies. These are a valuable source of clean white paper: well, not quite white and that's exactly what we need. The clean paper is the endpapers of the book, so these are what we strip out; the rest of the book can go for recycling. How do we use this paper? Well, most photocopiers have a 'sheet feed' option where you can use your own choice of special paper instead of the bulk supply in the hopper. And that's what we use. Take an original label (or a photocopy of one) and photocopy it onto the period paper. You now have a label on old paper that's indistinguishable from the real thing. It's amazing what modern technology can do!

DISPLAYS AND CONTESTS

Many collectors society meetings encourage members to display items from their collections and there's an increasing trend to hold *concours d'élégance* contests at these exhibitions.

These *concours* are borrowed from the world of classic cars and are a kind of contest in which the judging reflects the best restored item. Normally there is a trophy awarded to the outright winner and plaques for subsidiary winners in each class. There is a fair amount of prestige attached to the winning entries (and their owners), whilst the contests also help focus attention on successful restoration techniques.

Here are some of the factors which influence the judges' decision; they are written with radios in mind but with a little imagination they can be applied to other objects too.

Cabinet. Does it look attractive or is it grubby, cracked or even over-restored? Nobody is expecting perfection, particularly on very early sets or those with plastic cases that are prone to cracking. But there should be no grubby knobs or smears of polish. Old materials should be restored sympathetically, so I deducted marks for over-shiny polyurethane varnish finishes on wood. To quote Gordon Bussey: 'A well restored radio need not look as if it has just been bought, merely as if it had stood on the sideboard for the last few decades and been lovingly dusted and polished now and then'.

Chassis. Yes, judges do take the backs off and look inside. Rust and cobwebs lose marks, although nobody is expecting to see plating so shiny you can see your own reflection. Paint finishes are acceptable if they represent the only way of making the chassis look presentable.

Performance. Judges check RF performance, lack of mains hum, adequate volume and so on. This is not a problem with mains sets but if you offer a battery set and no batteries, you will inevitably lose all marks in this section. Sloppy drive cords, failed lamp bulbs and wobbly control knobs can lose vital points too.

Quality of restoration. Marks are awarded for craftsmanship but not for over-restoration. Marks should not be lost for the unavoidable ravages of time but an unrealistically fine finish will not win extra points and may detract in the judging.

Originality. The set's overall appearance should be authentic. Replica transfers and modern fabric-covered mains leads are fine . . . as long as they fool the judges. But please don't use modern looking rub-down lettering on 1920s breadboard sets; the alphabets look all wrong and the whiteness is dazzling. Where a wooden cabinet originally had a cellulose lacquer finish, try and replicate this rather than using French polish. Getting all these details just right involves effort, but surely that's the only way of setting your heirloom ahead of the next person's?

A judge should not mark someone down who had used current production high-voltage capacitors but he might award extra points to someone who had gone to the trouble of encasing them inside the original waxed-paper cylinders. PVC-covered wire or sleeving looks wrong in old sets; you can still get the old-fashioned spaghetti at swap-meets, whilst fabric-covered wire can be robbed from junker sets sold cheaply at swap-meets and jumble sales.

Scale of project. Large-scale projects that must have involved many hours of work deserve to score more points than a simple bakelite bedside radio. So a compensating factor should be applied here.

Overall appearance and attention to detail. This speaks for itself; you can usually tell how much work went into a project.

Presentation and descriptive information, explanation of work done and how found. Some people go to a lot of effort to present an information sheet giving a description of the item in question, how they found it, how much work they had to do on it, its performance today as compared to when made, photos as found, etc. etc. You don't have to provide this but anyone who goes to this trouble definitely deserves extra marks!

Historical significance. This may or may not have an influence, depending on the specific conditions of entry for the contest. Is the item a 'milestone' set? Does it have a famous owner or some special association? Alternatively, is it unique in some way or at least uncommon?

Bribes. It's always a fix who wins these contests, everyone knows that. Working out how much to bribe the judges (and how much other entrants have offered) is a skilled art, however. The golden rule is that you can never offer too much and guineas are more acceptable than pounds.

Seriously, ignore the last paragraph. But if you take the rest to heart, you might stand a better chance of winning next year. I hope so and I certainly look forward to seeing a dazzling array of beautifully restored sets next time round, whoever is judging.

CHAPTER 20

Conservation techniques

Conservation may sound like a strange topic for a hobbyist's book – more for museum curators or professional wildlife keepers. Well, not really, because the treasures we enjoy are also at risk. For this reason the serious collector needs to be aware of the ways in which objects deteriorate, and also how these processes can be arrested and reversed.

Permanence is for ever but that's about all. Certain substances we imagine to be stable for all time just aren't! Some deteriorate visibly: the forked cradles of 1950s telephones are prone to 'deliquescing'; that is, they attract moisture from the atmosphere and slowly disintegrate. Other plastics, such as the clear celluloid dial covers of old radio receivers, go yellow and clouded, sometimes distorting into crazy shapes. Anyone who has wrapped a PVC mains lead around an ABS plastic-cased object will be familiar with the way the mains cord 'imprints' itself in the ABS, cutting significant 'grooves'. Metal is not always stable: mazak gears and pulleys in dial cord assemblies contort and crack, eventually crumbling into brittle fragments (see Chapter 18, 'Mechanical matters'). Steel rusts and the lead-out wires on valves and tubes can work-harden and break off right at the metal-to-glass seal – so frustrating!

Other deterioration is invisible, such as the germanium transistors which grow internal 'whiskers' and cease to work. The later plastic silicon transistors are just as prone to change; their cases are microscopically porous and can absorb atmospheric moisture, causing distorted operation. Leaky electrolytic and wax capacitors and out-of-tolerance resistors are also well known. Sometimes these components can be replaced, although 'new old stock' ones may exhibit the same problems as the faulty ones.

Telephone and computer collectors have discovered that even 'modern' plastic materials can change colour badly. Photocopier repairmen are issued with a special xylene-based cleaning fluid which can correct this effect but the product is hazardous and not widely available. Edward Then, a senior conservator with the Science Museum in London, considers that xylene is not the best solution to the problem of yellowing in plastics and that ABS plastic would be damaged by a xylene-based solvent. In a paper he writes that the yellowing or discoloration of plastic cases is a common phenomenon, comparable to the discoloration of an apple skin, and is similarly irreversible. The only reassurance is that the damage associated with discolouring affects only the surface of the artefact.

According to Dr Then, different chemical reactions occur along the polymer chain during degradation. These can result in discoloration of the polymer and the chief factors initiating or accelerating this degradation are ultraviolet rays and light, in proportion to the light intensity. Although stabilizers are added during manufacturing, most of these additives are consumed as the polymer ages. Once used up, the polymer is often left unprotected and will deteriorate very rapidly. Little can be done to prevent this at present and the best advice is, perhaps, to do nothing, leaving discoloured surfaces untreated.

Dr Then adds that maybe, one day, the discoloration will be seen as desirable or inevitable, like the patina on metals! Dirt and grime are a separate problem, and may be cleaned with distilled or deionized water. Until there is a solution to this problem, he says the only prudent strategy is preventive conservation, keeping objects away from strong light, especially direct sunlight and other strong UV sources, and from any heat source.

When it comes to the preservation of other plastics, which for us generally means dials, knobs, insulators, face-plates and sometimes cases, Dr Then has some valuable recommendations. Remember these are for museum pieces where the items are intended for very long-term storage.

1. Separate plastics by polymer type and isolate degraded plastics and metal parts.
2. Keep temperature and humidity constant, preferably low.
3. Keep lighting low, avoiding ultraviolet (sunlight, fluorescent lights).
4. Keep potential gas-emitting plastics, such as cellulose nitrates and acetates and polyvinyl chloride, in well ventilated areas and leave space between objects to avoid cross-contamination.
5. Handle objects with gloves.
6. Identify the polymer before doing any treatment, especially with solvents.
7. Use inert padding to support objects liable to harden in a distorted shape.

Number 4 may surprise you. It seems this is a common problem in museums because deterioration is often 'contagious'. Nitrogen dioxide from the denitrification of cellulose acetate is particularly bad since it forms nitric acid which can attack other objects and especially metals. Cellulose nitrate, whilst not often seen today because of its fire hazard, was once the most common plastic, having been commercially available since the 1870s. Off-gassing of plasticizers in polyvinyl chloride plastics is also a problem as it leaves the remaining plastic brittle but with a sticky surface. The 'sweating' of vinyl can stain surrounding items and attack some metals such as aluminium. The other common cellulosic ester plastic is cellulose acetate. It is not nearly as damaging as cellulose nitrate when it degrades and, while it degrades much more slowly, the same ventilation rules apply. The article goes on to suggest the use of a special indicator paper that can be placed near plastic pieces to give warning of acid fumes being given off by degradation.

So, those of you collectors who carefully wrap up museum-piece treasures such as early transistor radios, place them in a box and put them on a shelf never to be opened until the time comes to collect on your 'investment' – you may be in for a big surprise a few years from now!

Industrial chemist Barry Ornitz, who passed on this last item, adds that ageing is inevitable and we cannot totally prevent the process in our radios, much less in ourselves. Our collectables may not be of museum quality and most of us would prefer to enjoy them, but there's no harm in taking the best possible care of them.

Given the disappointment of the ageing process just described, perhaps it would make more sense to enjoy these treasures in their original state – as illustrated in period books and magazines. These need care in their conservation too.

CONSERVING PAPERWORK

So what about our paper treasures? Old newspapers and magazines are prone to yellowing and can become very brittle. Pre-war copies of the *Radio Times*, *Radio Pictorial*, *Practical and Popular Wireless*, even *Wireless World*, as well as many catalogues, are all prone to this problem, largely because they were printed on cheap paper and were not intended for long-term preservation.

It makes eminent sense to take care of paper exhibits, simply because they are vulnerable. You can lengthen the life of all paper items by keeping them out of direct sunlight and by filing them carefully. Purists preserve them by keeping them in a dry, cold atmosphere but not all of us have temperature and humidity-controlled vaults at our disposal.

Magazines deserve to be filed in binding cases or filing boxes (you can buy low-cost cardboard magazine filers at larger stationers and office supply stores). Individual documents and loose papers can be protected inside individual pockets of clear plastic (A4 and A5 size). As these wallets are punched, they can then be filed in a ring-binder. The low-cost pockets sold at W.H. Smith and the like are adequate for most collectors, although at greater cost you can also buy conservation-grade wallets made of clear melamine.

When it comes to repairing old documents, be very careful. Inexpert repairs damage books and their value permanently. Normal clear tape is taboo and must not be used at any cost; it shrinks, oozes glue and goes yellow with age. The so-called magic tapes, with their matt surface, may not be so harmful but they are still to be avoided because they are plastic and do not look right on paper. Music tape, gummed transparent paper sold by sheet music specialists, is a paper-friendly material, as is gummed brown parcel tape (for repairing covers and reinforcing the spine of tattered magazines). There is also a specialist kind of tissue-paper gummed with a special low-tack adhesive and this will reinforce tears, yet can be peeled off at any time later. More details on this below.

Serious collectors will make greater efforts towards paper conservation and fortunately this is not difficult if you use the proper conservation-quality materials. You won't find them at your local stationer's but a specialist supplier is listed in the 'Buyer's guide' section of this book.

Museums and Record Offices use what they know to be made of archival-quality materials when it comes to looking after the paper in their care. This means that all the materials used – pockets, paper mounts and boxes – are all known by chemical analysis to be inert, acid-free or neutral, and so unable to have any adverse chemical reaction with the paper they are intended to protect. The cleaners they use are not abrasive so don't remove paper fibre, nor do they leave microscopic fragments in the paper to cause damage in the future. The albums used are designed to exclude light and airborne pollutants. The same materials are available to you to protect your collection.

Polyester is the material acclaimed and approved by the British Museum, the National Postal Museum and all other major archives for long-term storage and protection, as being 100 per cent archivally safe. It is inert, resistant to most acids, solvents and moisture, does not embrittle or yellow with age and has glass-like clarity. In the UK it is known as Melinex, and in the US it's called Mylar. Manufactured in sheet form, it's available in rolls to make jackets for dust-wrappers or books, and is turned into pockets to suit single sheets of paper as well as pamphlets or magazines.

When buying pockets to store your collection in, ask what they are made of. If the answer is not Melinex, you are not protecting your collection. If you're not satisfied with the answer you get, or the vendor hasn't a clue what you are talking about, don't buy. The value of your collection is always greater than the price of any kind of pocket.

Dust-wrappers are best made from 50 micron Melinex; this protects pamphlets, comics, magazines and books from damage caused by normal handling. It needs no tapes to hold it in place, since the folds you make in it will not come undone. Some collectors of single sheet items tend to mount their items on card holding a written description, while other collections require visibility of both sides of the item. Whichever is your choice, inside a Melinex pocket the item is protected from migrant acidity for storage and for handling, and their clarity makes them ideal for exhibiting. Any size pocket can be made, but many sizes are already held in stock, from post-cards to posters, and are sold singly or in bulk as required.

For removing pencil marks and dirty finger marks an excellent remedy is either an art gum eraser or an artist's putty eraser. Both of these are 'paper-friendly' and can be had at any good artist's supplier. For repairing torn pages the professional method is called Filmoplast P. It is a self-adhesive paper repair tape but the adhesive is intentionally just tacky, so it can be removed again later. It's made of paper and is so fine that you can read through it as if it wasn't there. A variant is slightly tinted to suit yellowed paper such as old newspapers. It is the safest alternative to cellulose tape, which should never be used on anything of value, not even the so-called 'magic' tapes.

Where the repair will not be seen, such as on the back of dust-jackets of books and posters, you can use gummed parcel tape – the old fashioned brown stuff, still available at stationers. As it dries it might cause the paper to cockle and there is a thicker version of Filmoplast (P90) that can be used if necessary. Finally, Filmoplast R is a heat-seal tissue that you can apply to large areas using a household iron.

The gummy residue of self-adhesive tape and labels is easily removed with lighter fuel. Other solvents may cause staining or might even dissolve the surface. Specialist paper dealers also stock a label lifter, which is a solvent for animal and vegetable-based glues and the 'lick and stick' variety of gums.

CHAPTER 21

Entering the spirit of the age

There are some people who have a remarkable capacity to enjoy things in a totally detached way and follow a pursuit without getting fully involved . . . such as the railway photographers who take superb line-side pictures of the fire-breathing steeds. They do this by car, however, and never set foot on trains as a means of travel. In fact they seem to have a very strange, arm's-length relationship with the objects of their passion.

To my way of thinking, that's a shame. I have met the same thing with people who collect telephones but do not try to make them work or connect them to a real telephone line. Then there are people with rooms full of ancient television sets, all totally inanimate like a row of tombstones. In the same way, I have met radio collectors with marvellous collections of sets, all dead, incapable of uttering another sound. By this I mean the sets, not the collectors, but I feel the collectors have something dead in their soul as well if they have never let their imagination run a little further and embrace the lore – and joys – of old radio (or television or whatever) in its broadest context.

All these products were made for a specific purpose, to fulfil a real-world function, and I feel people who preserve items without keeping them in working order are cheating the generations to come. Surely the whole point of keeping something beyond its rational working life is to allow our successors in future times to see how things worked and to experience life (and enjoy entertainment) the way we once did.

This applies to us already because the treasures in our collections are already relics of an era gone by. The first thing to do therefore is to make sure your favourite relics work – have them restored and make them function as they were intended. You can then take this a stage further and argue that the only way of truly appreciating an artefact of the past is to put it in its proper context – in the past. And if we are to recapture a golden age of times gone by, then that means putting yourself into the mind-set of that era. Try and enjoy the music and entertainment of that era, read the technical literature of the period in old magazines and books, try soaking it up! Eventually you'll get to understand much better why a particular radio, television or telephone turned out to be the shape it was, why it was made of that material and why it had a particular feature-set.

NEW SOUNDS FOR OLD SETS

What do I mean? How do you go about it? For a start, old radio sets give more pleasure if they work. Why, I'm not sure. It may be the enforced delay and anticipation while you wait for the valves to warm up, perhaps the mingled smell of burning dust and wax capacitors, or else the mellow sound of the speaker in a decent-sized cabinet. Most transmitting bands are unchanged, so a set of fifty, sixty or seventy years ago will still pick up today's programmes, and if you are unhappy with most of the programmes on the medium wave, there are the Radaptor converters that will convert FM programmes on VHF so they can be heard on a medium wave set (enquiries to Vytek-Radiocraft Ltd, 56 Main Street, Sedgeberrow, Evesham, Worcs. WR11 6UF; telephone/fax 01386-882280).

You can go one stage further if your radio has a GRAM or AUX input; it can play vintage programmes for you! As explained in an earlier chapter, playing period programmes on vintage sets completes the recreation process; it can also be a lot of fun and an introduction into a rewarding and fascinating interest area – almost a whole new hobby, generally known to its devotees as old time radio or OTR. OTR has its own clubs, for pre-war and early post-war programming on the one hand, whilst on the other hand there are also a whole raft of organizations devoted to the offshore radio phenomenon of the 1960s and 70s. Large chunks of the Internet are also devoted to the OTR hobby. Television nostalgia has its devotees as well, again with clubs and Internet areas.

TAP(P)ING THE ARCHIVES

Archive programme material need not be inaccessible. If you don't feel like joining an organization yet-a-while, perhaps all you need do is scour the *Radio Times* more closely than usual. In recent years the BBC has made a good job of repeating archive programme material, both on radio and tele-

Figure 21.1 The *Nine O'Clock News* (above) and *Children's Hour* (below) are powerful memories for many collectors. Listening to today's programmes (or tape recordings of older ones) is more pleasurable on older sets, with their warm glow and mellow tone. A wide variety of vintage programmes is available on cassette tape from commercial suppliers (see 'Buyer's guide' chapter) and on loan through tape clubs (see 'Clubs and societies' chapter).

vision, although it can be extremely frustrating if you missed the programme you wanted. There's no simple mechanism for buying copies of past programmes from the BBC (the charges are high and the criteria exclude most enquirers), and other than advertising in the newsletters of the societies listed in this book, there's no proper forum for swapping tapes with other enthusiasts.

You can of course buy a number of old shows on tape and CD. For radio the BBC has released a few old programmes (look in your local branch of W.H. Smith for these), whilst other material is available abroad, especially in the USA. ORCA, the Oldtime Radioshow Collectors Association has a big lending library of old radio programmes and their £5 membership fee is modest enough to deter nobody from joining.

The same applies to television. There are groups both in the UK and in America devoted to trading old programmes, whilst there are now a wide range of old programmes rereleased on sell-through videotape. Purists of course run these through a standards converter and view them in soothing 405 lines and restful black-and-white. The selection of tapes available commercially in this country is not desperately broad, however, whereas in America there is a huge range (and they don't have the problem of changed line standards either; the 525-line system has been in use there since 1941).

OTHER NOSTALGIA ITEMS

The field of radio and television nostalgia goes wider than archive programmes, though. It starts with old copies of the *Radio Times* and other programme papers and runs (in the printed matter field) through photo albums of the stars, Christmas annuals and cigarette card albums to pre-war radio licences and the large colour advertising placards on cardboard which used to hang up in radio and TV dealers' shops. Then we find picture postcards with radio-related cartoons or photographs of radio station buildings, advertising novelties such as china models of the His Master's Voice 'Nipper' dog, cigarette lighters in the shape of a radio valve and all manner of other gimmicks. Add to this the hundreds of disc and tape recordings, both original and reissues, and you have a huge potential field to hunt down and collect.

Some collectors take this a stage further and recreate 'retro' room settings around their period radios, televisions, hi-fi or phones. Table or TV lights can be very characteristic of a period, whilst another essential is a television trolley or table, the latter in the obligatory 'contemporary' style of rich glossy brown wood and tapering splayed legs in black, capped with lacquered brass ferrules.

Other accessories are decorated leather covers for your copies of the *Radio Times* and *TV Times* and stick-on screens that gave a 'life-like'

Figure 21.2 Advertising postcards such as this are cheap to buy and easy to find, yet they make a compact and desirable collection which complements and enhances an accumulation of larger items.

colour impression to TV programmes ('winner of the Brussels Inventors Fair', no less). It's worth recalling that at a time when colour television was still no more than a laboratory novelty in this country, some people were so desperate to have colour television that they bought these garish affairs of translucent multi-coloured plastic to give their monochrome screens a 'coloured' effect!

None of this stuff need be expensive unless you want to pay top-notch prices at auction. But because it's one of the less well organized collecting hobbies, you'll have to track down your quarry at boot fairs, general antiques and bygones fairs, ephemera swap-meets, second-hand bookshops . . . and not just specialist vintage wireless events, where prices are inevitably higher.

THEME MUSIC AND MORE

There is a growing interest in all kinds of radio, film and TV music, not just the classic main themes but in sound-tracks, interludes, library and production music (used as incidental music in films and TV programmes, and also for the background to newsreels and the background music to the test card on television – joy of joy, there are now a few CDs of Test Card Classics). Sound-tracks are only part of the story, therefore; the other music used in films and broadcasts is categorized either as 'presentation' or 'production'

Figure 21.3 A powerful part of the imagery of broadcasting in days gone by is the Marconi 'coffin' microphone so widely used by the BBC. Its Florentine bronze finish and restrained good looks make it a desirable – but relatively unattainable – addition to any collection. Examples can fetch a couple of thousands of pounds at auction and do not appear frequently.

Figure 21.4 As the supply of affordable 'hardware' slowly dries up some collectors are turning to paper items, such as pamphlets, books and magazines illustrating their subject. This attractive football pools advertisement is an example of the kind of ephemera which is found at specialist auctions and collectors' fairs.

music, depending on whether it identifies the production or is merely incidental to it.

Sound-track albums have always been fairly accessible (so long as you bought them whilst still available) but the production or 'library' music used incidentally in programmes was not to be had for love or money since such music was made available only for professional users. Gradually the specialist music libraries recognized there was a market for this as well, and many of these original performances are now being made available to the public too. You can now find them on commercial recordings in the form of limited edition albums, available either through specialist labels or by special arrangement through organizations such as the Robert Farnon Society (which is allowed to sell library music CDs to members only – a good enough reason for joining!).

TIMES CHANGE . . .

A final note emphasizes that it's easy to take this business too seriously, and what may be icons today were not venerated in their day. To prove it, here's what sixty years can do to change people's perception of 'good taste'.

There's a delightful book called *Interior Decorating* by Duncan Miller. The title is misleadingly understated in a delightfully British way, for this is no DIY handyman's book; no, it's a complete aesthete's guide on how to have *other* people fix up your dwelling for you, you know . . . workmen, tradespeople and the like. It is published by that guide to style, *The Studio* magazine, and has a publication date of 1937, issued simultaneously in London and New York.

The illustrations look like real photographs and are pasted in on to captioned pages, rather as in a photo album; you can now see this is no ordinary person's book. Those of you who are devotees of the *art déco* masterpiece the Round Ekco will be delighted to know that this book is dripping with pictures of room settings by Wells Coates, including one with an *à la mode* stainless steel and chrome electric fire/fireplace combination. There is a strong implication that, as well as endorsing this design house's creations, the book also embodies some of their line of thought.

That's why I was so excited to see what the style gurus thought of radio sets for the modern home. Hold on to your hat . . .

'*Radios.* As radios are bought in the manufacturers' cases and so few are properly designed, they can do great damage to a scheme of decoration. It is nearly always possible to build a radio into a cupboard or a bookshelf.'

Absolutely, couldn't agree more. Pass me that hammer and screwdriver, I'm just going to remove this frightful two-tone catalin case from this bijou radio I picked up . . .

PART 3

Resources

CHAPTER 22

Buyer's guide – a comprehensive directory of dealers, auctions and swap-meets

This chapter is another of my secret weapons! The information isn't secret really but you'd be surprised how many people don't know where to go, in particular for specialist and hard-to-find parts and services.

Note that valve prices are really starting to heat up, largely on account of high-end audio enthusiasts and professional guitar players. These people have deep pockets and can out-bid most other users, whilst a lot of stock is moving out to Italy and Japan, where collectors' interest is allegedly at fever pitch. All this means that if you need valves, you ought to lay in stock now while prices are still reasonable.

Dealers are the obvious first port of call; although bargains can sometimes be had in general antique shops, their merchandise is frequently suspect in condition or authenticity and often overpriced. Specialist dealers, who rely on repeat business, cannot afford to be so lax. Swap-meets are another excellent source of plunder, as is the National Vintage Communications Fair held twice yearly at the National Exhibition Centre, Birmingham. Details of all these are given below.

For specialist military and avionics radio equipment the specialist swap-meets devoted to these hobbies (aerojumbles and the like are well worth visiting – read *Flypast* magazine for dates and locations). Autojumbles are good for car radios and enamel signs from the time when automotive garages were also wireless dealers, whilst some interesting items also turn up at toy fairs (don't ask me why but they do!). Your local paper and the specialist hobby publications have details.

Don't be afraid to do some lateral thinking when you are trying to source uncommon parts. The high-voltage wire used in television sets and some transmitters is readily available from electronics parts suppliers such as Maplin but it does of course look modern. If you are seeking something that looks older, how about trying auto ignition wire? Not the kind with carbon filament but the old-fashioned type used in earlier cars up to the 1980s. This stuff was designed to handle voltages of 15 000-plus without breaking down and you might try car breakers, autojumbles or a long-established car repairer (or marinas or airports where they do repairs). Rubber feet for telephones . . . today's neoprene ones look too fresh and shiny? Then what about the black rubber bumpers for protecting doors at your local hardware shop? A good hardware store or a specialist furniture shop is where you will find felt feet – just what you need for the older type of wireless or to add to a telephone whose original rubber feet have dried rock-solid.

Ignore the fact that several of the dealers listed in the pages which follow are in the USA, not in the British Isles. Apart from the novelty of phoning, faxing or writing to the States, there's no reason why you too shouldn't buy from the USA. They have a lot of products not available here and often the price is still low, even after accounting for the postage. Small shipments seldom attract the attention of H.M. Customs, by the way, so don't worry about having to pay duty. Most of the businesses mentioned accept British credit cards, so you don't have the problem of getting hold of American money, and if they don't, you can always buy dollar bills from your bank and send these in a letter – it's a lot cheaper than buying international money orders.

Another thing. Because the market is larger over there, many of the American dealers issue free or low-cost catalogues that are a revelation to European eyes. These catalogues contain useful information and are well worth requesting.

Lastly, there is a notion that dealers are parasites and genuine collectors should not patronise them. Ah yes. The American author Mark V. Stein has some very apt words on this subject.

'A word about dealers and dealer prices: expect to pay a premium when purchasing from a dealer. The dealer offers you the luxury of eliminating the time-consuming hunt through yard and estate sales, flea markets, antique shows and the like. It is he who goes through the trouble of rooting out those hard-to-find items – ones you might not happen upon except after years of hunting yourself. Dealers' inventories represent long hours and related expenses, and so their prices must reflect those additional costs.'

Do please note that inclusion in this list does not imply any endorsement of the firms concerned. Please be understanding, too. It is inevitable that this information will go out of date – and there's not a thing the publisher can do about that! New suppliers will come along, others will change address or call it a day. Many of these traders are in the business more for the love of it than just to make money; if the latter was their aim, they'd probably be better advised to put their money in the bank. What this means is that some businesses go as suddenly as they come, and some of these 'firms' may in fact be part-time businesses working from home (so turning up unannounced may not bring the welcome you might expect). This is also the reason why some entries have only a telephone number or a P.O. Box number. Even proper shops may have idiosyncratic opening times, so it's always worth making a phone call before travelling long distances. Finally, if you do find errors or omissions, please feel free to write in with your corrections so that future editions can be updated.

ANTIQUE FAIRS (ALL-SORTS)

ARDINGLY SHOWGROUND, Sussex. One of the largest in the South of England. Information from IACF (01636-702326).
MALVERN SHOWGROUND, Worcs. Another large fair. Information from IACF (01636-702326).
NEWARK SHOWGROUND, near Newark-on-Trent. Probably the largest event of its kind in the UK – no more need be said. Information from IACF (01636-702326).
SWINDERBY, held at the former RAF camp, on the A46 road, between Lincoln and Newark-on-Trent. Regular three-day events, with 250-plus inside stalls under canvas marquees and 500-plus outside pitches. Details from Arthur Swallow (01298-27493 or 0860-797200, fax 01298-37493).

ANTIQUE SHOPS

Hank van Cleef reminds us that antique stores often have a few old radios and telephones, for atmosphere if nothing else. Don't expect any bargains for the most part. These dealers are in business to make money. They typically price radios at or above the 'book' values given in the collector guides. And since the dealer is not necessarily knowledgeable about electronics, the set is usually offered 'as is', without any form of guarantee.

AUCTIONEERS

ACADEMY AUCTIONEERS & VALUERS, Northcote House, Northcote Avenue, Ealing, London W5 3UR (0181-579 7466, fax 0181-579 0511). Regular sales of old telephones, radio and TV equipment.
AUCTION TEAM KÖLN, Box 501119, 50971, Köln, Germany (00 49 221

387049, fax 00 49 221 374878). Twice-yearly sales of 'old technology' with detailed illustrated catalogues captioned in German and English. UK agent (for catalogues and information): Robin Kahan, The Mill, Rimpton Road, Marston Magna BA22 8DH (tel/fax 01935-851203).
CHRISTIES, 85 Old Brompton Road, London SW7 3LD (0171-581 7611, fax 0171-321 3321). Regular sales of important early radios, TV receivers and gramophones. Catalogues available.
DOROTHEUM, Dorotheergasse 17, Postfach 528, A-1011 Wien, Austria (00 43 1-515 60, fax 00 43 1-515 60 443). Regular auctions of quality wireless, television and telephone apparatus. Web site *http://www.dorotheum.com*.
ESTES AUCTIONS, 7404 Ryan Road, Medina, OH 44256, USA (00 1 216-769 4992, fax 00 1 216-769 4116). Regular sales of vintage radio equipment.
Mes LIÈVRE, MAICHE, PARIS, Chartres, France (00 33 2-3784 0433). Regular auction sales of radios, telephones and telegraph instruments.
PHILLIPS AUCTIONEERS (Mechanical Music department), 10 Salem Road, Bayswater, London W2 4DL (0171-229 9090 ext. 214). Tony Jones is head of mechanical music and radio/TV; he arranges regular sales. Catalogues issued.
TWENTIETH CENTURY ANTIQUE PROMOTIONS, Tanners Farm, Balcombe Road, Horley RH6 9EF (tel/fax 01293-822469). Auctions held at Copthorne Hotel, Copthorne, Sussex of juke-boxes, records, pin-balls, fruit machines, bakelite radios, art deco, Americana, ephemera, pulps, poster, comics, fairground arcadia, tin toys etc.

BOOKS AND MAGAZINES (OLD): GENERAL SOURCES

BOOKLAND, The Fosse, Fosse Way, Radford Semele, Leamington Spa CV31 1XN (01926-614101, fax 01926-614293). Huge stock of old books and magazines on technology subjects; will search for back issues of magazines (send SAE for form).
OLD TIME SUPPLIES, Box 209, Banbury OX16 7GR (0973-144041). Old radios, books and components.

BOOKS AND MAGAZINES (OLD AND NEW): SPECIALIST SUPPLIERS

ANTIQUE ELECTRONIC SUPPLY (for details see Electrical and electronic components section). New collectors' guides on radio and TV. Their mail order service is first-class and they have a beautiful free colour catalogue (or is it color catalog?).
ARC, Box 2, Carlisle MA 01741, USA (00 1 508-371 0512, fax 00 1 508-371

7129). Large stock of vintage radio books from around the world. Mail order, credit cards.
CHEVET BOOKS, 157 Dickson Road, Blackpool FY1 2EU (01253 751858). Books, valves, some receivers.
Len KELLY BOOKS, 6 Redlands, Blundell's Road, Tiverton EX16 6BU (01884-256170, fax 01884-242550). Books on broadcasting; catalogues issued.
LAKE ELECTRONICS, 7 Middleton Close, Nuthall, Nottingham NG16 1BX (0115-938 2509). Vintage radio textbooks and data books. Visitors by appointment only please.
L-W BOOK SALES, Box 69, Gas City, IN 46933, USA (00 1 800-777 6450). Ten colour picture books/price guides of old radio sets for collectors – wooden, catalin etc. Informative and good value for money.
NEW WIRELESS PIONEERS (James and Felicia Kreuzer), Box 398, Elma, New York, NY 14059, USA (00 1 716-681 3186, fax 00 1 716-681 4540). Rare books on early radio. Catalogues issued.

BOOKS AND MAGAZINES FOUND TO ORDER

The following book search services are recommended for finding specific out-of-print books (no charge to yourself for the finding service).
ASHWORTH BOOKS, 39 West Market Street, Red Hook, NY 12571, USA.
BOOK TRACERS, 1 Temple Gardens, Glasgow G13 1JJ.
BROWN'S BOOKSEARCH, Trenance, Round Ring, Penryn TR10 9LA.
CARCOSA BOOKS (C.D. Pollard), 3 Arundel Grove, Perton, Wolverhampton WV6 7RF (tel/fax 01902-758977).
DC BOOK SERVICE, 11 Kirkby Close, Milton Road, Cambridge CB4 1XP
CULPIN'S BOOKSEARCH INTERNATIONAL, Box 9324, Denver, CO 80209-9324, USA (+1 303-733 5700, fax +1 303-871 0172).
D.C. BOOKS, 5 Exeter Drive, Spalding, Lincs. PE11 2DY.
FRONTISPICE, B.P. 177, f-75224, Paris Cedex 05, France (00 33 1-4012 0536, fax 00 33 1-4012 0604).
GLYN'S BOOKS, 6 The Avenue, Lyneal, Ellesmere SY12 0QJ.
INPRINT, 31 High Street, Stroud, Glos. GL5.
Derek MOFFETT, Bookfinder General, 3 Thornvale, Abram, Wigan WN2 5YF.
NEWGATE BOOKS (John Dwywer, MA), 89 North Seatin Road, Newbiggin-By-the-Sea NE64 6XR (01670-855381).
Malcolm and Christine NOBLE, 1 Stuart Crescent, Lubenham, Market Harborough LE16 7RL (tel/fax 01858-434671).
OUT-OF-PRINT BOOK SERVICE, 13 Pantbach Road, Birchgrove, Cardiff CF4 1TU.

Jerry SIMKIN, 10 Avalon Lane, Matawan, NJ 07747, USA (*gsimkin@monmouth.com*).
VENTURA PACIFIC BOOKFINDERS (*books@west.net*)
William WILSON, 3a Derwentwater Terrace, Leeds LS6 3JL.

BOOKS AND MAGAZINES (OLD): ON-LINE SELLERS

Many second-hand booksellers from around the world have their books listed at one or more of the following sites:

http://www.interloc.com
http://www.abebooks.com
http://www.bibliofind.com
http://www.bibliocity.com

Some of these also allow you to enter titles you want into their data base. There is also a Bibliophile mailing list; subscription instructions are at:

http://www.auldbooks.com/biblio/

After a two-week trial subscription the cost is $30 a year. Biblio has well over 1000 subscribers, mostly book dealers, and has a high 'traffic level', which means you should find a wanted book quickly.

CAR RADIO SPECIALISTS

ANTIQUE AUTOMOBILE RADIO, 700 Tampa Road, Palm Harbor, FL 34683, USA (00 1 800-933 4926). Free catalogue of repair parts.
A.C. JAMES, 10 Westview, Paulton, Bristol BS18 5XJ (01761-413933). Car radio repair/restoration service, ancient or modern, polarity changing etc. etc.
TORQUAY AUTOPARTS, 43 Ellacombe Road, Torquay TQ1 3AT (01803-200436). Supply period car radio sets from 1950s and 60s.
VINTAGE RADIO SERVICES (Lester Moon) 37 Court Road, Frampton Cotterell, Bristol BS17 2DE (01454-772814). Repairs, restoration and supply of all radios from late 1920s onward. Car radio specialists, particularly Radiomobile. FM conversions. No connection with Vintage Wireless Company of Bristol or Manchester.
VINTAGE WIRELESS COMPANY (Manchester) Ltd, Britannia Garage, 8 Britannia Road, Sale M33 2AA (0161-973 0438). Domestic radios from 1930s to 1960s plus period car radio sales and restoration. No connection with former Vintage Wireless Company of Bristol.

ANGELA INSTRUMENTS, 10830 Guildford Road, Suite 309, Annapolis Junction, MD 20701, USA (00 1 301-725 0451, fax 00 1 301-776 2892). Tube-type hi-fi gear, test equipment, books and components.

ANTIQUE AUDIO, 5555 N. Lamar, H-105, Austin, TX 78751, USA (00 1 512-467 0304, fax 00 1 512-467 2944). Components, kits and books.

ANTIQUE AUDIO OF MICHIGAN, 41560 Schoolcraft, Plymouth, MI 48170, USA. Components for rebuilding old amplifiers.

ARMSTRONG HI-FI VIDEO SERVICES Ltd, 32a Blackhorse Lane, Walthamstow, London E17 6HJ (0181-523 0051 or 0023, fax 0181-523 4395). Service for all Amstrong equipment, including valve-era products.

AUDIO CITY USA, Box 802, Northridge, CA 91328-0802, USA (00 1 818-701 5633). Buy/sell high-end consumer and professional tube-type amplifiers and speaker units.

AUDIO EMPORIUM (John Dempsey) at DAVE MANN MUSIC, 123/125 Mansfield Road, Nottingham NG1 3FQ (0115-947 6911). Large shop concentrating on musical instruments but goodly selection of valve-era and other classic hi-fi (Quad, Leak etc).

AUDIO LABORATORIES (LEEDS), 0113-244 0378. Service specialist, from the oldest valves to the latest digital laser disc.

B&W AUDIO SERVICES, 135 High Street North, Dunstable LU6 1JN (01582-663391). Record player stylus and cartridge specialists. Mail order.

CEDAR ELECTRONICS, Cedar House, 12 Isbourne Way, Broadway Road, Winchcombe, Cheltenham GL54 5NS (01242-602402). Heathkit spares and service.

CHURCHILL'S VINTAGE TECHNOLOGY (0118-966 6595, 0421-887559). Early radios, television, phonographs, gramophones, vintage hi-fi.

CLASSIQUE SOUNDS, 61 Aylestone Drive, Leicester LE2 8QE (0116-283 5821). Refurbished vintage products plus restoration service.

CONCORDANT AUDIO, 14 Glyn Close, Barwell LE9 8GL (01455-43752). Rebuilds Leak Troughline tuners.

CRYSTAL CLEAR SYSTEMS, Flat 2, 18 Canfield Gardens, London NW6 3JY (0171-328 9275, fax 0171-833 3008, mobile 0831-860499). Interesting early valves and hi-fi bought and sold.

CARTRIDGE MAN (0181-688 6565). Cartridge and stylus specialist, retipping service.

DIAMOND STYLUS COMPANY (01492-860881). Cartridge and stylus specialist, re-tipping service.

EKCOS OF THE PAST (David Barrow), 32A Netherhall Road, Doncaster (01302-368882, evenings 01302-366094). Vintage wireless, radio-grams, gramophones, classic audio, early technology.

EMPORIUM, 28 St Nicholas Street, Diss, Norfolk (01379-650744, fax 01379-641392). Vintage hi-fi equipment.

FALCON ACOUSTICS (01508-78272). Publications on valve hi-fi.
FERROGRAPH Ltd, Suite 7, Cookson House, River Drive, South Shields NE33 1JX (0191-427 7774). A full repair and documentation service is available on all Ferrograph products, also Revox and Uher. Refurbished products bought and sold, also a users/enthusiasts group. Ring up for details.
GRAMEX AUDIO, 25 Lower Marsh, London SE1 7AB (0171-401 3830, fax 0171-928 3999). Valve amplifiers.
GT AUDIO, 5 Upper Road, Higher Denham UB9 5EJ (01895-833099). Repair, restoration and sale of classic hi-fi equipment and valves.
HADEN BOARDMAN AUDIO CLASSICS/MELLOTONE (01942-257525, fax 01942-525861). Supplier (and purchaser) of classic audio equipment. Vintage hi-fi, modern quality component audio, absolutely anything of interest to the audiophile. Repairs and restorations but no spare parts or circuit diagrams sold.
HEATHKIT – see Cedar Electronics.
HEAD TECHNOLOGY, 11 Britannia Way, Stanwell, Staines, Middx (01784-256046). Tape-head relapping, reprofiling and new heads supplied.
HERTS. HI-FI (01923-893711). Technical support, repairs and spare parts for all Revox equipment from the 1950s onwards.
HI-FI CURIOSITY SHOP, 23 Bell Street, London NW1 5BY (0171-723 8545). Valve equipment sales. Repairs, rebuilds.
Graham HINDE, 4 Kyles Way, Bartley Green, Birmingham B32 4JW (0121-478 3308 evenings and weekends). Vintage audio bought, sold and exchanged.
Ward KREMER, 5807 Papaya Drive, Ft. Pierce, FL 34982, USA (phone/fax 00 1 561-468 7732, e-mail WKremer798@aol.com). Buy, sell and repair wire recorders. Webcor spares. Transfers of old wire recordings.
LOCKWOOD AUDIO (0181-207 4472, fax 0181-207 5283). New, used and vintage speakers bought, sold and serviced. Tannoy, Lockwood etc.
LORICRAFT AUDIO, 4 Big Lane, Goose Green, Lambourn RG16 7SQ (01488-72267). Remanufacture and restoration of Garrard 301 and 401 turntables to original standards of performance.
LOUGHBOROUGH HI-FI, 43a Market Street, Loughborough, Leics. (01509-239521, 236113). Valve to modern hi-fi bought, exotic or not, working or not.
MANTICORE AUDIO VISUAL, The Courtyard, 56c Shortmead Street, Biggleswade SG18 0AP (01767-318437). Rebuilding and rewiring tonearms plus turntable spares.
MIDLAND AUDIO EXCHANGE, 181 Franche Road, Kidderminster DY11 5AD (01562-822236). Hard-to-find items plus rebuild and repair service.
MIDLAND VALVE HIFI, Solihull, West Midlands (tel/fax 0121-722 2317). Suppliers of new valves, components and hi-fi amplifier parts, also vintage literature, classic equipment.

H.J. MORGAN-SMITH, Unit 3, Vernon Bldg., Westbourne Street, High Wycombe HP11 2PX (01494-532421). Stock item and custom instrument cases, 19 in rack cases and other metalwork. Spraying.
MUSONIC (UK) Ltd, Unit 3, Second way, Wembley HA9 0YS (0181-903 7067, fax 0181-903 4607). Original and replacement styli, with many obsolete and 78 rpm versions.
OLD COLONY SOUND LABORATORY, Box 243, Peterborough, NH 03458, USA (00 1 603-924 6371, 00 1 603-924 6526, fax 00 1 603-924 9467). Comprehensive stock of current and classic hi-fi equipment, spares, books and magazines, even software for designing speaker drivers. Definitely for the serious enthusiast. Mail order, catalogue available.
PRESENCE AUDIO, Haywards Heath (01444-461611, fax 01444-461510). Service for all Decca and London cartridges and tone-arms, plus replacement ribbons for Kelly speakers.
QUAD ELECTROACOUSTICS Ltd, Huntingdon, PE18 7DB (01480-52561, fax 01480-413403). It is worth noting that this long-established company still offers a first-class after-sales service for all its classic-era products. Articles on refurbishing the Quad II power amplifiers appeared in the February 1994 issue of *Everyday with Practical Electronics* and the special Quad supplement to the February 1994 issue of *Hi-fi World*.
RATA Ltd, Edge Bank House, Skelsmergh, Kendal LA8 9AS (01539-823247, fax 01539-823317). High grade audiophile components and cables, also some valves, valve bases and coolers.
REVOX Ltd – see Herts. Hi-Fi and Ferrograph.
SAVOY HILL PUBLICATIONS, 50 Meddon Street, Bideford EX39 2EQ (01237-424280, fax 01237-424280). Old gramophone cartridges and styli.
SOUND AFFECTS, Lower Rainham, Kent (01634-377339). Classic and modern hi-fi.
SOWTER TRANSFORMERS (E.A. Sowter Ltd), Box 36, Ipswich IP1 2EL (01473-219390, 252794, fax 01473-236188). Audio and mains transformers, chokes. Established over 50 years.
STATION SOUNDS, Down Platform, Worthing Station, Worthing, Sussex (01903-39980, evenings 01903-213192). Vintage audio equipment sold, repaired and overhauled. Parts and spares.
STUDIO SAV, 17 Bell Street, London NW1 (0171-258 3448). 1930s to 1960s vintage hi-fi and wireless, buy/sell, repairs.
SUNSHINE SOUNDS, 1413 Magnolia Lane, Midwest City, OK 73110, USA (00 1 405-737 3312). Hi-fi tube and commercial amps, speakers and horns.
SUPERIOR SOUND AUDIO, 25 Montefiore Road, Hove, Sussex (01273-206327). Valves galore; everything from complete amplifiers built to order, to output transformers.
TANDBERG (RDE TANDBERG Ltd), 47 Wensleydale Rise, Baildon, Bradford BD17 6TA (01274-588053). This company, run by a former

Farnell-Tandberg engineer, bought up the remaining spares in the UK. Spares and service.
TECHNICAL & GENERAL, Box 53, Crowborough TN6 2BY (01892-654534). Turntable specialist offering remanufacturing, spares, arms, cartridges, styli etc. for all the classic brands.
UHER – see Ferrograph.
VINTAGE RADIO RESTORATION, 30 Melbourne Avenue, Worthing BN12 4RT (01903-501158). Accessories and spares for restoration projects.
WATTS RADIO (Jim Badman), 1 West Street, Somerton, Somerset (01458-272440). Quad main dealer, repairs to most classic audio equipment.
WEST-TECH SERVICES (Steve Karpiak), 3725 Hollywood Boulevard, Suite 129, Hollywood, FL 33021, USA (00 1 305-929 3007). Cartridges rebuilt, idlers resurfaced. 78 rpm styli fabricated and rebuilt. Stock 1500 types of stereo/phono styli and needles.
WORLD DESIGNS (Hi-Fi World), 64 Castellain Road, London W9 1EX (0171-289 3533, fax 0171-289 5620). Selection of kits for brand-new valve amplifiers and other items.

CUSTOM-MADE CRYSTALS FOR ELECTRONIC EQUIPMENT

C-W CRYSTALS (formerly Phoenix Crystals), 1714 North Ash Street, Nevada, MO 64772, USA (00 1 417-667 6179, fax 00 1 417-667 6169, e-mail *cwxtal@u-n-i.net*). Supplying custom crystals for vintage equipment, amateurs and experimenters since 1933.
GOLLEDGE ELECTRONICS, Merriott, TA16 5NS (01460-73718, fax 01460-76340).
McKNIGHT-FORDAHL CRYSTAL COMPANY, Hardley Industrial Estate, Hythe, Southampton SO4 6ZY (01703-848961, fax 01703-846532).
QUARTSLAB MARKETING Ltd (QSL Crystals), Box 19, Erith DA8 1LH (01322-330830, fax 01322-334904).

ELECTRICAL AND ELECTRONIC COMPONENTS

AGILE TOOLS, Box 1408, London NW10 9ES (0181-452 6724). Early ICs and transistors.
ANTIQUE AUDIO, 253 Blanche, Plymouth, MI 48170, USA (00 1 313-455 4169). High-voltage capacitors and other components for valve circuitry.
ANTIQUE ELECTRONIC SUPPLY, Box 27468, Tempe, AZ 85285-7468, USA (00 1 602-820 5411, fax 00 1 602-820 4643). Comprehensive source of tubes, parts and supplies for restoring old radios. Elaborate 32-page catalogue free, mail order a speciality, credit cards taken.

ANTIQUE TRIODE, 653 Commercial Street, Farnham, NY 14061, USA (fax 00 1 716-549 3823). More than 1000 types of new old stock valves, including scarce types.
AQL TECHNOLOGY (01252-341711). For hard-to-find transistors.
ART AUDIO, 130 Main Street, Calverton, Nottingham NG1 6LU (0115-965 3604). Supplies Vaic valves.
AUDIO NOTE (UK) Ltd, Unit 1, Block C, Hove Business Centre, Fonthill Road, Hove BN3 6HA (01273-885511, 220511, fax 01273-731498). Audio valves.
BILLINGTON EXPORT, Unit 1E, Gillman's Industrial Estate, Billingshurst RH14 9EZ (01403-784961) Valves and CRTs, £50 minimum order.
Michel BLANCHARD, 21 Lot Miailhe, F-33720 Barsac, France (00 33 5-5627 0523 after 20.30 local time). Supplies cotton-covered wire.
W. BURCHER, 676 Foxhall Road, Ipswich IP3 8NQ. Valves and some other valve-era spares.
CENTRE ELECTRONICS, 345 Stockfield Road, Yardley, Birmingham B25 8JP (0121-706 0261). Valves and other components.
CHEVET BOOKS, 157 Dickson Road, Blackpool FY1 2EU (01253-751858). Books, valves, some receivers.
COLOMOR ELECTRONICS Ltd, 170 Goldhawk Road, London W12 8HN (0181-743 0899). Valves at economic prices, prompt mail order service but no credit cards.
CONNECTOR WORLD SUPPLY, Inc., Seattle, USA (00 1 206-789 7525). Huge selection of plugs and sockets.
CROWTHORNE TUBES (Gerald Horrox), 65 Greenwood Road, Crowthorne RG11 6JS (tel/fax 01344-776542). Wide range of old valves, obscure CRTs.
DIVERSE DEVICES, 75 Priory Road, St Denys, Southampton SO17 2JQ (tel/fax 01703-584680). Large quantity of manuals and hard-to-find components, including some germanium transistors.
DOUGLAS ELECTRONICS INDUSTRIES Ltd, 55 Eastfield Road, Louth LN11 7AL (01507-603643). Wide range of transformers for 'valve' type HT voltages.
ELECTRON TUBE ENTERPRISES, Box 8311, Essex, VT 05451-8311, USA (00 1 802-879 1844, fax 00 1 802-879 7764). New and used valves.
ELECTRONIQUE DIFFUSION, 15 rue de Rome, 59100 Roubaix, France (00 33 3-2070 2342). RF components including old-pattern valves.
Jean-Luc FRADET, 9 rue de l'Égalité, 36130 Déois, France (00 33 2-5434 4951). Components for old radio and hi-fi. Repairs.
W.W. GRAINGER, USA (00 1 800-323 0620). Acclaimed as a source of unusual bulbs such as lumiline tubes for Collins speakers and 656s for Johnson transmitters. The phone number is their order number.
HOT ROX UK, 10 Avondale Road, Carlton, Nottingham NG4 1AF (tel/fax 0115-987 3163). Sovtek valves supplied.

ICP, 63 rue de Coulommes, BP 12, 77860 Quincy Voisins, France (00 33 1-6004 0424). Old valves.
Leslie HINE, 9 Well Street, Ulverston LA12 7EG (01229-582557). Rare 1950s CRTs.
INTERSTUDIO Ltd, Priors Corner, Barnet Lane, London N20 8AL (0181-446 5298, fax 0181-446 6035). Wartime German valves, EL12, LS50 and RV12P2000 types.
KAL UK Ltd, Box 84, Herne Bay CT6 7GA; KAL Germany, Kraichbachstr 1a, D-68766 Hockenheim, Germany (tel/fax 00 49 6205-16504). Buy/sell vintage valves and audio equipment.
KENZEN, Unit 9, 16-20 George Street, Balsall Heath, Birmingham B12 9RG. (0121-446 4346, fax 0121-446 4245). Valves and valve-era components. Catalogue issued. (Please ring for an appointment if you intend visiting the warehouse.) This business also has an extensive range of new production high-voltage electrolytics at advantageous prices. These are good quality, British-made caps, in fact the same brand as the Vintage Wireless Company of Bristol used to sell. They are ideal for replacing problematic caps in old valve equipment. Information sheet/price list will be sent in return for SAE.
LANGREX SUPPLIES Ltd, 1 Mayo Road, Croydon CR20 2QP (0181-684 1166, fax 0181-684 3056). Valves, semiconductors and CRTs, especially industrial.
Bernard LOQUEN, 2 place de la Mairie, 22130 Plancoet, France (00 33 2-9684 1037). Valves and industrial tubes, magnetrons, klystrons.
LUKE SYSTEMS INTERNATIONAL, 27827 Via Amistosa, Suite 101, Agoura Hills, CA 91301, USA (00 1 818-991 9373). Specializes in obsolete integrated circuits.
MAJESTIC TRANSFORMER COMPANY, 245 Rossmore Road, Parkstone, Poole BH12 2HQ (01202-734463). Rewinds old transformers.
MB ELECTRONICS (01438-720222, fax 01438-720333). For hard-to-find semiconductors.
OLD TIME SUPPLIES, Box 209, Banbury OX15 5DP. Old radios, books and components.
PM COMPONENTS, Selectron House, Springhead Road, Gravesend, DA11 8HD (01474-560521, fax 01474-333762). Valves and semiconductors. Credit cards, 24 hr number.
PREMIUM PARTS, Whitewater, WI, USA (00 1 800-558 9572, fax 00 1 800-887 2727). Comprehensive source for all kinds of drive belts and wheels for tape recorders. They can custom-make stuff, too.
PROBUS ELECTRONICS Ltd, Findon, Southill Lane, Eastcote, Pinner HA5 2EQ (0181-866 7272, fax 0181-866 2999). UK distributor of production and servicing products made by Caig Laboratories (USA).
RATA Ltd, Edge Bank House, Skelsmergh, Kendal LA8 9AS (01539-823247, fax 01539-823317). Some valves, valve bases and coolers, also high grade audiophile components and cables.

ROCHESTER ELECTRONICS, 10 Malcolm Hoyt Drive, Newburyport, MA 01950, USA (00 1 508-462 9332). Specializes in long-obsolete integrated circuits.
SAVOY HILL PUBLICATIONS, 50 Meddon Street, Bideford EX39 2EQ (01237-424280, fax 01237-424280). Old parts from time to time.
The SEMICONDUCTOR ARCHIVES (0181-691 7908). For hard-to-find transistors.
STIFFKEY LAMP SHOP, Stiffkey, Wells-Next-Sea NR23 1AJ (01328-830460). Fabric-covered mains cable (looks better than PVC-sheathed flex on vintage receivers but has modern blue, brown, yellow/green wires inside). Available in two colours, gold and dark brown; the conductors are twisted and each has its own fabric covering.
T&J ENTERPRISES, Box 1963, Hall Green, Birmingham B28 9LP (0121-777 3386). Specialist suppliers of Lucas electrical spares 1930 to date. Mainly automotive but many parts used in other low-voltage applications (e.g. festoon bulbs in radios).
UNIVERSAL SEMICONDUCTOR DEVICES (01494-791289). For hard-to-find transistors.
VAIC VALVE, Ing. Alesa Vaic, Jilovska 1164, 142 00 Praha 4, Czech Republic (tel/fax 00 42 2-471 8524). Replica valves: this constructor in the Czech Republic is now making new productions of classic radio and audio valves of the period 1915 to 1937. A catalogue (in German) is available and prices of individual valves range from 85 to 300 Deutschmarks. UK agent: Art Audio, 130 Main Street, Calverton, Nottingham NG1 6LU (0115-965 3604).
VALVE & TUBE SUPPLIES, Woodlands Vale House, Calthorpe Road, Ryde PO33 1PR (01983-811386, fax 01983-564708). Large stock of transmitting, output and receiving valves, tuning indicators, valve-holders.
VECTIS COMPONENTS Ltd (01705-669885). For hard-to-find transistors, including obscure Japanese types.
VINTAGE AUDIO, The Wall, West Drive, Lodge Road, Hurst, Reading RG10 0SG (0118-932 1612, fax 0118-934 0717). Hard-to-find valves plus reprints of vintage USA valve data books.
VINTAGE TV & RADIO SUPPLY, 3498 West 105th Street, Cleveland, OH 44111, USA (00 1 216-671 6712). Components, refinishing supplies, reproduction parts, books and literature. Free catalogue.
WATFORD VALVES, 3 Ryall Close, Bricket Wood, St Albans, Herts. (01923-893270). Huge range of unused old and new-production valves.
WILSON VALVES (Jim Fish, G4MH), 28 Banks Avenue, Golcar, Huddersfield HD7 4LZ. (01484-654650, 420774, fax 01484-655699). Valves.

ENGINEER'S SUPPLIES (TOOLS, SHEET METAL, NUTS, SCREWS AND BOLTS)

AGILE TOOLS, Box 1408, London NW10 9ES (0181-452 6724). BA taps and dies, nuts and bolts.

J.A. CREW & CO., Waterygate Farm, Dovers Hill, Chipping Camden GL55 6QU (01386-841979). Broad range of engineering supplies by mail order at sub-retail prices.
CARR'S MODELLING PRODUCTS, Unit 5, Centre 88, Elm Grove, Wimbledon, London SW19 4HE (0181-946 5038, fax 0181-947 7801). General purpose and special solders (even for aluminium and mazak!), soldering tools, metal blackening chemicals.
CETEM POLISHING SUPPLIES, 1 Valepits Road, Garretts Green Industrial Estate, Birmingham B33 0TD (0121-786 1840, fax 0121-789 9656). Polishing materials and abrasives, mail order catalogue.
M. FARRANT, Steinkirk Building, 93 Dunkirk Road, Lincoln LN1 3JU (01522-542366). Polishing tools and materials for metals, glass and Perspex.
GFM ENGINEERING, Clayfields, Tickhill Road, Balby, Doncaster DN4 8QG (01302-859556). Whitworth BSF nuts and bolts.
HARRIS & WALEY, 17 Lothair Road, Aylestone Park, Leicester LE2 7QE (0116-283 2970, fax 0116-2440482). Nuts and bolts, screws, fasteners of all kinds.
NAMRICK – THE NUT & BOLT STORE, 124 Portland Road, Hove BN3 5QL (01273-779864, fax 01273-726708). Fasteners, tools, washers, sealants, pipes, clips, taps and dies. Mail order catalogue.
RESTORATION MATERIALS, Proctor Street, Bury BL8 2NY (0161-764 2741). Full range of clockmaker's supplies, including chemicals, sealants, adhesives, metal finishes and polishes, solvents, wire wool and tools.
A. & R. SHELDON, 33 Bramhall Park Road, Bramhall, Stockport SK7 3JN (tel/fax 0161-440 0821). Comprehensive mail order catalogue of engineer's and automotive tools.
NB: for square section wire try the better hobby shops, some of which sell brass profile sections for model boat and train constructors. Sheet copper of various thicknesses is available in art/craft stores where they sell copper enamelling supplies. Another source is any place where they make copper gutters or roofs – you should be able to get scraps for little or nothing. Sheet brass and nickel silver can be had from good model railway and model boat-building shops; see your local Yellow Pages or the advertisements in a specialist hobby magazine.

A source of loose-bearing balls (for ball races in tuning dials and so on) is your local bicycle repair shop; they are typically stocked in 1/8, 3/32 and 1/4-in sizes. Failing this, there are specialist ball-bearing supply houses in larger towns (see 'bearings' in *Yellow Pages*).

GARAGE SALES, FLEA MARKETS, CHARITY SHOPS AND THRIFT STORES

As Hank van Cleef says, if you're keen for bargains, these sources can yield some amazingly cheap goods. But you never know what you'll find, and you'll have to be willing to slog through acres of junk, as well. If you enjoy this kind of hunting, sifting through the junk is all part of the fun. If you're looking for a specific item, however, you're better off trying more specific sources.

GLASS AND MIRRORS

BEAVER GLASS RESTORATION, 36 Marchmont Road, Edinburgh EH9 1HX (0131-228 9966). Specializes in grinding and polishing chips in glass surfaces. Glass bevelled for dial glasses, carriage clocks etc. Glassware declouded. Glass tubes made for crystal sets.
GLASS SERVICE (0181-520 5353). Glass and mirrors manufactured, coated and finished.

GRAMOPHONES

ALL OUR YESTERDAYS, 3 Cattle Market, Sandwich CT13 9EA (01304-614756). Gramophones, phonographs and electrical antiques.
ANTIQUE PHONOGRAPH SUPPLY COMPANY, Route 23, Box 123, Davenport Center, NY 13751, USA.
AVANTIQUE, 7 The Crofters, Sale M33 2LE (0161-973 8857). Phonographs, radios, books and magazines.
CAMERA CENTRE, 53 High Street, Hailsham, Sussex (01323-840559, fax 01323-442295). Wind-up gramophones bought, sold and repaired.
CHURCHILL'S VINTAGE TECHNOLOGY (0118-966 6595, 0421-887559). Early radios, television, phonographs, gramophones, vintage hi-fi.
Reg DYKES (0181-363 7494). Phonographs, gramophones, 78 rpm records, bygones.
EKCOS OF THE PAST (David Barrow), 32A Netherhall Road, Doncaster (01302-368882, evenings 01302-366094). Vintage wireless, radiograms, gramophones, classic audio, early technology.
FRANK'S ANTIQUES, 10 Town Hall Place, Bovey Tracey, Devon (01626-833325). Gramophones and vintage radios.
GRAMOPHONALIA, 397 High Road, Woodford Green IG8 0XG (0181-559 2592). Shop is on A104 road, next to The Castle public house. Gramophones, phonographs bought and sold, needles, main-springs stocked, selection of cylinders and 78s. Closed Wednesdays and Sundays.
GRAMOPHONE SHOP at two locations: Wednesdays and Thursdays 11.00–19.00 at 31 John Street, Luton, Beds., Saturdays 09.00–18.00 at the Gramophone Stall, under Westway Flyover, Portobello Road Market, London (near Ladbroke Grove underground station) (01234-340829). Selling phonographs, gramophones, needles, springs, spares, repairs, HMV and 'Nipper' memorabilia etc.

Graham GREEN (0181-423 2658). Vintage radios, gramophones and record players.

GREENING'S VINTAGE GRAMOPHONES (N.C. Philips), Oxford Antiques Centre (The Old Marmalade Factory), 27 Park End Street, Oxford (opposite railway stations) (home tel 01993-881740). Open Monday to Saturday. Old gramophones bought, sold and repaired, also radios.

HOLMFIRTH ANTIQUES, 28 Roaine Drive, Cinderhills, Holmfirth HD7 1EX. Repairs, restorations and spare parts for old gramophones and phonographs. Send A4-size SAE for catalogue.

Howard HOPE, 21 Bridge Road, East Molesey KT8 9EU (0181-941 2472 shop hours, 0181-398 7130 other times, fax 0181-398 7630). Buy and sell gramophones, musical boxes, organettes, optical toys, early sewing machines, primitive typewriters and other mechanical/musical devices. Open 10.00–17.00 Fridays and Saturdays and by appointment at other times.

J&M WHOLESALE (BEDFORD) Ltd, 16 Hardwick Road, Bedford MK42 9LF (02314-340829). Gramophone springs and needles by mail order.

Philip KNIGHTON, 1c South Street, Wellington, Som. (01823-662647). Vintage radio sales, repairs and restoration. Radios and gramophones always wanted.

Ruth LAMBERT, 24 Churchway, Weston Favell, Northampton NN3 3BT (01604-405184). Gramophone needle tins bought and sold, steel needles always on sale. A 15-page sale catalogue available.

NAUCK'S VINTAGE RECORDS, 6323 Inway Drive, Spring, TX 77389-3643, USA. Rare cylinders and 78 rpm records. Catalogue issued.

PHONOGRAPHIC RESTORATIONS (Richard Taylor) (01952-814854). Repairs and spares.

RETRO RENAISSANCE, 1096 Nichols Drive, Raleigh, NC 27605-1108, USA. High-quality new steel needles (loud or soft tone) for Victrola gramophones.

SAVOY HILL PUBLICATIONS, 50 Meddon Street, Bideford EX39 2EQ (01237-424280, fax 01237-424280). Old gramophone cartridges and styli.

Ronald SITKO, 26 Tekakwitha Court, Clifton Park, NY 12065, USA (00 1 518-371 8549). Has needles, horns, reproducer parts, gears, belts, even some complete mechanisms, cases and so forth.

Marie-Claude STEGER, 76 avenue Michelet, 93400 Saint-Ouen, France (00 33 1-4012 5378, fax 00 33 1-4874 1404). Restoration of phonographs, gramophones, all spare parts.

TALKING POINT ANTIQUES, 66 West Street, Sowerby Bridge HX6 3AP (01422-834126). Phonographs and gramophones, records, accessories, music books, repairs and restorations.

Brian TAYLOR, 24 Molesworth Road, Plymouth PL1 5LZ (01752-569061). Gramophones, phonographs, polyphons, radio, advertising bought and sold.

VICTROLA REPAIR SERVICE (Rod Lauman), 19 Cliff Street, St Johnsbury, VT 05819, USA (00 1 800-239 4188 evenings). Wind-up phonograph repairs, parts, springs, needles etc. by mail order.
WAVES, Suite 1005, 110 West 25th Street, Manhattan, New York, USA (00 1 212-989 9284, fax 00 1 201-461 7121). Good selection of radios, gramophones, microphones and ephemera.
WYATT'S MUSICAL AMERICANA TALKING MACHINE COMPANY, Box 601, Lakeport, CA 95453, USA (00 1 707-279 2824). Open Monday to Thursday 10.00–16.00, CA time. Almost any part for any model (both cylinder and disk) of Edison, Columbia or Victor machine can be obtained from here. ('Dwayne and Donna Wyatt own this company – they are very nice people.')

HARDWARE AND CABINET SUNDRIES

AGILE TOOLS, Box 1408, London NW10 9ES (0181-452 6724). Panel locks, tubular keys and locks.
ANTIQUE ELECTRONIC SUPPLY, 6221 S. Maple Avenue, Tempe, AZ 85283, USA (00 1 602-820 5411, fax 00 1 602-820 4643). Comprehensive source of tubes, parts and supplies for restoring old radios. Elaborate 32-page catalogue free, mail order a speciality, credit cards taken.
ANTIQUE RADIO HARDWARE (Guy Frederick), 1121 Powers NW, Grand Rapids, MI 49504, USA (00 1 616-456 9378). Knobs, handles, pendants, pulls etc. Original dies for hundreds of different patterns dating back to 1920. Single piece or multiple orders.
ANTIQUE RADIO LABS, Route 1, Box 41, Cutler, IN 46920, USA (00 1 317-268 2214). Wide range of radio restoration parts and custom replica parts service, also copies of technical manuals.
DAVIES MOLDING Co. (NAPCO), Houston and Dallas, Texas, USA (00 1 800-580 2521 in Houston, 00 1 800-580 9515 in Austin and 00 1 214-241 9450 in Dallas). The company lists the old fashioned general-purpose phenolic black instrument cases, ABS cases etc.
Rita HUTCHINGS, Box 382331, Germantown, TN 3818302331, USA (00 1 901-754 5565). Plastic dial covers remade.
OLD TIME REPLICATIONS (Larry Bordonaro), 574 Tobias, Van Nuys, CA 91411, USA (00 1 818-786 2500). Knobs, push buttons, escutcheons, plastic grilles and handles remade.
Dick OLIVER, 28604 Schwalm Drive, Elkhart, IN 46517, USA (00 1 219-522 4516). Reproduction Philco cabinet parts, clock finials etc.
Nigel POLLICOTT (0181-840 1075). Tough insulation board for making replica back panels for various sets. The material is 2.3 mm thick and comes in sheets 800 × 1200 mm for £9.50 each.

RESTORATION MATERIALS, Proctor Street, Bury BL8 2NY (0161-764 2741). Full range of clock-maker's supplies, including chemicals, lacquers, wood stopping, sealants, adhesives, pigments, metal finishes, polishes, shellac and cellulose polishes, solvents, stains, waxes, gums, wire wool and tools.
ROCK-SEA ENTERPRISES, 323 E. Matilija St. #110-241, Ojai, CA 93023, USA (00 1 805-646 7362). Reproduction dial scales for old radios.
SAVOY HILL PUBLICATIONS, 50 Meddon Street, Bideford EX39 2EQ (01237-424280, fax 01237-424280). Old and reproduction radio transfers, speaker cloth.
SMALL PARTS, Inc., USA (00 1 800-220 4242). Phenolic (i.e. bakelized paper) sheet and rods.
L.W. TERRELL, 7109 E. Arbor Avenue, Mesa, AZ 85208, USA (00 1 602-830 7849). Custom replicas of old knobs and dial lenses.
William TURNER, 1117 Pike, Saint Charles, MO 63301, USA (00 1 314-949 2210). Remanufactured plastic dial and clock covers, any style.
VINTAGE TV & RADIO SUPPLY, 3498 West 105th Street, Cleveland, OH 44111, USA (00 1 216-671 6712). Components, refinishing supplies, reproduction parts, books and literature. Free catalogue.
WADE'S WORLD OF KNOBS, 7109 E. Arbor Avenue, Mesa, AZ, 85208, USA (00 1 602-830 7849). This organization makes resin castings of old knobs.

JUKE-BOXES

ANTIQUE APPARATUS, 2335 208th Street, Torrance, CA 90501, USA (00 1 310-328 1306).
AMERICAN JUKEBOX COMPANY (Tony Rolf) (01273-475083). Home breakdown service with full workshop back-up. Specialist in Wurlitzer 78 mechs, restoration and overhaul.
BEYST JUKEBOX COMPANY, Muckton Bottom, Muckton, Louth, LN 8NT (01507-480385). Original juke-boxes, neon signs, advertising signs, slot machines and pin-ball tables. Ring for colour brochure.
CHICAGO SOUND COMPANY, Northmoor House, Colesbrook, Gillingham SP8 4HH (01747-824338). Always a large selection of machines for sale as is or painstakingly restored. Visitors welcome by appointment.
DORIAN'S JUKEBOX RESTORATION (Dorian Walker), 21 Springmeadow Road, Dudley DY2 6DW (01384-828001). Complete or part restoration service for private and trade customers.
ELECTRONS PAST, 851 Enchanted Way, Midvale, UT 84047, USA (00 1 810-262 3903).
FUNTIQUES, Box 825, Tucker, GA 30085, USA (00 1 404-564 1775).
GAMES UNLIMITED, Units 4/5 Lion Park, Holbrook Industrial Estate,

Halfway, Sheffield S19 5GZ (01742-470242, fax 01742-510727). Juke-boxes bought and sold. New and used AMI spares.
Max HOLIDAY, Sittingbourne (01795-473340). Wurlitzer silver age boxes and parts.
HOME ARCADE, 1108 Front Street, Lisle, IL 60532, USA (00 1 708 964 2555).
J&J CLASSIC JUKEBOXES (J. Lang), Oakridge, Shootersway, Berkhamsted HP4 3ND (01442-865871). Vintage juke-boxes, amusement machines, Americana.
JET LEISURE, Unit G5, Chadwell Heath Industrial Park, Kemp Road, Dagenham RM8 1SL (0181-590 7236). All Rockola juke-boxes and Rockola spares.
JOOKS, 10 Hamilton Road, Thame OX9 3XY (01844-214743). Restored juke-boxes for sale, also rebuilds and restoration, even roving repair service.
JUKE JOINT (Tony Holmes), 634 Staniforth Road, Sheffield S9 4LP (01742-446367). Good stock of interesting machines, restored and unrestored. DIY projects with back-up information.
JUKEBOX CENTRAL, 5540 River Road, Madison, OH 44057, USA (00 1 216-428 3523 p.m.).
JUKEBOX DELUXE, Stome Cottage, Beenham RG7 5NN (01734-713341). Restored juke-boxes for sale.
JUKEBOX HIRE (0181-318 2852). Hire a juke-box for a party or to see how it would fit into your home. Delivery and collection included, choose from over 1000 hits 1950–1993. Prices from £95, also sales.
JUKEBOX JUNKYARD, Box 181, Lizella, GA 31052, USA (00 1 912-935 2721).
JUKEBOX MAN (Geoff Young), Unit 19, Monks Way, Monks Road, Lincoln LN2 5LN (01522-513083). The largest parts specialist in the UK, also service manuals and amplifier repairs. Loads of glass, title strips, fully restored machines, you name it.
JUKEBOX MEDIA, 8795A Corvus Street, San Diego, CA 92126, USA (00 1 619-271 8294).
JUKEBOX SERVICES, 31 Wick Road, Teddington TW11 9DN (0181-943 1700). Unrestored juke-boxes sold, also Rockola glass.
JUKEBOX SHOP, 14 High Street, Lye, Stourbridge DY9 8UT (01384-424325, fax 01384-424661). Open six days a week for juke-boxes, always a large selection in stock. Also restoration, spares and repairs.
JUKEBOX WORKSHOP (Warwick McGill), 53 Glenafeoch Road, Carluke ML8 4DR (01555-72175). Juke-box spares and repairs, Seeburg specialist.
JUKEBOXES ETCETERA, 223A North Glendora Avenue, Glendora, CA 91740, USA (00 1 818-914 9434).
JUKEVILLE (Stuart Coleman), Laburnum Fram, Chiltern Green, Luton

LU2 9PW (01438 832-522). Sells machines for all ages. Reliable call-out repair service. Full restorations. Fully stocked juke-boxes available for hire.
JUST SLOTS, 141 Askew Avenue, Hull (01482-54666). Over 300 pull-handle bandits, juke-boxes, pin-tables and wall machines. Viewing by appointment.
MADDY'S JIVING SCHOOL, Box 2149, London W13 9XF (0181-566 5226, 0181-568 4900). Ten-week courses in authentic 1940s/50s jive start regularly in Hammersmith, Slough, Ealing and Richmond.
NEW ENGLAND JUKEBOX, 77 Tolland Turnpike, Manchester, CT 06040, USA (00 1 203-646 1533).
NORMAN TERRY JUKEBOXES, The Beeches, 5 Westfield Road, Dereham NR19 1JB (01362-694547). Import, buy, sell and renovate juke-boxes.
OSCARMATICS (01766-85221). Deal in fully refurbished 1960s/70s/80s juke-boxes.
POWELL'S AUTOMATICS, 8 Marine Gardens, Whitley Bay NT26 1EQ (0191-253 1985). Juke-boxes and slot machines.
Terry PRICE, Box 1964, Selly Oak, Birmingham B29 4BZ (0121-475 6540). Wurlitzers and other classic juke-boxes bought, sold and exchanged. Fabulous showroom, neons, books, video, friendly advice.
Tim RAND, 12027 Saint Vrain Road, Longmont, CO 80501, USA (00 1 303-772 5158).
R.G. EDWARDS AMUSEMENT MACHINES, 108 Thornton Road, Thornton Heath CR4 6BB (0181-684 6239). One of the longest serving and most highly qualified professionals in the juke-box fraternity.
SPECIALIST AMPLIFIER SERVICES, Mike Nicholls, 8 Byron Way, Romford RM3 7PS (01708-341934). All makes of juke-box valve amplifiers and sound systems repaired, renovated and serviced.
STATESIDE MEMORABILIA, Graham MacDonald, 357 Finchampstead Road, Wokingham RG11 3JU (01734-732633). Metal signs, juke-boxes and Americana.
Vern TINSDALE, 8402 North 18th Avenue, Phoenix, AZ 85021, USA (00 1 602-944 8444).
VICTORY GLASS COMPANY, 3260 Ute Avenue, Waukee, IA 50263, USA (00 1 515-987 5765).
WILLIAMS AMUSEMENTS, Bluebird House, Poveycross Road, Horley RH6 0AG (01293-782222). Juke-boxes and other coin-operated equipment for sale.
The WIRELESS WORKSHOP (Jim Cookson), Fen Hill, Hall Common Road, Ludham NR29 5NU (01692-630285). Meticulous restoration of military and communications receivers, vintage radio, gramophones and juke-boxes.
WURCO, 908 Niagara Falls Boulevard, North Tonawanda, NY 14120, USA (00 1 716-694 6247).

LOUDSPEAKER CLOTH AND REPAIR

S.W. CHAPLIN, 43 Lime Avenue, Leigh-on-Sea SS9 3PA (01702-73740, fax 01702-435864). Loudspeaker cloth for period sets.
Barry McDANIEL, 6 Pecos Lane, Little Rock, AK 72212, USA. Reproduction grille cloth.
JACKSON SPEAKER SERVICE Antique Loudspeaker Repairs (Ron McGee), 217 Crestbrook Drive, Jackson, MI 49203, USA (00 1 517-789 6400). Thirty years' experience, speaker parts and repair for virtually any item.
Michael KATZ, 3987 Daleview Avenue, Seaford, NY 11783, USA (00 1 516-781 6202 evenings). Authentic 1930s speaker grille cloth, twelve patterns available.
LOCKWOOD AUDIO (0181-207 4472, fax 0181-207 5283). New, used and vintage speakers bought, sold and serviced. Tannoy, Lockwood etc.
John MAYFIELD, Unit 15, 95 Bath Street, Birmingham B4 6HG (0121-236 1114). Loudspeaker reconing and repair.
PRESENCE AUDIO, Haywards Heath (01444-461611, fax 01444-461510). Replacement ribbons for Kelly speakers.
SAVOY HILL PUBLICATIONS, 50 Meddon Street, Bideford EX39 2EQ (01237-424280, fax 01237-424280). Old and reproduction radio transfers, speaker cloth.
SCAVENGER SOUND INC., Baltimore, USA (00 1 301-636 1951). Reconing loudspeakers.
SOUND REMEDY Loudspeaker Repair Service (Richard Stamer), 331 Virginia Avenue, Collingswood, NJ 08108, USA (00 1 609-869 0238).
SPEAKERWORLD, 2000 Warm Springs Court #6, Fremont, CA 94539, USA (00 1 510-490 5842, fax 00 1 510-490 1961). Expert speaker refoaming and reconing, Visa/MC.
TEMPO RADIO (Nicolo Rubini), Via Mazzini 7, 35040 Masi (PD), Italy (tel/fax 00 39 245-52150). Grille cloths including some original Philips patterns.

MILITARY RADIO, RADAR AND AVIONICS

A.H SUPPLIES, Unit 12, Bankside Works, Darnall Road, Sheffield S9 5HA (0114-244 4278). Some military surplus radio and radar items. Send two first-class stamps for latest list.
AMERICAN CONNECTORS, Box 5397, Hialeah, FL 33014-1397, USA (00 1 305-362 5388, fax 00 1 305-362 5540). All kinds of military surplus connectors.
ANCHOR SURPLUS, The Cattle Market, Nottingham NG2 3GY (0115-986 4902, 864041, fax 0115-986 4667). Some ex-military radio equipment

from time to time, lots of other genuine military (non-radio) surplus equipment. Open seven days a week.
BACO, Kromhoutstraat 36–38, Ijmuiden, Netherlands (00 31 255-511612). General military surplus place with a significant electronics and communications section. Three stars, worth a visit.
JOHN BIRKETT, 25 The Strait, Lincoln LN2 1JF (01522-520767). Occasional military surplus radio and telephone items.
CENTRE ELECTRONICS, 345 Stockfield Road, Yardley, Birmingham B25 8JP (0121-706 0261). Communications receivers and other military equipment, valves, manuals.
COLOMOR ELECTRONICS LTD, 170 Goldhawk Road, London W12 8HN (0181-743 0899, fax 0181-749 3934).
CORKE AERO, Boston, Lincs. (tel/fax 01205-750101, mobile 0831 553271). Multipole connectors, aircraft radio equipment.
E2 ELECTRONIC ENTERPRISES, 131 Gibson Street, Parry Sound, Ontario, Canada, P2A 1X8 (00 1 705-746-1376, fax 00 1 705-746-1570). Advertises as specializing in military surplus radios (buy, sell, trade, service) and in No. 19 sets in particular.
FAIR RADIO SALES CO., Box 1105, 1016 E. Eureka Street, Lima, OH 45802, USA. Commercial and military radio, also surplus equipment, handbooks, magazines.
G.W.M. RADIO, 40/42 Portland Road, Worthing, Sussex (01903-34897). Some military radio equipment and test-gear.
Samuel HEVENER, 3583 Everett Road, Richfield, OH 44286, USA (00 1 330-659 3244 before 18.00 hours EST). World War II military radio equipment.
Bob McKOWEN, 215 S. Ave. C., Washington, IA 52353, USA (00 1 319-653 5776, fax 00 1 319-653 2548). GF/RU rigs, accessories, mounts (some NOS), lots of IFF, radar, intercom etc. His prices are reasonable/negotiable.
M&B RADIO, 86 Bishopsgate Street, Leeds LS1 4BB (0113-243 5649, fax 0113-242 6881). Military and ex-broadcast equipment. The company has a warehouse nearby for larger equipment.
METHODICAL ENGINEERS Ltd, Manor Trading Estate, 4/6 Armstrong Road, Benfleet SS7 4PW (01268-792681, fax 01268-795375). Large stocks of military surplus connectors and other vital oddments.
MILITARY MARKETING Inc., USA (00 1 770-729-9315 or 00 1 800-619 0900 for ordering). Old military radio equipment.
MILITARY MARKETPLACE, 3643 East River Road, Bainbridge, GA 31717, USA (00 1 912-243 0430). Radio equipment, manuals and general militaria.
William A. PERRY, 92 Beechwood Road (Rear), Louisville, KY 40207, USA (00 1 502-893 8724). Amphenol and Bendix connectors by mail.
P.M. QUAKKELSTEIN, Westhavenplaats 28, 3131 BT Vlaardingen (near

Rotterdam), Netherlands (00 31 10-4344523). 'Several US military sets of post-WW II vintage in the window boded well. Inside, the friendly proprietor, one assumes Mr Quakkelstein himself, presided over one of those places crowded with shelves and bays with almost no place to walk. Contents ranged from RT-68s to a 'modern' 1 kW SSB ship's transceiver, but mostly odd parts, components, tank coils, tubes etc., etc. Once again, three stars, worth the visit.'

RADIO-RESEARCH INSTRUMENT Co., Inc., 584 North Main Street, Waterbury, CT 06704-3506, USA (00 1 203-753 5840). 'Their catalog is a wonder to behold. They have all kinds of radar stuff including manuals for the SCR-584.'

The SIGNAL CORPS (Sam Heverer, W8KBF), 3583 Everett Road, Richfield, OH 44286-9723, USA (00 1 216-659 3244). Illustrated 50-page catalogue of WW II military radio sets for $5 foreign post paid, refunded on first purchase.

VIDEO DISPLAY, Inc., Atlanta, GA, USA (00 1 800-241 5005). CRT rebuilders. Most of their business is the CRTs in airport displays, ATM machines and bowling alley displays. The division here does the above listed tubes, but there is a division in Pennsylvania that does 'entertainment' type CRTs. This includes colour, black and white, and the high intensity lamps for big-screen 'projection' TV sets. They stock many CRTs, but are more than willing to rebuild one that they do not have on hand.

Jim WILLIAMS AND SONS SURPLUS, Rear 330 Main Street, Dickson City, PA 18519, USA (00 1 717-383-1295). 'We specialize in military radios and related communications equipment, both used and government reconditioned.'

VTH, Postfach 2274, D-76492 Baden Baden, Germany (00 49 7221-50870, fax 00 49 7221-508752). Publishes acclaimed series of books on East German and Russian military radio equipment.

The WIRELESS WORKSHOP (Jim Cookson), Fen Hill, Hall Common Road, Ludham NR29 5NU (01692-630285). Meticulous restoration of military and communications receivers, vintage radio, gramophones and juke-boxes.

OIL FOR REFILLING TRANSFORMERS, DUMMY LOADS AND THE LIKE

What you need is a light mineral oil (naphthenic oil) such as medicinal mineral oil (not castor oil or cooking oil as this will go rancid). A veterinary surgeon would be a good person to try. If you feel you must have the proper transformer oil, it would be worth trying the public relations department of your local electric power supply company, pleading that transformer oil distributors will sell only 55-gallon barrels of the stuff. If you are lucky they'll contact the engineering department, and spring loose a stray gallon. Do not use motor oil; it has poor heat transfer characteristics and detergent additives may cause unwanted reaction.

PAINT AND LACQUERS FOR RESTORATION PROJECTS

ANTIQUE ELECTRONIC SUPPLY, 6221 S. Maple Avenue, Tempe, AZ 85283, USA (00 1 602-820 5411, fax 00 1 602-820 4643). Comprehensive source of tubes, parts and supplies for restoring old radios. Elaborate 32-page catalogue free, mail order a speciality, credit cards taken.
CASTLE CUSTOMS, Unit 6, Honeyholes Lane, Dunholme, Lincoln LN2 3SH (01673-861685). Paint specialists, computer-cut vinyl lettering, signwriting.
EAGLE INDUSTRIAL PAINTS Ltd, Bridgnorth WV16 5JL (01746-764242). Old paint shades matched. Access, Visa.
The EASTWOOD Company, Box 1729, Yate, Bristol BS17 5PB (01454-329900, fax 01454-329988). As well as selling wrinkle finish paint, this enterprising firm sells a set of chemicals for blackening ferrous metals and a kit of materials for tin-zinc electroplating almost any metal parts, including brass, copper and steel, and another kit for copper plating. They also sell an aerosol can for spraying a reflective aluminium finish, which they claim is the closest paint finish to chrome plating and polished aluminium. This and much more can be found in their free mail order catalogue of car restoration tools and materials.
FOXELL & JAMES Ltd, 57 Farringdon Road, London EC1M 3JB (0171-405 0152). Comprehensive selection of lacquers, varnishes, stains, French polishes and waxes for all types of cabinet restoration.
LIBERON WAXES Ltd, Mountfield Industrial Estate, Learoyd Road, New Romney TN28 8XU (01797-367555).
MATCO, 138 Walton Road, Liverpool L4 4AY (0151-207 2858). Paint for classic vehicles and other projects, custom matching and aerosols.
Don PATTERSON, 636 Cambridge Rd, Augusta, GA 30909, USA. Can supply crackle finish paints.
The PINBALL RESOURCE, Steve Young, 37 Velie Road, La Grangeville, NY 12540, USA (00 1 914-223 5613, fax 00 1 914-223 7365). Supplies CYG Lite acrylic preservation fluid for glass designs.
R&R DESIGNS, 202 Midvale Drive, Marshall, WI 53559, USA (orders 00 1 800-372 4287, technical support 00 1 608-255 0400). Specialist paints matched for refurbishing Collins, Eico, Hallicrafters, Gonset, Hammarlund, Heathkit, Johnson, National, Western Electric and Yaesu equipment. Also lacquers, primers, reducers and fine-point brushes.
RESTORATION MATERIALS, Proctor Street, Bury BL8 2NY (0161-764 2741). Full range of clock-maker's supplies, including chemicals, lacquers, wood stopping, sealants, adhesives, pigments, metal finishes, polishes, shel-

lac and cellulose polishes, solvents, stains, waxes, gums, wire wool and tools.
VERNOL S.A. (attention M. Sousquet), 56, Rue d'Epinay, F-95100 Argenteuil, France (00 33 5-3410 0671). This is one of the very few firms that can manufacture crackle paints to the old formulae. It is willing to manufacture them, even in small quantities and in the desired shades. The paints are applied in three coats and have proved entirely satisfactory under museum test. They even allow variations in the depth and width of the cracks.
VINTAGE TV & RADIO SUPPLY, 3498 West 105th Street, Cleveland, OH 44111, USA (00 1 216-671 6712). Components, refinishing supplies, reproduction parts, books and literature. Free catalogue.
WOODWORKERS SUPPLY, USA (00 1 800-645 9292). Cellulose toning lacquer sold by the can under the name of Behlen, which is Mohawk's non-commercial brand; there is no minimum order requirement.

PAPER CONSERVATION MATERIALS

PAPERSAFE, The North Lodge, Adderley, Market Drayton TF9 3TF (01630-655342, fax 01630-658751). Wide range of hard-to-find, archival-quality paper care and repair materials.

PLASTIC POLISH

Despite the boring-sounding name, Paste Polishing No. 5 is known to connoisseurs as the ideal material for cleaning bakelite and other plastics (even plastic baths!). Unlike Brasso and other liquid polishes, it leaves no active residue and, as it also contains a waxy agent, it also gives a gloss finish. Paste Polishing No. 5 is the stuff the Post Office used to polish up their old bakelite phones and is marvellous stuff – ask any user! Unfortunately the demand for it is reduced nowadays (BT doesn't need it now!), so it is only manufactured at intervals. You can buy it in small quantities under the name Baykobrite from The Radiophile, 'Larkhill', Newport Road, Woodseaves, Stafford ST20 0NP at £1.95 a tube plus postage. Larger quantities and trade supplies from Greygate Chemical Company, Fir Tree Lane, Groby, Leicester LE6 0FH (0116-287 7777). This company also makes special liquid polish for Perspex-type plastics.

PRO AUDIO, MICROPHONES ETC. (FREQUENTLY AT PRO PRICE LEVELS!)

AUDIO TOYSHOP, Bristol (0117-946 7711, fax 0117-973 0503). Sells classic valve studio microphones (at pro prices). Specialist in all tube and esoteric equipment.

ASTROMAN, USA (00 1-213 477 2108, fax 00 1-213 477 2292). Vintage and recent recording equipment.
CLASSIC ELECTRONICS, Los Angeles, CA, USA (00 1 818-992 0999, fax 00 1 818-340 4331). Claims huge selection of vintage microphones and recording equipment.
HARBORSOUND, 180a Central Street, Saugus, MA 01906, USA (00 1 617-231 0095, fax 00 1 617-231 0295). Vintage microphones, consoles, recorders.
Tony LARKING PROFESSIONAL SALES Ltd (01462-490600, fax 01462-490700). England's largest stockist of new and used pro audio equipment. Commercial prices.
LOCATION TECHNICAL FACILITIES Ltd, 11 The Green, Brill, Aylesbury HP18 9RU (01844-238444, fax, 01844-238106). Classic studio microphones when available plus lots of other hard-to-find professional plugs, sockets, leads and other fittings through associated company Future Film Developments.
M&B RADIO, 86 Bishopsgate Street, Leeds LS1 4BB (0113-243 5649, fax 0113-242 6881). Military and ex-broadcast equipment. The company has a warehouse nearby for larger equipment.
The MIC SHOP, USA (00 1 615-595 1717, fax 00 1 615-595 1718). Tube microphones bought, sold and serviced.
MISTY HILL AUDIO, USA (00 1 512-338 6777). Vintage Neumann microphones, Neve and Studer equipment.
Olaf PRINZ, Kurfürstenring 27, D-46483 Wesel, Germany (00 49 282-31187, fax 00 49 281-71724). Buy/sell microphones and recording equipment 1920s to 1970s.
STUDIO SAV, 17 Bell Street, London NW1 (0171-258 3448, fax 0171-258 3449). Current and some older pro audio equipment.
STUDIOSPARES, 61/63 Rochester Place, London NW1 9JU (0171-485 4908, order line 0171-482 1692, fax 0171-485 4168). Professional equipment supplier (not vintage), handy for plugs, sockets, leads and other fittings.
UHER. Servicing carried out by Ferrograph Ltd, Suite 7, Cookson House, River Drive, South Shields NE33 1JX (0191-427 7774).
VINTAGE KING AUDIO, USA (00 1 313-965 0645, fax 00 1 313-964 3338). Vintage microphones and studio equipment.
WESTERN VINTAGE MICROPHONES, Russia (tel/fax 00 7 095-299 1161). Vintage Russian valve microphones.

RADIO AND TV NOSTALGIA

A&R BOOKSEARCH, High Close, Winnick Cross, Lanreath, Looe PL13 2PF (01503-220246, 01503-220965). Huge 160-page catalogue of books including radio and TV programmes and nostalgia.

ADVENTURES IN CASSETTES, 5353 Nathan Lane, Plymouth, MN 55442, USA (1-800-328-0108). American old-time radio shows on tape.

BBC RADIO TAPES. A wide selection of old BBC radio programmes is available at outlets such as W.H. Smith. Catalogue from Video Plus Direct, Box 190, Peterborough PE2 6UW (0800-258259). Publisher is BBC Enterprises, Woodlands, 80 Wood Lane, London W12 0TT (0181-576 2000).

CHEVET BOOKS, 157 Dickson Road, Blackpool FY1 2EU (01253-751858). Copies of the *Radio Times* and other old magazines available from time to time; check stocks before visiting.

EAST ANGLIAN PRODUCTIONS, 21–23 Walton Road, Frinton-on-Sea CO13 0AA (01255-676252, fax 01255-850528). Comprehensive colour catalogue of books, CDs and videotapes.

ERSTWHILE RADIO, Box 2284, Peabody, MA 01960, USA. Nearly 5000 old American radio shows.

EUROSOUND, 41 Swanmore Road, Boscombe East, Bournemouth BH7 6PD. Recordings of wartime German radio broadcasts on cassette tape and some videotapes of newsreels etc. List available.

HAZELDENE ENTERTAINMENT, Hazeldene, George Street, Wordsley, Stourbridge DY8 5YN. Mail order supplier of American classic radio programmes, catalogue free.

INTERSTATE MUSIC Ltd, Box 74, Crawley RH11 0LX. Transcriptions to CD of German songs from wartime propaganda broadcasts. Sold only through retailers, such as Red Lick Records (see below).

KEN'S (Joyce and Ken Graham), 29 High Street, Newport Pagnell, Milton Keynes MK16 8AR (01908-610003). Ephemera specialist, with some old radio/TV magazines and books. Closed Thursdays.

METRO GOLDEN MEMORIES, 5425 W. Addison, Chicago, IL 60641, USA (00 1 312 736 4133). Shop devoted to old films, radio and TV shows, tapes, books, photos and so on. Mail order.

MUSIC INN, 7–11 West End Arcade, Nottingham NG1 6JP (0115-947 0754). Old-established record store with an extremely wide range of nostalgia titles on CD. Mail order specialists.

Doug POULTNEY, 219 Lynmouth Avenue, Morden SM4 4RX (0181-330 3472). Old copies of the *Radio Times* from time to time. Mail order business.

RADIO SHOWCASE, P.O. Box 4357, Santa Rosa, CA 95402, USA. Vendor of American old-time radio tapes.

RADIO SPIRITS, Box 2141, Schiller Park, IL 60176, USA (00 1 800 723 4648). Very commercial marketeer of old-time American radio shows. Mail order, thick colour catalogue.

RADIO YESTERYEAR, Box C, Sandy Hook, CT 06482, USA (00 1 203 426 2574). Very commercial marketeer of old-time American radio and television shows. Mail order, large colour catalogue.

RADIO WAVES, Box 515, Alvin, TX 77512, USA. Rare radio shows, horror and science fiction.
RED LICK RECORDS, Box 3, Portmadoc, Gwynedd LL48 6AQ. (01766-770990, fax 01776-771010). Retailer for Interstate Records (see above).
SCREENTHEMES, 3 Newland Close, Toton, Nottingham NG9 6EQ (tel/fax 0115-973 2431). Thousands of current and deleted items by the world's top composers of television and film music.
SHOKUS VIDEO, Box 3125, Chatsworth, CA 91313, USA (00 1 818-704 0400, fax 00 1 818-701 0560). Mail order supplier of old TV programmes on tape, PAL as well as NTSC formats. Several shows that were shown on British TV are available, such as Robin Hood (different intro sequence from UK presentation), Highway Patrol and Amos & Andy.
SKYMAG, 8/9 Henley Business Park, Trident Close, Strood ME2 4ER. Huge stocks of pre- and post-war copies of the *Radio Times*.
STICHTING MEDIA COMMUNICATIE (SMC), Box 53121, 1007 RC, Amsterdam, Netherlands (00 31 20-662 1141 between 19.00 and 21.00 only). Rare radio recordings from 1950s, 60s and 70s.
R. PARR, 65 Bere Lane, Glastonbury BA6 8BE. Recordings of wartime German radio broadcasts on cassette tape. List available.
RADIO SHOWCASE, Box 4357, Santa Rosa, CA 95402, USA. This company boasts over 9000 titles, with any odd audio problems and edited materials listed with each cassette. Catalogue is also on the World Wide Web (*http://www.crl.com/~lgenco/otr.html*).
SYMPOSIUM RECORDS, 110 Derwent Avenue, East Barnet EN4 8LZ. *Reminiscences of Radio, 1895–1995* is a sixty minute cassette of archive recordings (price around £10, send SAE to confirm current price).
VINTAGE MAGAZINE COMPANY, 39/43 Brewer Street, London W1R 3FD (0171-439 8525, fax 0171-439 8527). Some radio and TV weekly magazines and ephemera.
Dean WENTWORTH, Box 2523, Vancouver, WA 98668, USA. 'His new tapes catalog is loaded with cassettes at $3 each.'

REPAIR SERVICES: RADIO AND TV

Bruce ADAMS, Birmingham (0121-550 0019). Repairs and overhauls of old sets. Collect/deliver throughout West Midlands.
GREENING'S VINTAGE GRAMOPHONES, (N.C. Philips), Oxford Antiques Centre (The Old Marmalade Factory), 27 Park End Street, Oxford (opposite railway stations) (home tel 01993-881740). Open Monday to Saturday. Old radios bought, sold and repaired.
David HIGGINSON, 28 High Street, Misterton, Doncaster DN10 4BU (01427-890768). Repairs and overhauls. Will collect from many areas.
Philip KNIGHTON, 1c South Street, Wellington, Som. (01823-662647).

Vintage radio sales, repairs and restoration. Radios and gramophones always wanted.
Brian MALEY, 203 Tankerton Road, Whitstable CT5 2AT (01227-262491). Valve radios repaired but not radiograms or tape recorders.
NOSTALGIA RADIO repair service, 157 Darnley Road, Gravesend DA11 0SR (01474-352981). Estimates £25.
RADIOCRAFT Ltd, 56 Main Street, Sedgeberrow, Evesham WR11 6UF (tel/fax 01386-882280). A select range of vintage radio receivers restored electrically and cosmetically to the highest standard, also bespoke restorations. Visitors by appointment.
RETRORADIO (0181-670 3302). Spares, repairs and restorations for vintage radios, also buy/sell/trade/rent. Free estimates, personal service.
VALVE EQUIPMENT REVIVAL, Unit 18, Grays Farm Production Village, Grays Farm Road, St Paul's Cray, Orpington BR5 3BD (0181-302 2102). Repairs to valve amplifiers, radios, tape recorders, test gear etc.
VINTAGE RADIO RESTORATION, 30 Melbourne Avenue, Worthing BN12 4RT (01903-501158). Accessories and spares for restoration projects.
John L. WILSON, 23 Malting Close, Stoke Goldington, Milton Keynes MK16 8NX (01908-551363).
The WIRELESS WORKSHOP (Jim Cookson), Fen Hill, Hall Common Road, Ludham NR29 5NU (01692-630285). Meticulous restoration of military and communications receivers, vintage radio, gramophones and jukeboxes.
Andrew ZIMMER, Surrey (01342-834363). Valve radio repairs with coil and transformer winding capability.

SPECIALIST WIRE SUPPLIERS

COASTAL CONCEPTS, 407 Ridge Court, Kohler, WI 53044, USA (00 1 920-458 8825). Litz wire in various gauges. Credit card orders taken.
THE COAX CONNECTION, Joe Cassata KA9CAI 10S226 Meadow Ln, Naperville, IL 60564, USA (00 1 708-420-0342). Range includes magnet wire.
KERRIGAN LEWIS WIRE PRODUCTS, 4421 West Rice Street, Chicago, IL 60651, USA (00 1312-772 7208). The company catalogue lists a wide range of sizes and also has some interesting technical information on Litz wire. If you ask for a catalogue be sure to also ask for the technical tips bulletin 'Soldering of Litz wire'.
MIDCO, PO Box 2288, Hollywood, FL 33022, USA. Cotton-covered wire in small gauges.
MODERN RADIO LABS, Box 14902, Minneapolis, MN 55414-0902, USA. Cotton-covered wire in small gauges.
SCIENTIFIC WIRE COMPANY, 18 Raven Road, London E18 1HB

(0181-505 0002, fax 0181-554 1114). Wide range of bare, enamelled and plated wires, also Litz and manganin wires.
WIRECOM Inc., Box 2569, Asheville, NC 28802, USA (00 1 888-298 9473, fax 00 1 704-628-9473). Coax, solid and stranded 'antenna wire', related connectors, insulators, magnet wire.

TELEGRAPH INSTRUMENTS

ANTIQUES OF SCIENCE & TECHNOLOGY (Jim and Rhoda Morris), Box 1852, Wakefield, MA 01880, USA (00 1 617-245 2897, fax 00 1 617-245 3572). Old telegraph instruments.
GALERIE ALTE TECHNIK, Moltkestrasse 48, D-12202, Berlin, Germany (00 49 172-924 4013, fax 00 49 5421-1272). Telephones and telegraph items, also German radios.
Dr Joe JACOBS, 60 Seaview Terrace, Northport, NY 11768, USA (00 1 516-261 1576). Possibly the best known manufacturer of telegraph apparatus in the US was J.H. Bunnell & Co. in New York. When the company finally closed its doors many years ago, most (if not all) of its remaining inventory was purchased by Dr Joe Jacobs. Among the items that are available from him are bottles of Bunnell Golden Lacquer, used to protect brass on Morse keys.
Bruce MORRIS, 62 Gerllan, Tywyn, Gwynedd LL36 9DE. Sells tape entitled *500 kHz – The End Is Nigh!*, a unique record of the final Morse transmissions by coast radio stations world-wide.
SCIENTIFIC & MEDICAL ANTIQUES, P.O. Box 4252, London SW20 0XS (0181-946 1470, fax 0181-944 7961). Sales stand at Portobello Antiques Market (Units 12–13 downstairs), 282 Westbourne Grove, London W11. Open Saturday 09.00–13.00 or by appointment. Induction coils, telegraphs, galvanometers, X-ray tubes, Wimshurst and electrostatic apparatus, early wireless, books and other electrical antiques.

TELEPHONES

ANTIQUE TELEPHONE COMPANY, 12 Caxton Hill, Hertford SG13 7NE (01992-584650). Antique and reproduction phones.
BAKELITE TELEPHONES, Creech Farm, Southwick, Fareham PO17 6HJ (01705-254275). Reconditioned phones.
William BATH (shop 01202-434003, home 01929-425338, London base 0181-941 1687). Dealing in and collecting telephones and other artefacts of the first half of the twentieth century. Bakelite phones and kiosk equipment always wanted.
BILLARD'S OLD TELEPHONES, 21710 Regnart Road, Cupertino, CA 95014, USA (00 1 408-252 2104).

Figure 22.1 Telephone heaven! Swap-meets organized by phone collectors are a good place to find something of interest, while other sources are antique markets and the National Vintage Communications Fair held twice a year at the National Exhibition Centre.

Ken BUSHELL, Box 572, Baulkham Hills, NSW 2153, Australia. Wide range of accurate reproduction parts for repairing old British, Australian, Swedish and USA telephones, including handset and fixed transmitter mouthpieces, bell clapper covers and decorative transfers.

CANDLESTICK & BAKELITE (Martin Barnett), Box 308, Orpington BR5 1TB (0181-467 3743). Selection of original old phones.

CARGILL TELEPHONE COMPANY (Cartelco), Box 1000, Cargill, Ontario N0G 1J0, Canada (00 1 519-366 2249). Warehouse full of old phones and spares, mail order.

CHICAGO OLD TELEPHONE Co., 327 Carthage Street, Sanford, NC 27330, USA (00 1 919-774 6625 or 00 1 800-843 1320).

CLASSICFONES, Park Hall Exhibition Centre, Charnock Richard, Lancs. every Sunday (01254-263119 any time). Bakelite telephones.

DEJA-VU ANTIQUES, Hatters Row, Horsemarket Street, Warrington, Lancs. (01925-232677). Reconditioned old phones.

Tony FALZON, 7 Kilbreck Street, Benowa Waters, Qld 4217, Australia (00 61 7-5539 3097). Reproduction transfers for Ericsson and British Western Electric telephones.

John FERRIS (01429-223922). Kiosk backboards, notice frames, notices and adverts, also cash tray locks.
GALERIE ALTE TECHNIK, Moltkestrasse 48, D-12202, Berlin, Germany (00 49 172-924 4013, fax 00 49 5421-1272). Telephones and telegraph items, also German radios.
Roger HARVEY, Stone Lodge, Deer Park, Milton Abbas, Blandford DT11 0AY. Telephones bought and sold, also invisible mending and restoration service for coloured bakelite telephones.
HOUSE OF TELEPHONES, 2677 East Valley Drive, San Angelo, TX 76903, USA (00 1 915-482 0101, fax 00 1 915-655 4177). Free catalogue of antique and reproduction phones, with large range of spare parts, cords etc. Visa, Mastercard.
Geoff JULL, 20 Valley Road, Browns Bay, Auckland, New Zealand (tel/fax 00 64 9-478 5161). Accurate reproduction metal parts finished in nickel plate or polished brass: nuts, cradles, tear-drops and pegs for repairing Ericsson skeleton telephones. Also turned wooden handles for early Ericsson spoon receivers.
Marco LAUDANI, 716 Elgar Road, Doncaster, Vic 3108, Australia (00 61 3-9848 7518). VHS videotapes (PAL and NTSC) on telephone collecting.
Patrik LAUNO, Turin, Italy (tel/fax 00 39 11-669 2172). Specialist in rare European telephones.
Jill LEE, 27 Crawford Road, Chelmer, Qld 4068, Australia (00 61 7-379 5834). Reproduction twisted receiver cords and three-, four- and six-conductor plaited cords for telephones in tan, black, green, white and red.
MAHANTANGO MANOR Inc., Box 170, Dalmatia, PA 17017, USA. Quality reproduction telephones and replacement parts.
Laurie MANGLESON, 82 Griffin Parade, Illawong, NSW 2234, Australia (fax 00 61 2-9543 0658). Accurate reproduction parts for British, Australian and American telephones, including transmitters, receivers, mouthpieces, switch-hook forks and transmitter mounts.
METRO RETRO, 21 Canonbury Lane, London N1 2AS (shop 0171-288 1086, enquiries tel/fax 01245-442047). 'Deco' phones and electric fans.
NEW WIRELESS PIONEERS (James and Felicia Kreuzer), Box 398, Elma, New York, NY 14059, USA (00 1 716-681 3186). Rare books on early telephones and telegraphs. Catalogues issued.
THE OLD TELEPHONE COMPANY, Battlesbridge Antiques Centre, The Old Granary, Battlesbridge, Wickford SS11 7RE (01245-400601). Superbly restored phones at antique shop prices.
ON THE AIR, 42 Bridge Street Row, Chester (01244-348468). Chester's unique vintage technology shop and museum. Open Monday to Saturday 10.00–17.30.
OTTERSWICK TELEPHONES, Flat 2F2, 6 Lady Lawson Street, Edinburgh EH3 9DS (0131-228 3690). Specialists in original 1890s to 1950s metal, wooden and bakelite telephones.

PHONECO, PO Box 70, 207 East Mill Road, Galesville, WI 54630, USA (00 1 608-582 4124). Legendary assembly of old telephones, more than the eye can possibly count. A 'must visit' place by all accounts.
PT SUPPLY, Box 38, Pillow, PA 17080, USA (00 1 717-758 1706).
F.G. RICHARD, Marché Vernaison, Allée no. 3, Stand 107 bis, 94300 St-Ouen, Paris (stall 00 33 1-4011 2913, home 00 33 1-6459 9776). This stall is in the famous Vernaison flea market, close to the Porte de Clignancourt station on the Paris Metro. The market is lively on Saturdays and Sundays, pretty dead on other days; it is also the only flea market in Paris worth trawling for this kind of merchandise. You will also find other stalls with similar sorts of plunder.
L. RUDOLF, London (0181-466 4983). Telephones and red telephone kiosk spares for sale. Items also purchased.
RUPERT'S VINTAGE WIRELESS, 151 Northfield Avenue, Ealing, London W13 9QT (0181-576 1368). Mainly radio but usually some telephones as well.
SCIENTIFIC & MEDICAL ANTIQUES, P.O. Box 4252, London SW20 0XS (0181-946 1470, fax 0181-944 7961). Sales stand at Portobello Antiques Market (units 12–13 downstairs), 282 Westbourne Grove, London W11. Open Saturday 09.00–13.00 or by appointment. Induction coils, telegraphs, galvanometers, X-ray tubes, Wimshurst and electrostatic apparatus, early wireless, books and other electrical antiques.
STUDIO SAV, 17 Bell Street, London NW1 (0171-258 3448). Some telephones, also 1930s to1960s vintage hi-fi and wireless, buy/sell, repairs. Not the lowest prices.
TELEPHONE ENGINEERING, Simpson, PA 18407, USA (00 1 717-282 5100).
TELEPHONE LINES Ltd, 304 High Street, Cheltenham GL50 3JF (01242-583699, fax 01242-690033). Wide range of antique and reproduction phones, plus plaited cords sometimes.
Rick UNWIN, 18 Ulaka Road, Ingle Farm, SA 5098, Australia (00 61 8-396 1740). Reproduction parts for Swedish Ericsson phones. Walnut crowns and metal keyhole escutcheons.
Arthur WILLIAMS, 26 Centre Street, Invercargill, New Zealand (00 64 3-216 8985). Reproduction wood parts for old telephones including scroll-top crests, handset grips, ear caps, button feet and writing slopes.
Ernie WOODWARD (01509-672860). Telephones 706/746, dials, cords, PMBXs, jack-strips, indicators and many other spares.

VINTAGE RADIO, TELEVISION AND TAPE RECORDERS

ALL OUR YESTERDAYS, 3 Cattle Market, Sandwich CT13 9EA (01304-614756). Radios and electrical antiques.

ANTIK 'ART, Marché Vernaison, Allée no. 2, Stand 157, 94300 St-Ouen, Paris (00 33 1-4945 5688). This stall is in the famous Vernaison flea market, close to the Porte de Clignancourt station on the Paris Metro. The market is lively on Saturdays and Sundays, pretty dead on other days; it is also the only flea market in Paris worth trawling for this kind of merchandise. You will also find other stalls with similar sorts of plunder.

ANTIKVA RADIO PRAHA, U. Paliarky/Graficka 39, 150 00 Praha 5-Smichov, Czech Republic (00 42 2-643 7444, -537221, -43672).

Roland ASENSI, 4 rue Laënec, 35400 St Malo, France (00 33 2-9956 2844). Domestic and military radio sets.

AVANTIQUE, 7 The Crofters, Sale M33 2LE (0161-973 8857). Phonographs, radios, books and magazines.

CHEVET BOOKS, 157 Dickson Road, Blackpool FY1 2EU (01253 751858). Books, valves, some receivers.

CHURCHILL'S VINTAGE TECHNOLOGY (0118-966 6595, 0421-887559). Early radios, television, phonographs, gramophones, vintage hi-fi.

CORKE AERO, Boston, Lincs. (tel/fax 01205-750101, mobile 0831 553271). Multipole connectors, aircraft radio equipment.

DECODENCE, 13 The Mall, Camden Passage, Islington, London N1 (0171-354 4473 or 0181-458 4665, 24 hrs). American vintage radios bought, sold and exchanged. Open Wednesday, Friday, Saturday 10.00–17.00 or by appointment.

DINOSAUR CONVERTER (David Grant), 4 Kemble Drive, Bromley BR2 8PZ (01869-857086). A first-class 625-to-405 line standards converter (send SAE for details).

DOCKS DE LA RADIO, 34 Rue Jules Vallés, F-93400 Saint-Ouen, France (00 33 1-4011 0990, fax 00 33 1-4011 3365). Closed Wednesdays and Thursday but open rest of week, including Sunday. Close to the main Paris flea market.

Reg DYKES (0181-363 7494). Early radios sold.

EKCOS OF THE PAST (David Barrow), 32A Netherhall Road, Doncaster (01302-368882, evenings 01302-366094). Vintage wireless, radio-grams, gramophones, classic audio, early technology.

FAIR RADIO SALES Co., Box 1105, 1016 E. Eureka Street, Lima, OH 45802, USA. Commercial and military radio, also surplus equipment, hand-books, magazines.

FRANK'S ANTIQUES, 10 Town Hall Place, Bovey Tracey, Devon (01626-833325). Gramophones and vintage radios.

GALERIE ALTE TECHNIK, Moltkestrasse 48, D-12202, Berlin, Germany (00 49 172-924 4013, fax 00 49 5421-1272). Telephones and tele-graph items, also German radios.

Graham GREEN (0181-423 2658). Vintage radios, gramophones and record players.

GREENING'S VINTAGE GRAMOPHONES (N.C. Philips), Oxford

Antiques Centre (The Old Marmalade Factory), 27 Park End Street, Oxford (opposite railway stations) (home tel 01993-881740). Open Monday to Saturday. Old radios bought, sold and repaired.

HEAD TECHNOLOGY, 11 Britannia Way, Stanwell, Staines, Middx (01784-256046). New and remanufactured tape-heads.

Philip KNIGHTON, 1c South Street, Wellington, Som. (01823-662647). Vintage radio sales, repairs and restoration. Radios and gramophones always wanted.

L-W BOOK SALES, Box 69, Gas City, IN 46933 (00 1 800-777 6450). Publishes the book *Classic TVs, Pre-war Through 1950s*. Colour picture book/price guide of old sets for collectors.

MAC'S RADIOS, 19 Stanley Road, Hornchurch RM12 4JS. Large stock, mainly pre-war. Sets, valves and components.

MAD (MILITARY & DOMESTIC), 12 Long Street, Great Gonerby, Grantham NG31 8LN (01476-74047, 0831-790639). Domestic and military radios bought and sold.

MAITLAND RADIO, 4 West Maitland, Edinburgh EH12 5OS (0131-221 0404). Sell radios and TVs, and specialize in mostly 1945–1960 table-top AM radios.

MAJESTIC TRANSFORMER COMPANY, 245 Rossmore Road, Parkstone, Poole BH12 2HQ (01202-734463). Rewind old transformers, including TV line output transformers.

MARTIN BROWN ANTIQUES, Birmingham (0121-585 5758). 1950s-style cabinets (similar to Bush TV22) in modern materials for housing security monitors, TV sets, fish tanks etc.

METHODICAL ENGINEERS Ltd, Manor Trading Estate, 4/6 Armstrong Road, Benfleet, SS7 4PW (01268-792681, fax 01268-795375). Large stocks of military surplus connectors and other vital oddments.

NOSTALGIE RADIO, Rue Grande, 04230 St Etienne Les Orgues, France (00 33 4-92 73 17 25, fax 00 33 4-92 73 17 26). Sales and repairs of old radios and TVs.

OLD TIME SUPPLIES, Box 209, Banbury OX15 5DP. Old radios, books and components.

ON THE AIR, located in Melody's Antique Galleries, 42 Bridge Street Row, Chester (01244-328968 or 661062). Open Monday to Saturday 10.00–17.30.

Harry POSTER, Box 1883, South Hackensack, NJ 07606, USA (00 1 201-410 7525, fax 00 1 201-794 9553, e-mail *hposter@worldnet.att.net*). Vintage television, buy/sell/repair/parts.

RADIO DAYS, 87 Lower Marsh, London SE1 7AB (0171-928 0800). Open 10.30–17.00 Monday to Friday and 11.00–16.00 Saturday. Collectables and memorabilia from 1920s to 1960s.

RADIO TIMES (Martin Pratt), Home Farm, Staunton Harrold, Ashby de la Zouch LE6 5RU (01332-864586). Vintage radios and collectables.

RADIO TUBES, 40 Boulevard du Temple, 75011, Paris, France (00 33 1-4700 5645).
RADIOCRAFT Ltd, 56 Main Street, Sedgeberrow, Evesham WR11 6UF (tel/fax 01386-882280). A select range of vintage radio receivers restored electrically and cosmetically to the highest standard, also bespoke restorations. Visitors by appointment.
RETRORADIO (0181-670 3302). Spares, repairs and restorations for vintage radios, also buy/sell/trade/rent. Free estimates, personal service.
RETROVISOR modern colour TV in vintage 'bakelite' cabinet from selected radio/TV dealers. Details from DeBray Ltd, Eagle House, Middleholme Lane, Sutton on Trent, Newark NG23 6PG (01636-821197 or 0181-841 9983, fax 0181-841 8631).
F.G. RICHARD, Marché Vernaison, Allée no. 3, Stand 107 bis, 94300 St-Ouen, Paris (stall 00 33 1-4011 2913, home 00 33 1-6459 9776). Another stall in Paris's famous Vernaison flea market (see Antik 'art above).
RIN TIN TIN (Rick Irvine), 37 North Road, Brighton, BN1 1YB (01273-672424). Ephemera specialist, with many advertising novelties and framed advertisements taken from old magazines. Also promotional items, magazines, toys, games, early plastics, radios and other domestic fittings.
RUPERT'S VINTAGE WIRELESS, 151 Northfield Avenue, Ealing, London W13 9QT (0181-576 1368).
SAVOY HILL PUBLICATIONS, 50 Meddon Street, Bideford EX39 2EQ (01237-424280, fax 01237-424280). Old sets and parts from time to time, also cabinet transfers.
SCIENTIFIC & MEDICAL ANTIQUES, P.O. Box 4252, London SW20 0XS (0181-946 1470, fax 0181-944 7961). Sales stand in one of the Portobello Antiques Markets, London, W11. Open Saturday 09.00–13.00 or by appointment (ring first for location). Induction coils, telegraphs, galvanometers, X-ray tubes, Wimshurst and electrostatic apparatus, early wireless, books and other electrical antiques.
SCOT TRANSFORMERS Ltd, Hanley Swan, Worcs. (01684-569104, 563113). EHT transformers rewound.
The SOUNDBOX (Ian Maxted), 10 High Street, Hampton Hill TW12 1PD (0181-977 4802, fax 0181-943 5595). Buy and sell gramophones, musical boxes, polyphons, early sewing machines, typewriters and radios. Open Saturdays and by appointment at other times
SPATZ ART DECO CENTRE, Waterloo Antiques Centre, Leeds (0113-244 4187). Specialists in pre-war radios.
Brian TAYLOR, 24 Molesworth Road, Plymouth PL1 5LZ (01752-569061). Gramophones, phonographs, polyphons, radio, advertising bought and sold.
Joseph URBAN, 44 High Street, Anstruther, Fife (01333-310471). Valve radio repairs.
VINTAGE RADIO SERVICES (Lester Moon) 57 Court Road, Frampton

Cotterell, Bristol BS17 2DE (01454-772814). Supplies, repairs old car radios.
VINTAGE WIRELESS COMPANY (Bristol). Ceased trading in 1993.
VINTAGE WIRELESS COMPANY (Manchester) Ltd, Britannia Garage, 8 Britannia Road, Sale, Ches. M33 2AA (0161-973 0438). Domestic radios from 1930s to 1960s plus period car radio sales and restoration. No connection with former Vintage Wireless Company of Bristol.
WAVES, Suite 1005, 110 West 25th Street, Manhattan, New York, USA (00 1 212-989 9284, fax 00 1 201-461 7121). Good selection of radios, gramophones, microphones and ephemera.
WEST ELEVEN WIRELESS (Traylings), 215 Westbourne Park Road, London W11 1EA (0171-221 2653). Valve and early transistor sets bought, sold and hired. Open Saturdays 10.00–18.00. Near Portobello Road market.
The WIRELESS WORKS, 48a Fore Street, Bugle PL26 8PP (01726-852284). Good stock of restored radios plus spare parts to rummage through.
Dennis YATES, The Hewarths, Sandiacre, Nottingham NG10 5NQ (0115-939 3139, fax 0115-949 0180). Quality radio sets and occasional pre-war television.

SWAP-MEETS

Last, but not least, we come to what is often the most fruitful source of treasures – the swap-meet. Most of these are organized by societies and that means you'll have to be a member to find out about these meetings. But that's the best bit, the fellowship you get from being part of the crowd.

GREAT BRITAIN

Radio/TV swap-meets are organized several times a year by the British Vintage Wireless Society (members only) and *The Radiophile*; telephone swap-meets are held by the Telecomms Heritage Group (see 'Clubs and societies' chapter for addresses).

For classic hi-fi, the best source is the AUDIOJUMBLE, Victoria Hall, London Road, Southborough, Kent. Quality vintage hi-fi. For dates contact John Howes, 11 Crendon Park, Southborough, Tunbridge Wells, Kent (01892-540022 evenings) or see advertisements in *Hi-Fi World* magazine.

Best of all, for all of our interests is the NATIONAL VINTAGE COMMUNICATIONS FAIR, held at the National Exhibition Centre, Birmingham, in May and December. Details from the organizer, Jonathan Hill, 13 Belmont Road, Exeter EX1 2HF (01392-411565).

Also worth considering are: the BAKELITE FAIR, organized by the Bakelite Museum Society (0181-852 3492 and 01374-126670), held in May at the De La Warr Pavilion, Bexhill, Sussex; and the MEMORABILIA ROADSHOW AND FAIRS, held at Birmingham (NEC), Glasgow (City Halls) and Manchester (at Bowler's, Trafford Park), organized by Made in Heaven, 216 Kristiansand Way, Letchworth SG6 1TU (01462-683965, 0860-355620).

FRANCE

Paris. The magnet here is the famous Vernaison flea market, close to the Porte de Clignancourt station on the Paris Metro. The market is lively on Saturdays and Sundays, pretty dead on other days; it is also the only flea market in Paris worth trawling for our kind of merchandise (radio/TV/telephones). You will also find old electrical goods there and books but not much hi-fi. Expect to pay quite high prices; there are few bargains here. Many of the stalls are in the older *allées* with their quaint wooden shanty-town huts, whilst opposite in a modern concrete hall is a collection of dealers with equally fine old radios and telephones for sale.

On Saturday and Sunday mornings (the guide books say all day but the stall-holders haven't read them) you will find another, more down-to-earth market in the streets avenue Georges Lafenestre and avenue Marc-Sangnier (nearest Metro station Porte de Vanves). This is a people's market and prices are substantially lower, but with equally tasty merchandise. Telephones, televisions, radios and cine films and projectors can all be found here. There is also plenty of paperwork to be had, including old catalogues and illustrated magazines, but you must be prepared to search. The French are keen collectors of ephemera, so old leaflets and directories are quite widely sold and easy to find.

Brocante fairs (glorified junk markets) in the street or in halls are another rich source to tap. See the inexpensive *Pariscope* events magazine (at all news kiosks) to find out where these are being held.

GERMANY

The vintage radio hobby is well organized in Germany, so if you are visiting the country it might be worth seeing if any of these swap-meets coincide with your trip.

Altenau (Harz). Swap-meet and radio market. Information from C. Timmermann (00 49 5328-8248).

Altensteig. Collectors meeting. Information from U. Lambertz (00 49 7453-8632).

Bad Laasphe. Radio market. Information from H. Necker (00 49 2752-9798).

Bielefeld/Altenhausen. Swap-meet and radio market. Information from K.-H. Lange (00 49 521-331061).

Datteln. Historic radio and TV market. Information from R. Berkenhoff (00 49 2363-65329) or W. Meier (00 49 2832-2544).

Ellerau. North German swap-meet. Information from A. Rieck (00 49 4106-72476 evenings) or K. Helt (00 49 4106-67408).

Erfurt. Thüringer radio flea market, September, in the Elektro Museum. Information from Stephan Hloucal, Hohe Str. 24, D-99094 Erfurt-Möbisburg (00 49 361-68448)

Eschborn. Radio swap-meet. Information from Dr R. Walz (00 49 6195-2506).

Friedrichshafen. Ham radio show plus flea market. Huge annual event in July; the flea market is recommended for all kinds of old things. Information from R. Suter (00 49 7553-7873).

Fürth. Radio flea market in September at the radio museum, Schlosshof 23. Information from G. Walther (00 49 911-756 8110, fax 00 49 911-752577).

Gronau/Leine. Swap-meet. Information from L.-D. Schmidt (00 49 30-3815881 or 00 49 5182-2894).

Inning/Ammersee. South German swap-meet and radio market. Information from M. Roggisch (00 49 89-870688) or C. Dingfeider (00 49 8143-390).

Lamstedt. Swap-meet. Information from S. Grüttner (00 49 4777-8156) or W. Tenschert (00 49 4773-7368).

Nürnberg. Flotronica computer and radio flea market. Information from H. Kammler (00 49 911-644863).

Ulm. GFGF meeting and radio market. Information from G. Bogner (00 49 731-719149) or W. Hauf (00 49 7346-2800).

NETHERLANDS

An international 'market for old technology' is held at regular intervals at **Nijkerk**. Information from M.P. Ritmeester (00 31 5910-13721).

RUSSIA

Moscow. The largest market is held on Sunday mornings at Ismailova Park. There is a Metro station nearby and the market stretches from there along a broad pathway for about a mile to a stadium and then around the corner some! All kinds of goods are sold there, and on my visit in 1992 old telephones were quite plentiful (even a pre-1914 set), although I saw no radio or TV items. There is also a daily market in the forecourt of the Shabolovskaya Metro station where you can find radio valves, integrated circuits and so on. Of course, this may have changed.

SWITZERLAND

The national vintage radio and sound society, CRGS, holds an international flea market in May at **Egerkingen**. Information from A.F. Egli, Kreuzbuchstr. 97, CH-6006 Luzern. RETRO-TECHNIKA is an annual swap-meet/market for all technical hobbies (including radio/TV, telephones and hi-fi) held in September at the Wankdorf Exhibition Centre in **Bern**. Details from RTB, Postfach, CH-2537 Vuaffelin, Switzerland (00 41 32-581810, fax 00 41 32-581910).

USA

Regional swap-meets for old radios, TVs and telephones are plentiful and are listed in the relevant magazines: *Old Timer's Bulletin* (published by the

Figure 22.2 Swap-meets for vintage electronics enthusiasts are not confined to the British Isles and for the largest affairs of this kind you must visit those held in the USA. Perhaps they are luckier with their climate, since many of these events are outdoor affairs.

AWA) and *Antique Radio Classified* for radio and TV, and the ATCA and TCI society newsletters for telephones (address details in the chapters on 'Clubs and societies' and 'Buyer's guide'). The really big show is the **Rochester** event held every September by the AWA; this has a huge flea market (quite apart from all the other displays, visits and presentations). Also worth a visit is the **Dayton Hamvention**, another national event held in early May at Dayton, Ohio. This is predominantly an amateur radio show but there is a ten-acre open-air flea market as well, which always has some vintage radio gear.

OTHER COUNTRIES

The best suggestion is to get in touch with the relevant national society, if there is one; see the chapter on 'Clubs and societies'.

If any readers would like to offer some other suggestions, I'd be pleased to use them in subsequent editions of this book.

CHAPTER 23

Service data and other information

Servicing information is rather like a road map; few can manage without it, and the same applies to other technical data. Finding this information, even for sets sixty years old, is generally not too difficult – if you look in the right place – and this chapter will show you the right places. You may also wish to research other information – technical information, customer literature and so on – and some pointers are given at the end of this chapter. Make no mistake, the information you need is out there; it's just a matter of tracking it down methodically (and being prepared to pay a bit of money to get hold of it; you get nothing for nothing).

OLD TELEPHONES AND TELEGRAPH INSTRUMENTS

All British Post Office telephone and telegraph instruments were wired to standard circuits, which were printed in reference books. Up to the first decade of this century these were in hardback pocket-books – first a combined telegraph and telephone book, then separate issues for telephones and telegraphs (red for telephones, black for telegraphs; up to 1912 the separate National Telephone Company also produced a similar book). As the number of diagrams proliferated, they were produced in loose-leaf form, the most common being the N series of telephone circuits (universally known as N diagrams). These are easy to obtain, either second-hand at swap-meets or as photocopies from the Telecomms Heritage Group (members only; see the 'Clubs and societies' chapter for membership details). A third source is the British Telecom Archives Centre in London (address at end of chapter), although the cost of photocopies is higher.

Another place to find circuits of telephones and telegraphs is in textbooks. The standard books on the subject, such as Herbert's *Telegraphy*, Herbert and Procter's *Telephony* and *Atkinson's Telephony* are all generously provided with diagrams of the most common instruments, whilst older books on the subject (and general electrical works too) frequently have circuit diagrams. Failing this, the Telecomms Heritage Group has good library resources and a letter to its magazine may well produce results, although it helps to be a member.

For telephones made outside the UK, books are the first place to look as well. Ralph Meyer's book *Old Time Telephones* (details in the chapter 'Recommended reading') covers all commonly used American telephones and many of these were exported to Europe and the rest of the world (or made under licence there). In other countries there are standard textbooks on telecommunications in which you will find circuit diagrams, and these books will be found in libraries there. There are also telephone museums and collectors' societies around the world which should be able to help if approached in the right fashion. They are all listed later in this book and you should not be too surprised if some say their resources are for members only; you will undoubtedly benefit from becoming a member.

RADIO, TELEVISION AND AUDIO EQUIPMENT

The most obvious producer of service information is the original manufacturer, who would have printed the genuine service manual. In addition, several trade magazines issued service information of their own and third-party publishers would also supply this information in book form. This applies not just to the United Kingdom but to Continental Europe and the United States as well.

As far as Britain is concerned, photocopies can be had of most official service manuals and of the so-called 'Trader sheets', which are summaries of the same information that was once published by the magazine *Electrical & Radio Trader*. Many sets were also described in the annual volumes of the book *Radio & TV Servicing*, which can be bought cheaply second-hand or borrowed from the library.

In the States there are several 'bibles' akin to *Radio & TV Servicing*. Howard W. Sams & Company started publishing their *Sams Photofacts* in 1946 and these are accepted as the standard for accuracy. Complete chassis photographs are included and each component is clearly labelled. Photofacts are still available, sold in numbered folders and volumes, and to obtain the correct Photofact, you need to find the reference number for your particular piece of equipment in the Photofact Index.

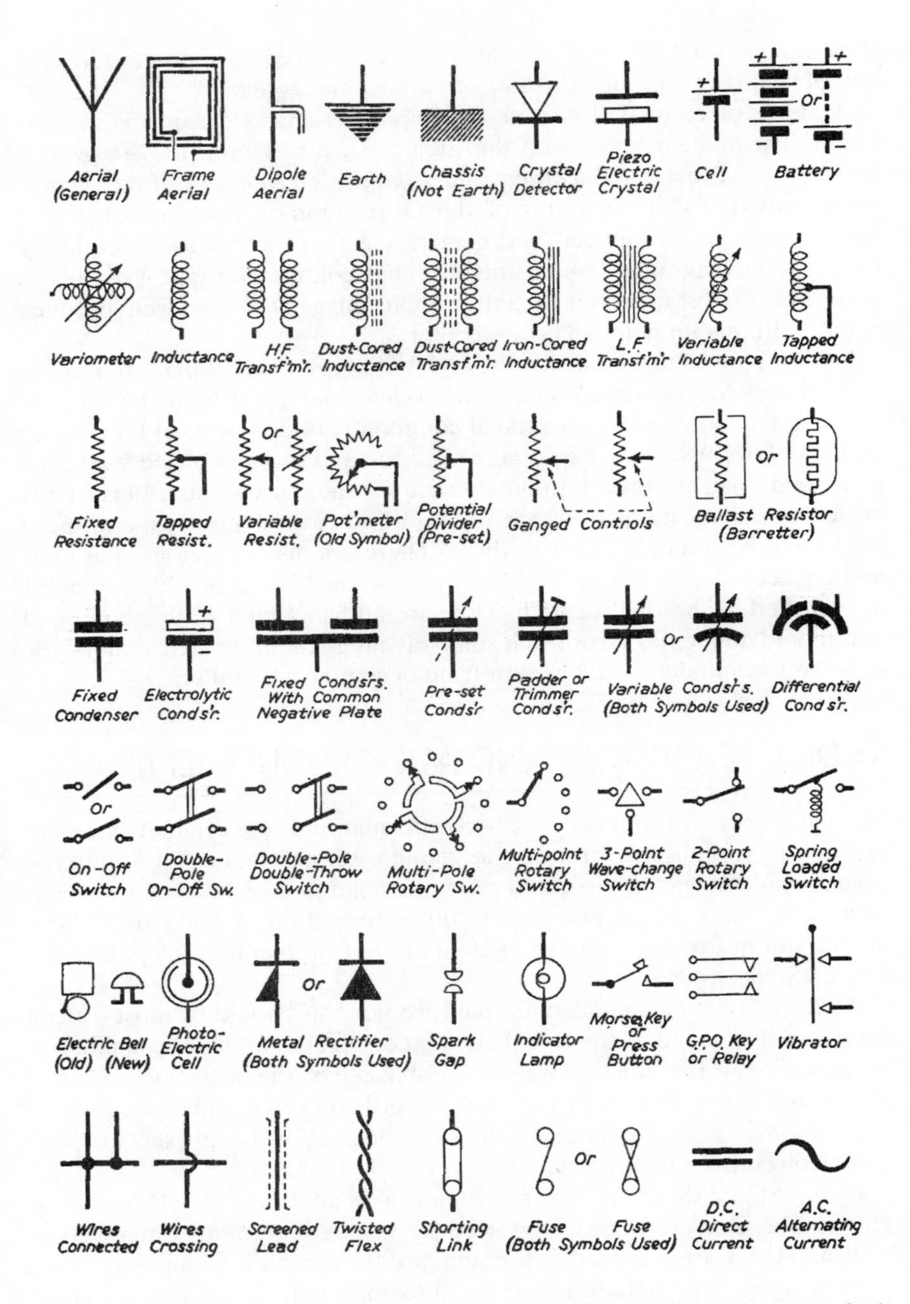

Figure 23.1 Although many of the symbols used in older circuit diagrams are fairly obvious in their meaning, not all of them accord with the ones we use today. Here is a selection.

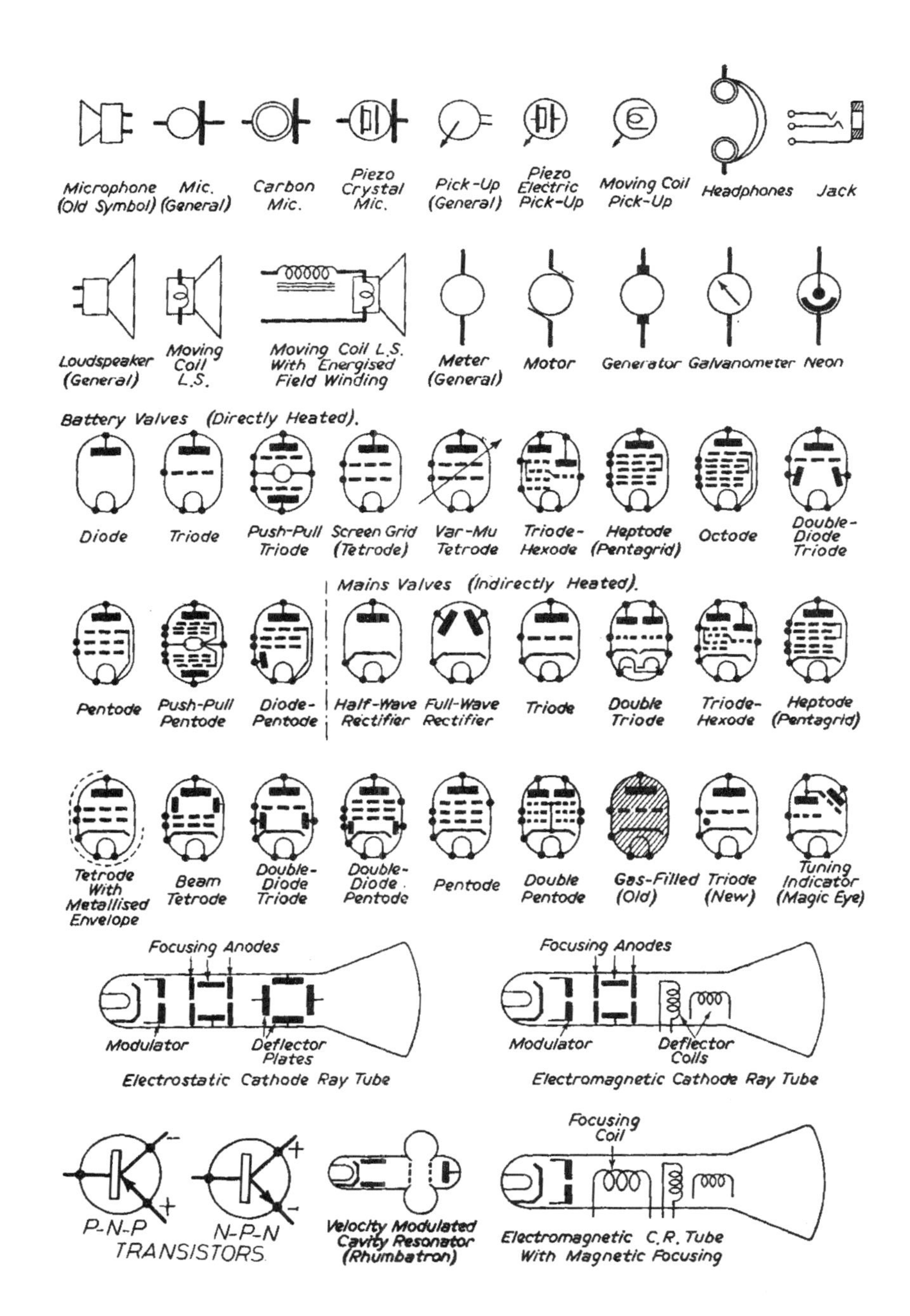
Microphone (Old Symbol)
Mic. (General)
Carbon Mic.
Piezo Crystal Mic.
Pick-Up (General)
Piezo Electric Pick-Up
Moving Coil Pick-Up
Headphones
Jack
Loudspeaker (General)
Moving Coil L.S.
Moving Coil L.S. With Energised Field Winding
Meter (General)
Motor
Generator
Galvanometer
Neon
Battery Valves (Directly Heated).
Diode
Triode
Push-Pull Triode
Screen Grid (Tetrode)
Var-Mu Tetrode
Triode-Hexode
Heptode (Pentagrid)
Octode
Double-Diode Triode
Mains Valves (Indirectly Heated).
Pentode
Push-Pull Pentode
Diode-Pentode
Half-Wave Rectifier
Full-Wave Rectifier
Triode
Double Triode
Triode-Hexode
Heptode (Pentagrid)
Tetrode With Metallised Envelope
Beam Tetrode
Double-Diode Triode
Double-Diode Pentode
Pentode
Double Pentode
Gas-Filled (Old)
Triode (New)
Tuning Indicator (Magic Eye)
Focusing Anodes
Modulator
Deflector Plates
Electrostatic Cathode Ray Tube
Focusing Anodes
Modulator
Deflector Coils
Electromagnetic Cathode Ray Tube
P-N-P
N-P-N
TRANSISTORS
Velocity Modulated Cavity Resonator (Rhumbatron)
Focusing Coil
Electromagnetic C.R. Tube With Magnetic Focusing

The second reference is 'Rider's' the *John F. Rider Perpetual Troubleshooters Manuals*. These cover radios from 1926 to 1953, and also televisions from 1948 to 1957. Rider's is indexed by a system that is very similar to the Photofacts.

'Red Books' is the name given to *RCA Victor Service Data Books*; these cover RCA receivers manufactured from 1923 to 1950.

Supreme Manuals cover radios made as early as 1926 until the company went out of business in the 1950s, generally containing schematics only.

Hugo Gernsback's Official Radio Service Manuals published six volumes from 1929 to 1936.

The *Mallory Radio Service Encyclopedia* was first published in the early 1930s and contained a very complete listing of radios by manufacturer and model number.

FINDING DATA TODAY

There are a number of firms around the country who supply photocopies of data sheets, generally at prices around £5 for a set of data (for a particular radio or TV). There are a few thoughtless people who think this price unreasonable and imagine these entrepreneurs are getting rich in this business. Sure, you can buy photocopies for 10p a sheet at the public library or the copy shop, but do you think you could run a business and support yourself by selling copies at that price? The people who offer this data service have spent years acquiring and storing the data, and also laid out large sums buying and maintaining their own photocopying equipment, so let's hear no more silly complaints about profiteering.

It's worth noting that in this information technology age there are alternative sources and a number of suppliers are turning to different distribution methods. Many North American schematics can be found on-line at: *http://www.nostalgiaair.org*, and Italian and other diagrams at: *http://www.radionostalgianet/schemi.html.* Resources for finding military radio equipment manuals on-line can be found at: *http://www.atsc-army.org/atdls.html.* And, no doubt, other such resources will appear in future.

Not everyone has connection to the Internet (yet) nor is this method ideal for everyone. But several suppliers are now offering radio schematics on CD-ROM disk as well and this promises to be an ideal solution. In the UK the Mauritron organization has compiled diagrams for 500 popular sets on to a CD, whilst in the USA the same system allows you to have access to 30 000 pages of Rider's manuals and other valuable publications. European sets are now represented on CD-ROM as well. Here is a list of all the suppliers we know of – give us a shout if you offer a commercial service and have not been included.

ALGRA FUNKHISTORISCHES LABOR, Kirchstrasse 15, D-38642 Grafhorst, Germany (00 49 5364-2593, fax 00 49 5364-8386). Schematics from USA, NATO, UK, Germany and the Eastern Bloc, specializing in military sets.

ANTIQUE AUDIO, 41560 Schoolcraft, Plymouth, MI 48170, USA. Older factory manuals for RCA, Zenith (including Transoceanics) and several others. Rider's Volumes 1–22, *Sams Photofacts* 1–600, Beitman's Volumes 1–10 and *Sams Transistor Service Manuals* Volumes 1–90.

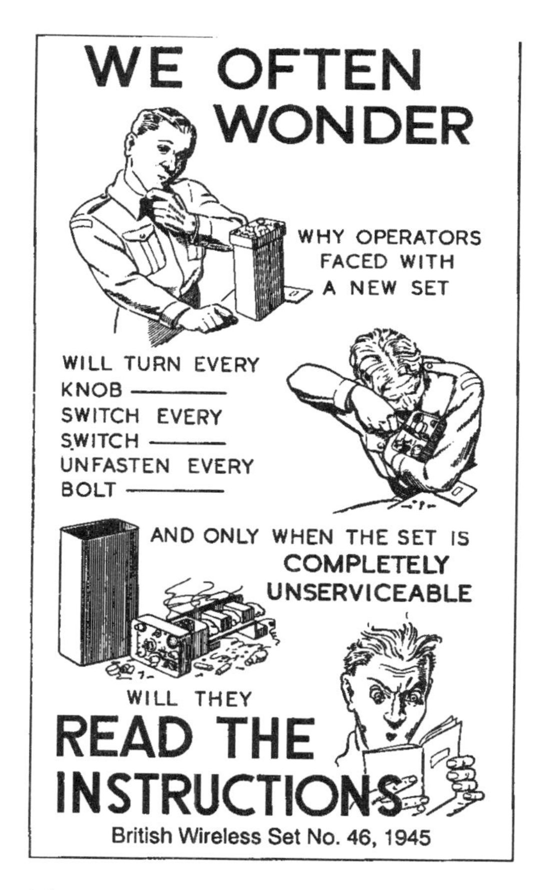

Figure 23.2 And there's no excuse for not having the manual either! Suppliers listed in this chapter will be able to source documentation for just about every communications device, even World War II radio sets.

ANTIQUE ELECTRONIC SUPPLY, 6221 South Maple Avenue, Tempe, AZ 85283, USA (00 1 602-820 5411). Pre-1960 service manuals for many electronic products, including communications receivers. The company offers reprints of some older service manuals, tube manuals, a tube substitution guide book and even a ballast-tube manual.
ARDCO ELECTRONICS, Box 95, Berwyn, IL 60402, USA. The source for copies of Hallicrafters manuals.
Walt BELSITO, 149 Southmayd Road, Waterbury, CT 06705, USA (00 1 203-756 6376). Originals and copies of American manuals and schematics.
Mr BENTLEY, 27 DeVere Gardens, Ilford IG1 3EB (0181-554 6631). Thousands of technical manuals and service sheets, top quality copies at modest prices.
Alton BOWMAN, 4172 East Avenue, Canadaigua, NY 14424-9564, USA. Schematics for all USA radio, TV, organ etc. equipment 1920–1970.
CARDWELL CONDENSER CORPORATION, 80 East Montauk Highway, Lindenhurst, NY 11757, USA (00 1 516-957 7200, fax 00 1 516-957 7203). Manuals for Hammarlund equipment.
CENTER FOR LEGISLATIVE ARCHIVES, National Archives, Washington, DC 20408, USA (00 1 202-501 5350). Record Group 287 supplies US Army Technical Manuals for radio equipment manufactured from 1940 to 1979. Photocopies can be obtained at a cost of 25 cents per page and a $6 minimum order. A 'reproduction service order' must first be completed by the National Archives to determine the cost of the specific manual you desire. The manual on this form must be identified by its proper Army Technical Manual number. A booklet entitled 'Indexes and Lists to Army Technical and Administrative Publications 1940–1979; The National Archives Microfilm Publications Pamphlet Describing M1641', also available from the Archives, describes this procedure.
CENTRE ELECTRONICS, 345 Stockfield Road, Yardley, Birmingham B25 8JP (0121-706 0261). Communications receivers and other military equipment, valves, manuals.
Jim COOKSON, Fen Hill, Hall Common Road, Ludham NR29 5NU (01692-630285). Specialist information for military radios.
DIVERSE DEVICES, 75 Priory Road, St Denys, Southampton SO17 2JQ (tel/fax 01703-584680). Large quantity of manuals and hard-to-find components.
FAIR RADIO SALES Co., Box 1105, 1016 E. Eureka Street, Lima, OH 45802, USA (00 1 419-223-2196). Manuals for surplus equipment. Also some Tektronix manuals and vacuum tube data. Catalogue available.
Jim FARAGO, 4017 42 Avenue South, Minneapolis, MN 55406-3528, USA. A complete set of Rider's, Volumes 1–23, covering radios from 1924 to 1954. The cost is $3 per copy plus SASE. The customer must provide the unit name, brand name and model number.
HI-MANUALS, Box 802, Council Bluffs, IA 51502, USA. Extensive col-

lection of manuals from the mid-1930s to the 1970s, including amateur radio equipment. You must order from a current catalogue and 'dropped list'. The cost is $3 for both. Many Heathkit manuals in stock. They also supply vacuum-tube data from the RCA Receiving and Air Cooled Transmitting Tube Manuals and the Rider #35 Tube Substitution Guide. The cost is $5 per tube plus a business-size SASE. Your cheque will be returned if the tube data is not available. Hi-Manuals cannot accommodate technical correspondence, quotation requests or telephone calls.

T. HULTERMANS PD0MHS, Postbus 4228, 5604 EE Eindhoven, Netherlands. Makes copies of old schematics for a small fee.

Dean K. KIDD, W7TYR, 27270 SW Ladd Hill Road, Sherwood, OR 97140, USA (00 1 503-625 7363). Manuals and data for older Tektronix test equipment.

LAND AIR COMMUNICATIONS, 95–15 108th Street, Richmond Hill, NY 11419, USA (00 1 718-847 3090). Carries Collins, Johnson, Hallicrafters, National and other manuals. Can provide schematics on most equipment dating back to 1945.

Richard A. LERCHE, 1561 Bluebell Court, Livermore, CA 94550, USA (00 1 510-447 9365, e-mail *lerche@aimnet.com*). Rider's and Sams schematics 1920–1970.

MANUAL MAN (Pete Markavage) (00 1 908 238-8964). 'The best source of high quality manual copies that I have found is Pete Markavage. I have purchased three from him, and his copies are first class. I have purchased copies from another popular service and was sent a lousy copy of a lousy copy. The pages were not centered, and no attempt was made to clean up the images. I received a refund, although I have to wonder why the product was sent out in the condition it was. On the other hand, Pete's copies were clearly off the original, with very clean and usable pictures. If you have a choice, call Pete.'

MANUALS PLUS, Box 549, T.A.D. #601, Tooele, UT 84074, USA (00 1 801-882 7188, fax 00 1 801-882-7195). Stocks 300 000 original manuals for test equipment, also military and amateur radio equipment.

MAURITRON TECHNICAL SERVICES, 8 Cherry Tree Road, Chinnor OX9 4QY (01844-351694, fax 01844-352554). Photocopies of old service sheets, other technical data.

MECCA, 1132 Conway Drive, Newark, OH 43055, USA (00 1 614-522-4944). Used manuals for test equipment, amateur equipment and early receivers. Prices range from $5 to $35. Send an SASE with your request for price and availability.

MILITARY MARKETPLACE, 3643 East River Road, Bainbridge, GA 31717, USA (00 1 912 243 0430). Radio equipment, manuals and general militaria.

Bernard MOTHERSILL photocopies (at cost) items from his own extensive collection of service sheets for 1950s and 60s TV sets. There are dozens and dozens, mainly Alba, Ekco, Bush, Ferguson/Thorn, GEC, Murphy, Perdio, Pilot, also a few Decca, Defiant, HMV, KB, McMichael, Peto Scott, Philco, Regentone and Ultra. Write with international reply coupon plus unstamped self-addressed envelope to him at 3 Cherrywood Close, Clonsilla, Dublin 15, Eire.

H. MUNIMUS, Jakobstr. 2, 70806 Kornwestheim, Germany (tel/fax 00 49 7154-182180). Circuits for radios from the USSR.

MUSTY MANUALS, 645 Wheeling Road, Wheeling, IL 60090, USA (00 1 708-634-6467). An extensive collection of owner's manuals, service and maintenance manuals, catalogues, magazines and other technical facts on all types of early and late model radio equipment, including ham gear. Manuals come with updates or articles, product reviews and other information. Catalogue available.

OLDE TYME RADIO COMPANY, Suite 317, 2445 Lyttonsville Road, Silver Spring, MD 20910, USA (00 1 301-585 8776). Some RCA Tube Manuals and Rider's for sale. Catalogue available.

PUETT ELECTRONICS, Box 28572, Dallas, TX 75228, USA (00 1 214-321-0927, 214-327 8721). Schematic diagrams 'for nearly any radio receiver'. Cost is $5 with manufacturer and model number; $15 plus drawing and description if the manufacturer or model number is unknown. Reproduction manuals, Rider's indexes, RCA Vacuum Tube Manuals, and many other service/instruction manuals for classic radios and related equipment are available. Some amateur equipment manuals including Collins. Test equipment manuals include Hewlett-Packard, Tektronix, Hickock and others. Catalogue available.

RADIO ROADMAP, Lance Wilson, 8429 Via Linda, Scottsdale, AZ 85258, USA (00 1 602 483 8993). Pre-1945 USA radios.

The RADIOPHILE, 'Larkhill', Newport Road, Woodseaves, Stafford ST20 0NP. Copies of radio and TV manuals.

RÉTRO-PHONIA, BP1462, 25008 Besancon cedex, France (tel/fax 00 33 3-8148 9398). More than 3000 circuits for European radios.

R. ROSENGARTEN, 1448 Lebanon Road, Clarksville, OH 45113-9711, USA (e-mail *brose@erinet.com*). Pre-1938 American schematics.

Howard W. SAMS & Co., 2647 Waterfront Parkway East Drive, Indianapolis, IN 46214-2041, USA (00 1 800-428 7267, 800-428-5331). Sams covers most consumer electronic items manufactured since 1945. A photocopy service for manufacturer's manuals and Rider's data is also available.

SAVOY HILL PUBLICATIONS (Tudor Gwilliam-Rees), 50 Meddon Street, Bideford EX39 2EQ (01237-424280). Large library of service data for photocopying. Historical matter such as wholesaler adverts or manufacturer sales brochures is often supplied as well with service data.

K. SCHOLTZ, Leisveldstraat 31, NL-5045 XS, Tilburg, Netherlands (00 31 13-571 2494 weekdays, 00 49 228-217321 weekends). Circuits for Czech radios 1945 onwards, Hungarian radios 1936 onwards.

Robert SCHRANTZ, 610 E. Juanita Avenue, San Dimas, CA 91773, USA (tel/fax 00 1 909-394 1194). American radio and TV schematics 1920–1960 mailed or faxed.
Chris SIMMONDS (01705-789320). Radio circuit diagrams for more than 4000 models 1930–1960.
SLEP ELECTRONICS Co., Box 100, Otto, NC 28763-0100, USA (00 1 704-524 7519). Operation and service manuals, schematic service for old and new amateur radios, commercial and military surplus test equipment, commercial equipment.
Philip TAYLOR, 3 Silver Lane, Billingshurst RH1 0RP.
TECHNICAL INFORMATION SERVICES/INFOTECH, 76 Church Street, Larkhall ML9 1HF (01698-883344/888343, fax 01698-884825), 'World's largest selection of manuals, 1930s to current date, British and foreign'.
TEST EQUIPMENT MANUALS (Ed Matsuda), Box 390613, San Diego, CA 92149, USA (00 1 619-479 0225, fax 00 1 619-479 1670). Good source of Hewlett-Packard and other manuals.
VINTAGE TECHNICAL SERVICES, 28 Welbourne Road, Liverpool L16 6AJ (0151-722 1178). Hundreds of technical manuals for WW2 radio and radar equipment, English, American and German.
Dave WILLIAMS, 16 Church Street, Owston Ferry, Doncaster, DN9 1RG (tel/fax 01427-728046). Circuits and service manuals 1930–1990s.

CD-ROMS

BPS Inc., 164 Winter Haven, Brownsville, TX 78521, USA. Complete Rider's collection of Volumes 1–23 (30 000 pages), priced $249.00.
MAURITRON TECHNICAL SERVICES, 8 Cherry Tree Road, Chinnor, Oxon OX9 4QY (01844-351694, fax 01844-352554). Disk contains Trader service sheets for 500 sets from the 1930s to the 1950s, priced £49.95.
Wilfried MEIER, Schepdonksweg 11, D-47625 Kevelaer-Wetten, Germany (00 49 2832-2544). CD1, Radio receiver circuits, comprising all 11 volumes of Lange/Nowisch; CD2, Complete Philips radio service documentation from 1927 to 1945; CD3, Philips radios BX series from 1947 to 1957; CD4, Philips test equipment (all GM types); CD5, Philips TV receivers from 1948 to 1968 (including British sets and first-generation colour sets); CD6, TV miscellany (British pre-war sets and post-war sets down to 1953, plus many German, Belgian and French sets and some Eastern Bloc examples). Each CD-ROM costs 98.00 Deutschmarks.
OLD COLONY SOUND LABORATORY, Box 243, Peterborough, NH 03458, USA (00 1 603-924 6371, 00 1 603-924 6526; fax 00 1 603-924 9467). Disk version of the fourth edition of *Radiotron Designer's Handbook*, $29.95.

RADIO ERA ARCHIVES, 2043 Empire Central, Dallas, Texas 75235, USA (00 1 214-358 5195, fax 00 1 214-357 4693). *Rider's Perpetual Troubleshooter's Manual* with additional new material and index, covering the original 23 volumes in six CDs, $85 per CD or $450 the set. Plus RCA *Radiotron Repair Handbook* and other old reference books on CD at $85 per volume (details at *http://www.electrosys.com*).

TELECOM DIGEST CD-ROM. Containing all back issues of the *Usenet Telecom Digest* from 1981 to the end of 1995, with 34 728 articles and messages, some on historical topics. Available in USA for $39.95 from Walnut Creek CD-ROM, Suite E, 4041 Pike Lane, Concord, CA 94520, USA (00 1 800-786 9907 or 001 510-674 0783 sales line open 24 hrs, fax 00 1 510-674 0821). On sale in the UK, price £23 + £2 p&p, from PDSL, Winscombe House, Beacon Road, Crowborough TN6 1UL (credit card orders 01892 663298, fax 01892 667473).

SCHEMATICS BY E-MAIL

RADIO ERA ARCHIVES (*http://www.electrosys.com*). Supplies circuit diagrams by e-mail (a charge is made for this service).

RÉTRO-PHONIA (*http://members.aol.com/retrophoni/bienvenue.htm*). A French collectors' organization which has several thousand schematics listed on its Web site (mainly French and other European sets) that can be e-mailed to members. Rétro-Phonia also has a list of the introduction dates of more than 6000 radio receivers; very handy for those wishing to know when their set dates from.

PROJECT ANUBIS

Project Anubis is an ambitious and praiseworthy idea to compile an international data base of circuit diagrams for old radio sets in digital form. The man behind it is Jean-Michel Bourque of the French organization Rétro-Phonia. In his own words, the objective of the project is to organize co-operation between wireless clubs and museums, for making inventories and exchanging radio schematics in digital form, so that these precious documents will be safeguarded.

He explains: 'Old radio clubs are very often approached by their members, asking for copies of radio schematics. Satisfying this requirement is very time-consuming; the clubs have to archive the documents, inventory them, and then retrieve and copy them every time a request is received. Making digital copies of radio schematics permits safeguarding of the original paper documents, speeds retrieval, and makes sending and receiving the documents easier. Moreover, "poor quality" documents can be enhanced digitally as they are archived. If each document is identified with a unique code, we can form a great European data bank.

'With this system, duplication of effort due to scanning schematics already digitized by someone else can be eliminated, and instant data retrieval is possible,' he continues. 'The format of files is not important as digital technology is rapidly evolving. However, it would be useful to adopt a standard naming convention. Even if you do not yet have a scanner, you can join in this effort by helping inventory documents and exchanging high-quality photocopies. On request, I can send you a list of available digitized schematics.'

READING CIRCUIT DIAGRAMS

A degree of understanding helps, especially when you find symbols and notations which don't match today's practice.

DON'T BE MISLED

When reading old diagrams bear in mind that some components may be denoted by different symbols from the ones we use today. Resistors were drawn as zig-zags instead of boxes, whilst lamps and fuses were also shown differently. What looks like a script lower case w (actually the small form of the Greek letter omega) denotes ohms and at this time the normal large omega symbol was reserved to denote megaohms. If you find both symbols together, the meaning of each is clear, otherwise you will have to use common sense!

In American diagrams another trap lurks. Mohms at one time denoted *thousands* of ohms, from the Latin word for a thousand (which also gives us 'milli-' for indicating thousandths. Usage ceased some time in the early 1930s, but not consistently (and mid-1930s ARRL Handbooks are very careful over this, which gives some idea of the confusion it caused). In really old US diagrams, you may find resistors drawn as 'square-wave' shapes, and inductors drawn with zig-zag lines, whilst condensers are shown as either overlapping squares or interlocked, square-cornered 'U' shapes (this appears in many US patent drawings but is not exclusive to these).

The designator VT may mean 'vacuum tube/valve, thermionic' or 'valve, transistor' and Q can be either a transistor or a capacitor. In all cases, however, the shape of the accompanying symbol will make clear which component is meant.

In wiring diagrams the colours are usually marked and on British drawings the convention is generally:

BK	black
BN	brown
R	red
O	orange
Y	yellow
G	green
B	blue
V	violet
S	slate (i.e. grey)
W	white
P	pink

DATA BOOKS

VALVE AND TRANSISTOR DATA

Data books for components are extremely valuable, not only for the characteristics of each type but the equivalents tables, showing substitutes for when the original type needed is not available. Each country has its own 'bibles' of data, and the bad news is that most of these are now long out of print. The good news is that a few have been reprinted and many more are easily found second-hand at swap-meets and in used bookshops. In Britain the many editions of the *Wireless World* guide (official title *Radio Valve Data* and later *Radio Valve and Transistor Data*, Iliffe Books) and Bernard Babani's *International Radio Tube Encyclopedia* are among the best books ever published.

Among the books currently in print (all inexpensive paperbacks) are:

Comprehensive Radio Valve Guide (five volumes) by Bernard Babani, reprinted by Radio Bygones, 9 Wetherby Close, Broadstone BH18 8JB (tel/fax 01202-658474).

Lampmètre Radio-Contrôle – Notice d'Emploi, 1941 (Valve tester Instructions, 1941). Photocopied reprint, 62 pages. Price 100 French francs post-paid, from J.-C. Montagné, 35 rue Salvador-Allende, 92220 Bagneux, France. The book covers all current (in 1941) French valves plus a sprinkling of British, RCA and German ones, with equivalence tables and base diagrams.

Radio Tubes and *TV Tubes*, published by SECG Éditions Radio, 3 rue de l' Éperon, F-75006, Paris, France. As well as basic data for each type, these books also show specimen application circuits.

Röhren-Codex, a reprint of the 1951 edition of this German valve data publication covering valves from all Western countries. Available from Verlag Historischer Technikliteratur, Nesselrodestrasse 7, D-45699 Herten, Germany.

Tableaux de Characteristiques et de Correspondances de Lampes (Valve characteristics and equivalents). Photocopied book, 48 pages. Price 100 French francs post-paid, from J.-C. Montagné, 35 rue Salvador-Allende, 92220 Bagneux, France. This useful book is a mixture of old publications and personal additions, made up into a kind of useful scrap-book. French valves come top of the list in this book but there are also plenty of American, British and German types. The data covers the pre-war period right down to 1964 and includes both tabular information and diagrams of valve bases.

Valves Electronic (GPO, 1947), *WD Valve Data, CV and Military Wartime Valve Equivalents, Services Radio Valve Manual* (1942), *Services Index of Valves, CRTs and Electronic Devices* (1944), *Admiralty Fleet Order* (1945, with RAF and USA valve types).

A specialist supplier of valve books worth mentioning is OLD COLONY SOUND LABORATORY, Box 243, Peterborough, NH 03458, USA (00 1 603-924 6371, 00 1 603-924 6526, fax 00 1 603-924 9467, e-mail *custserv@audioxpress.com*). The company offers *Operating Features of the Audion*, a small 32-page reprint of Edwin Armstrong's 1917 paper from the *Journal of the New York Academy of Sciences*. Old Colony also has many of the RCA tube books in reprint and some foreign tube books, including a 636-page book from Japan which has characteristics and specs for all major vacuum tubes (mainly in English), and there are other books in Japanese, German, Swedish and Italian.

In this day and age 'machine readable' resources are popular with some people (presumably those who cannot be bothered to buy books) and the good news for them is that valve data is available on the Internet and on floppy disk. Here are some pointers.

WEB SITES (ALL WITH LINKS TO OTHER SITES)

Audio tube data and spice models:
http://www.phy.ohoiu.edu/~cigna/tubes/tubes.html
Audiomatica (data on obscure and Russian tubes that you might not find elsewhere):
http://www.mclink.it/com/audiomatica/sofia
Peavey Electronics transistor cross-reference list:
http://www.peavey.com/technote/transbrk.htm
Robert Casey's pencil tube page:
http://www.netcom.com/~wa2ise
On-line tube substitution guides:
http://www.nostalgiaair.org
http://www.tri.net/radio
http://www.audioweb.com/newsstand/positivefeedback/6-1/tubesub.html

http://www.geocities.com/Paris/2356/tube.txt
http://www.audio.com.hk/vintage/ti.htm
Trevor Gale's European valve data site:
http://www.xs4all.nl/~tgale/valves.html
Vacuum tubes unlimited (general information):
http://www.cinternet.net/~vactubes

VALVE AND OTHER DATA ON DISK

Tube DataBase III, also *Transistor Substitution Database* and *Integrated Circuit Identification*. IBM format disk, $25 each from Jim Johnson, Box 6352, Kennewick, WA 99336-0352, USA.
Tubedata. Valve data and substitution information on more than 25 000 European, American and Russian valves. IBM format disk, $39.95 from ARC, Box 2, Carlisle, MA 01741, USA (00 1 508 371 0512, fax 00 1 508 371 7129).
VT Data. Interactive data search and inventory system. IBM format disk, $29 + postage from EPS/Solutions, Box 862, Broomall, PA 19008-0862, USA. Information at *http://members.aol.com/EPSweb/vtdata*.

OTHER SOURCES OF INFORMATION

As you research equipment more deeply you will find yourself needing more detailed information. In the past company archives were an excellent first port of call but these days many organizations have cut back on the material they keep. If you are lucky, you will find that these offices contain customer catalogues of the past, information bulletins and official photographs taken for use in company publications. In some cases, where the company is no longer in business, you may find that its documentation was deposited with the local public reference library nearest to where the business used to exist. This is the case, for instance, with Pye of Cambridge (documents at Cambridge Central Library local studies collection) and for the camera lens maker Dallmeyer (at Neasden Library). Obviously you will need to make enquiries but with a little effort you may well strike gold.

Reference libraries are a very useful resource (see chapter on 'Recommended reading'), since the larger ones of these often keep company house journals. Many communications equipment manufacturers used to issue (and still do, of course) regular glossy magazines for customers, and these frequently described new products, complete with photographs and circuit diagrams. The same applies to service organizations; the BBC issued its *Technical Monographs* for instance, whilst the BPO subsidized the production of the *Institute of Post Office Electrical Engineers Journal*. Both of these publications are full of valuable technical information and are typical of many similar kinds of publication.

Finally you may also find the information you require in the Public Record Office or the archives of the fighting services or other user organizations or of appropriate professional institutions.

BBC ENGINEERING INFORMATION, 201 Wood Lane, London, W12 7TS (0345-010313). Has a valuable library and photographic collection.

BBC PHOTOGRAPH LIBRARY, Centre House, 56 Wood Lane, London W12 7SB (0181-743 8000 ext 63314, fax 0181-746 0353).

BBC WRITTEN ARCHIVES CENTRE, Caversham Park, Reading RG4 8TZ (0118-947 2742). Open by appointment (two weeks' notice) from 09.45 to 13.00 and 14.00 to 17.15 Tuesday to Friday; fee payable.

BRITISH FILM INSTITUTE, 21 Stephen Street, London W1P 2LN (0171-255 1444, fax 0171-436 7950). The library has a number of resources for television studies, particularly of the Baird era. Telephone enquiries are handled 10.00–17.00 and an appointment is necessary for visits.

BRITISH TELECOM ARCHIVES, Third Floor, Holborn Telephone Exchange, 268–270 High Holborn, London WC1V 7ED (0171-492 8792, fax 0171-242 1967). Open 09.30–16.30 Monday to Friday except public holidays.

BROADCAST PIONEERS' LIBRARY OF AMERICAN BROADCASTING, located on the campus of the University of Maryland at College Park, about 20 miles from Washington, DC, USA. The library has a large collection of publications and sound recordings available for on-site research. See Web site at *http://www.lib.umd.edu/UMCP/LAB*.

IEEE CENTER FOR THE HISTORY OF ELECTRICAL ENGINEERING, Rutgers University, 39 Union Street, New Brunswick, NJ 08903, USA (00 1 908-932 1066, fax 00 1 908-932 1193).

IMPERIAL WAR MUSEUM, Lambeth Road, London SE1 6HZ (0171-416 5000). The Films and Photographs Department in the Austral Street Annexe (0171-416 5333, fax 0171-416 5379) has some images valuable to communications equipment enthusiasts.

INDEPENDENT TELEVISION COMMISSION, 33 Foley Street, London W1P 7LB (0171-255 3000, fax 0171-306 7800). Comprehensive library on broadcasting subjects, with public access by appointment.

INSTITUTION OF ELECTRICAL ENGINEERS, 2 Savoy Place, London WC2R 0BL (0171-240 1871, fax 0171-497 3557). Large library and document resource, with excellent bookshop. The archives contain several important collections relevant to the history of electrical science and technology.

NATIONAL SOUND ARCHIVE, 29 Exhibition Road, London SW2 2AS (0171-412 7440).

POST OFFICE ARCHIVES & RECORD CENTRE, Freeling House, Mount Pleasant Complex (Phoenix Place entrance), London EC1A 1BB

(0171-239 2570, fax 0171-239 2576). Open 09.00–16.15 Monday to Friday. Appointment unnecessary but proof of identity required.
PUBLIC RECORD OFFICE, Ruskin Avenue, Kew TW9 4DU (0181-876 3444, fax 0181-878 8905). Open 09.30–17.00; identity is necessary but no appointment is required. The PRO houses the national records of England and the United Kingdom; with a few exceptions, public records are available for inspection thirty years after the year of creation.

For photographs of apparatus the most obvious source would be the manufacturer's or user's own archive, as appropriate. Sometimes a local newspaper may have photographs or else one of the large press photography companies may have taken pictures. There are also several commercial picture libraries in London (see *Yellow Pages* directory), although their prices are generally pitched at commercial levels. Many museums also have photographic collections or else can take pictures of their exhibits (see Chapter 27, 'Museums and other places to visit').

CHAPTER 24

Professional services directory

This chapter is devoted to those back-up services you will probably need to call on sooner or later. Most of the names listed are commercial organizations but some are not, and you will find that these enthusiasts make more modest charges than full-time firms.

AUDIO RESTORATION, TRANSCRIPTION AND SOUND RECOVERY

AUDIO SERVICES (Philip Farlow) (0181-942 6788). Audio transfer bureau specializing in achieving the best possible transfer from mainly redundant formats such as open-reel tape of all sizes, speeds and formats plus acetate discs up to 17 in in diameter and any speed and groove format. The recovered audio can be copied to compact cassette, DAT or CD-R. Digital audio restoration and other enhancing processes can be used. Full information sheet on request. Professional service and charges.
Adrian COSENTINI, 3422 214th Place, Bayside, NY 11361, USA.
John DAVIES, Bryn Coleu, Plas Gwyn Road, Pwllheli LL53 6UT (01758-612932).
John R.T. DAVIES, 1 Walnut Tree Cottage, Burnham SL1 8DH (01628-604811).
Ted KENDALL, 1 Glan-yr-Afon, Gladestry, Kington HR5 3NX (01544-22211). Ted specializes in getting the best recovery from old acetates, 78 rpm records and 16 in transcriptions. He was responsible for restoring the Goon Shows rebroadcast by the BBC.

NATIONAL SOUND ARCHIVE, 29 Exhibition Road, London SW7 2AS (0171-589 6603). Particularly well placed to handle very early formats including wax cylinders and steel wire recordings. Sony Elcasets and other tapes in cartridges and cassettes can be removed from these and replayed on an open-reel machine.
Steve SMOLIAN, Smolian Sound Studios, 1 Wormans Mill Court #4, Frederick, MD 21701, USA.
Adrian TUDDENHAM, 88 Mount Road, Southdown, Bath BA2 1LH (01225-335974).
Scott TAYLOR, Audio Restoration Services, 13308 Enterprise Avenue, Cleveland, OH 44135-5106, USA (00 1 216-433 1220, fax 00 1 216-433 1223).
Seth B. WINNER, Sound Studios, Inc., 1296 East 48th Street, Brooklyn, NY 11234-2102, USA.
Jim WHEELER, Tape Archival and Restoration Service, 1763 Valley View, Belmont, CA 94002, USA (00 1 415-595 4090, fax 00 1 415-594 0951, e-mail *jimwheeler@aol.com*).

CRT REBUILDING

ABLE ELECTRONICS, 1-5742 Beresford, Burnaby, BC, Canada (00 1 604-438 8001). Established for many years rebuilding picture tubes for the video game industry.
DISPLAY ELECTRONICS, Unit 4, Swan Wharf, Waterloo Road, Uxbridge UB8 2RA (01895-255800). Tube rebuilding, mainly industrial CRTs.
HAWKEYE PICTURE TUBE Mfg, 724 Scott Ave, Des Moines, IA, USA (00 1 515-288 8567). CRT rebuilders.
RSC ELECTRONICS, Wichita, KS, USA (00 1 800-456 1545). CRT rebuilders.
SHERWOOD TUBES, Victoria Buildings, Peveril Street, Nottingham NG7 4AH (0115-978 6896). CRT rebuilders.
VIDEO DISPLAY, Inc., Atlanta, GA, USA (00 1 800-241 5005). CRT rebuilders. Most of their business is the CRTs in airport displays, ATM machines and bowling alley displays. The division here does the above-listed tubes, but there is a division in Pennsylvania that does 'entertainment' type CRTs. This includes colour, black and white, and the high-intensity lamps for big-screen 'projection' television sets. They stock many CRTs, but are more than willing to rebuild one that they do not have on hand.

ENAMEL BADGE REPAIR AND REPLICA ANODIZED PLATES

The cost of repairing a small enamel badge is likely to be around £20. These

are three specialists in the field; their main work is radiator badges for classic cars, so be prepared to explain that your job will be a little smaller!
AJP EMBLEMS LTF (Classic Services), Unit 23, Crawford House, West Avenue, Wigston, Leicester LE8 2FB (0116-257 0466, fax 0116-288 7429). Also general metal polishing, plating and chroming, no job too small.
REBUS BADGES & REGALIA, Stonehouse Farm, Bodenham, Hereford HR1 3HZ (01568-797401, fax 01568-797402). Enamel badge restoration, masonic regalia, competitive prices.
PD ENAMELS, 1 New Street, Chulmleigh EX18 7DB (01769-581122, fax 01769-520535). Specialist in enamel and paint-filled badge restoration. Replica badges and chassis plates can also be made in anodized aluminium.

ENGRAVED PLASTIC LABELS

NEWTON ENGRAVING, The Eyry, Newton St Petrock, Torrington EX38 8LU (01409-281342). Engraving in 'traditional' alphabets suitable for older equipment.

EXHIBITION DISPLAYS

HERITAGE SUPPORT SERVICES (Michael Powell), 4 Shorts Close, Burton, Christchurch BH23 7NG (01202-470126). High-quality, modestly priced displays and showcases for organizations on a budget.
SIGNSCAPE (Mark Clegg) (0976-942744, fax 0115-939 2423). Exhibition displays and settings for organizations on a budget.

FILM-TO-VIDEO TRANSFER (TELECINE)

There are of course many excellent telecine services. One of the most reliable and comprehensive is PLATO VIDEO in Bournemouth. They handle standard 8, super 8, 9.5 and 16 mm gauges (also 525/625 PAL/SECAM/NTSC standards conversion work). A professional multiplex and Sony broadcast camera are used, giving first-rate results at competitive prices. Twin 35 mm slide facilities with dissolve is also available. Output to domestic and broadcast (Betacam SP) tape formats. Speak to Lionel Fynn (01202-554382).

HIRE OF OLD TVS AND RADIOS

Working period sets for filming and TV productions.

Pat HILDRED, 18 Well House Avenue, Leeds LS8 4BY (0113-240 2841 evenings).
STUDIO & TV HIRE, 3 Ariel Way, London W12 (0181-749 3445).

HIRE OF 1950S/60S TELEVISION STUDIO EQUIPMENT

A variety of image orthicon and other cameras, tripods and monitors can be hired with a competent operator/technician. Contact GOLDEN AGE TELEVISION RECREATIONS in the persons of Dicky Howett (01245-441811) or Paul Marshall (01522-703348).

METER REPAIR

METERS & INSTRUMENTS, 9431 Winkler Street, Houston, TX 77017, USA (00 1 713-943 1211).
STANDARD METER LAB, Inc., 236 Rickenbacker Circle, Livermore, CA, USA (00 1 510-449 0220).
WHITE ELECTRICAL INSTRUMENT CO., Spring Lane North, Malvern Link, WR14 1BL (01684-892369, fax 01684-564920)
YE OLDE METER CELLAR (Leonard W. Cartwright), 879 Russet Drive, Sunnyvale, CA 94087, USA (00 1 408-739 6025). Repair and replacement of pre-1940 meters.

MICROPHONE AND BROADCAST EQUIPMENT REPAIR

CLASSIC ELECTRONICS, Los Angeles, CA, USA (00 1 818-992 0999, fax 00 1 818-340 4331). Specializes in rebuilding of vintage microphones.
Bill DAVIES, 5548 Elmer Avenue, North Hollywood, CA 91601, USA (00 1 818-761 9831). Repair of velocity microphones and type 70 transcription turntables. Worked for RCA Broadcast Division before they closed, factory-trained and stocks parts for the RCAs.
ENAK MICROPHONE REPAIR (Clarence Kane), 420 Carew Avenue, Pitman, NJ 08071 USA (00 1 609-589 6186, fax 00 1 609-589 7791, e-mail *enakmic@p3.net*).
The MIC WORKS (Tracy Korby), USA (00 1 412-937 1349). Dedicated to the art of restoring and repairing classic microphones.
MICROMIKE LABS, 1357 W. Cris Place, Anaheim, CA 92082, USA (00 1 714-774 0342). Repair and refurbishment of Western Electric and Altek microphones. Also has a supply of 442 connectors and 173A stand mounts.
QUAD ONE ELECTRONICS, located at Dubs Inc. in Hollywood, USA (+1 213-461 3726). Talk to Art Mairose for Quadruplex repairs and parts.

RIBBON MICS (Wes Dooley), USA (00 1 800-798 9127). Sales, parts and service for ribbon microphones such as RCA and Coles.

OBSOLETE VIDEOTAPE STOCK

People requiring cassette tapes for Technicolor 1/4 in and Philips 1500, 1700 and V2000 VCRs should contact Stephen Albrow, Globe Video Services, 192 Castelnau, London SW13 9DH (0181-748 1453). High density half-inch open-reel videotape can be had from Pat Hildred, 18 Well House Avenue, Leeds LS8 8BY (0113-240 2841 evenings).

OPTICAL REPAIRS

BALHAM OPTICAL INSTRUMENTS, 6 Weir Road, London SW12 0NA (0181-673 8513).
SCIENTIFIC OPTICS Ltd, Drury Lane, Ponswood Industrial Estate, Hastings TN34 1YA (01424-430371, fax 01424-441639).
VAN DIEMEN OPTICS CENTRE, 45–49 High Street, Earl Shilton, Leicester LE9 7DH (01455-847722, fax 01455-841224).
Each of these companies repairs, rebuilds and repolishes lenses.
SRB FILM SERVICE, 286 Leagrave Road, Luton LU3 1RB (01582-572471, fax 01582-573532). Custom-made filters, adapter rings and other metalwork for lenses.

PLATING SERVICE

CLASSIC SERVICES, Unit 23, Crawford House, West Avenue, Wigston, Leicester LE8 2FB (0116-257 0466, fax 0116-288 7429). Metal polishing, plating and chroming, no job too small.
PETERBOROUGH PLATING COMPANY, 12 Maxwell Road, Peterborough (01733-233250, fax 01733-371411). Precision chrome plating, nickel, zinc, anodizing, chemical blackening.
See also your local *Yellow Pages* directory under Electroplating. There are a large number of firms in this field and one local to you may be prepared to carry out small one-off jobs for a modest payment (offering cash helps). However, the two firms noted here are specialists and are used to handling enthusiasts' work.

RECORD CLEANING

Paul SAWTELL, Stourbridge, West Midlands (01384-896679). Paul has a Keith Monks Discwasher – marvellous for cleaning your cherished gramophone records. Conscientious handling, only 75p per disc. If you don't live close to Paul, try your local quality hi-fi dealer!

TAPE-HEAD RELAPPING AND REPLACEMENT

AMP SERVICES Inc., 224 Datura Street, Suite 614, West Palm Beach, FL 33401, USA (00 1 407-659 4805).
HEAD TECHNOLOGY, 11 Britannia Way, Stanwell, Staines, Middx (01784-256046). Tape head relapping, reprofiling, new heads supplied.
JRF MAGNETIC SCIENCES, 249 Kennedy Road, Greendell, NJ 07839, USA (00 1 201-579 5773, fax 00 1 201-579 6021).

TAPE MOTOR REBUILDING

AMP SERVICES Inc., 224 Datura Street, Suite 614, West Palm Beach, FL 33401, USA (00 1 407-659 4805).

TRANSFERS MADE TO YOUR SPECIFICATION

BOOST TRANSFERS, Wishanger, Churt, Farnham, Surrey (0125125-2772). Artwork made; small runs a speciality.
FOX TRANSFERS, Cranberry End Studios, Old School House, 12 Brougham Street, Leicester LE1 2BA (0116-262 6868).
KAYLEE TRANSFERS Ltd, New Tythe Street, Long Eaton NG10 2DP (0115-973 5247, fax 0115-946 0801). Over 40 years of specialist transfer manufacture. Oil-based and cellulose transfers, no minimum order. Full artwork service available.

TRANSFORMER REWINDING

Michel BLANCHARD, 21 Lot Miailhe, F-33720 Barsac, France (00 33 5-5627 0523 after 20.30 local time). Rewinds all manner of coils and transformers. Also supplies cotton-covered wire.
MAJESTIC TRANSFORMER COMPANY, 245 Rossmore Road, Parkstone, Poole BH12 2HQ (01202-734463). Rewinds old transformers.
Andrew ZIMMER, Surrey (01342-834363). Valve radio repairs with coil and transformer winding capability.
SCOT TRANSFORMERS Ltd, Hanley Swan, Worcs. (01684-569104, 563113). EHT transformers rewound.

UNUSUAL OR HARD-TO-FIND CONNECTORS OR LEADS

For unusual connectors or leads for your video camera, computer or VCR, NICAM, ATARI, SEGA etc., try A.R. COMMUNICATIONS (01302-321066). Open Monday to Friday, 10 till 6. Access and Visa accepted.

VIDEO FORMAT CONVERSION

The following people can offer tape format conversion, i.e. they can copy from one tape (not video) standard to another. Some are commercial organizations, others are not. In all cases, charges are by negotiation, often modest.

Pat Hildred, 18 Well House Avenue, Leeds L58 4BY (0113-240 2841 evenings). Pat has VHS, U-Matic, National cartridge and Technicolor/Funai Microvideo 1/4 in cassette.

Martin LOACH, 82 Honey Bottom Lane, Dry Sandford, Abingdon OX13 6BX (01865-735821 evenings). Martin has the following VTR formats.

One-inch reel to reel: Ampex VR7003 (A format), Philips/Peto-Scott EL3400, IVC 826. Two-thirds of an inch reel to reel: Ikegami TVR401. Half-inch reel to reel: Philips LDL, Rank-Nivico KU800, Sony CV-2000, Sony CV-2100, Shibaden, EIAJ-1 (e.g. high-density Hitachi, National, Sony, Sanyo). Quarter-inch reel to reel: Akai VT100. Cassette: Betamax, Grundig SVR, Philips N1500, Philips N1700, Sony U-Matic (low band), V2000, VHS.

Joe PAGANO, Obsolete Tape Transfer Service, RGB Broadcast Service Co., 1234 Southampton Road, Philadelphia, PA 19116, USA (00 1 215-4643158).

PMF VIDEO Ltd (Paul Farnsworth), Wallington, Surrey (0181-669 0011). Specializes in 2 in Quad and tape cleaning.

TELEVISION ARCHIVING Ltd, 10 Wivenhoe Business Centre, Wivenhoe, Colchester CO7 9DP (01206-823526, fax 01206-826751). Specialist in transferring from 2 in Quad and 1 in C-format to present-day formats.

THE VIDEO LAB (formerly Flintdown Channel 5 Television) (Donald Blakely, Chief Engineer), St Anne's on Sea (01253-725499). Collection of vintage VCRs so vast it confuses even the most ardent of format spotters. These include Philips EL3400, 1500, 1700 and 2000 including slow-motion; 2 in Quad in 405, PAL, SECAM and NTSC, with full tape cleaning facilities; Ampex A, B and C; EIAJ; Betacam and Betacam SP; Akai and Funai/Technicolor 1/4 in; Betamax, Video 8, Hi-8; VHS and S-VHS. Time-base correction available. In all cases NTSC variants are included where appropriate.

TWENTIETH-CENTURY VIDEO, Wembley (0181-904 6271). Also good for 2 in Quad format.

VIDEOTAPE RESTORATION

Peter BROTHERS, Specs Bros. L.L.C., P.O. Box 5, Ridgefield Park, NJ 07660, USA (00 1 800-852 7732, fax 00 1 201-440 6588).
Grace LAN, Bay Area Video Coalition, 1111 17th Street, San Francisco, CA 94107, USA.
Jim LINDNER, VidiPax, 920 Broadway, 16th Floor, New York, NY 10010, USA (00 1 800-653 8434).

CHAPTER 25

Clubs and societies

WHY CLUBS AND SOCIETIES ARE VITAL

Literally thousands of people are involved in hobby organizations, organizing events, producing newsletters, preserving collections and so on. They do all this purely for the love of it, not because they are paid to, which means they really put all of themselves into this work. It's marvellous and backs up the work of all the 'official' bodies, museums and so on. Some people feel they might not be welcome in organizations of this kind or that they are not the kind of person who joins clubs. What a shame!

Their newsletters and magazines in particular are treasure troves of information and the best start you can get into these specialist interests. So it's frustrating to read that so many of these publications are by subscription only. That means you can't just pick them up in W.H. Smith and see if they're your cup of tea or not. But by not finding out, you're cutting yourself off from some of the best magazines in the country. Most of these publications will be very happy to send a specimen copy to anyone interested, sometimes for a modest fee. So do write off: join a society too and get in the swim! You won't regret it . . .

Club subscriptions cost money, though. The £15 you spend on a year's membership might be better spent on a cheap radio. Many people think like that – is that your attitude too? If so, it's a shame. Joining a club doesn't cost money, it saves it! Virtually every newsletter is full of ads for sets, cheap spares and other goodies that you'd never find out about otherwise. Just one good deal you would otherwise never have stumbled upon makes signing up worth while. In each issue of these specialist newsletters you can read material you genuinely won't find anywhere else, and many members say they have made marvellous contacts with other enthusiasts through these club newsletters and magazines. In fact, the good deals and help available through the small ads are alone worth the subscription charged.

Members also play a very real and important role in the active conservation and preservation of radio and television history; without their membership, these things suffer to a degree. In Britain the premier society for vintage radio is the British Vintage Wireless Society. In the past it has not been without its critics and indeed there have been three breakaway groups. In my opinion (and it is no more than an opinion) the subjects of criticism have been attended to and what discontent that existed has now subsided. This is great; it means we are all one happy family again. I am further convinced of this because the three splinter groups appear to have folded, at least I have not seen any meetings publicized for some time now.

In the vintage radio field there are some very good 'single make' societies, devoted to the products of one particular manufacturer. This is not surprising for there is a strong following for the products of, say, Eddystone. Too bad that there is no organization yet for people similarly devoted to apparatus made by Racal or Marconi.

Finally, apart from all the other benefits of joining a group of kindred spirits, the small advertisements in their newsletters or journals are one of the best places of finding (or selling) scarce parts and components, data sheets, user manuals and so on. Another telling point: some swap-meets, such as those organized by the BVWS, are not open to non-members (unless guests of existing members) so by not joining, you are locking yourself out of a fruitful source of material and fellowship.

Please note: the information which follows is as complete and as up-to-date as possible. If you are in a position to add to it (or make corrections) please get in touch but we are not able to supply missing phone numbers, addresses and so on. If you believe a particular society or association exists but is not listed below, an admirable book to consult is the *Directory of British Associations and Associations in Ireland*, which is about as comprehensive as you can get. It is a substantial and expensive book but you should be able to find a copy in larger libraries.

VINTAGE WIRELESS SOCIETIES AND ALLIED ORGANIZATIONS (VIRTUALLY ALL OF THESE GROUPS HOLD MEETINGS AND ISSUE PUBLICATIONS)

ALBERTA VINTAGE RADIO SOCIETY. Renamed; see Canadian Vintage Radio Society.

ANTIQUE WIRELESS ASSOCIATION, Inc., Box E, Breesport, NY 14816, USA. The original and largest historical radio collector group in the USA. Monthly quality magazine.

ANTIQUE WIRELESS CLUB, JAPAN, Noriyoshi Tezuka, 1-11-2-403 Hiroo, Shibuyaku, Tokyo 150, Japan (00 81 3-3440 8392, fax 00 81 3 3440 8396).

ASSOCIACIO CULTURAL AMICS DEL LA RADIO, Rei Jaume 55, 08840 Cardedeu, Spain. Spanish vintage wireless club. Juan Julia is the president (93-867 1794) and Esteve Pujal the vice-president (tel/fax 93-879 4286).

ASSOCIATION DES AMIS DU MUSEE DE L'ELECTRO-ACOUSTIQUE (French Antique Radio Association – AEA), 135 avenue du President Wilson, 93100 Montreuil, France.

ASSOCIATION FOR RECORDED SOUND COLLECTIONS, Box 543, Annapolis, MD 21404, USA. Dues are $30 a year and include a subscription to the journal (published semi-annually) and the newsletter.

ASSOCIAZIONE ITALIANA RADIO D'EPOCA (AIRE – Fausto Casi is the President), Via Cavour 5, 52100 Arezzo, Italy.

BELGISCHE RADIO CLUB, François Marschang, Kromvendreef 53, B-2900, Schoten, Belgium.

BRITISH VINTAGE WIRELESS SOCIETY, Gerald Wells, Vintage Wireless Museum, 23 Rosendale Road, London SE21 8DS (081-670 3667). Informative, glossy magazine plus reprints of classic radio publications of the past. Regular swap-meets in several parts of the country and an annual garden party at the Vintage Wireless Museum in London.

CANADIAN VINTAGE RADIO SOCIETY, Box 43012, Metropolitan Place P.O., Edmonton, Alberta, T5J 4M8, Canada (00 1 403 449 5413).

CENTRE FOR THE HISTORY OF DEFENCE ELECTRONICS, Department of Conservation Sciences, Bournemouth University, Fern Barrow, Poole BH12 5BB. Impressive plans are afoot to create a research centre and museum here for the history of radar, sonar and electronic countermeasures. Major support is already pledged and the Centre is now looking for volunteers and members.

CLUB DER RADIO- UND GRAMMO-SAMMLER (Swiss Radio and Gramophone Collectors Club – CRGS), Armin Egli, Kreuzbuchstr. 97, CH 6006 Luzern, Switzerland.

CLUB HISTOIRE ET COLLECTION RADIO, Willy Ostertag, Secretariat CHCR, 11 rue Florival, F-67240 Bischwiller, France.

CRYSTAL SET SOCIETY; see XTAL SET SOCIETY.

EDDYSTONE USER GROUP, c/o Eddystone Radio Ltd., Alvechurch Road, West Heath, Birmingham B31 3PP. For collectors of Eddystone radios.

E.H. SCOTT HISTORICAL SOCIETY, John Meredith, Box 1070, Niceville, FL 32588-1070, USA. For collectors of Scott sets.

GESELLSCHAFT DER FREUNDE DER GESCHICHTE DES FUNKWESENS (GfGF, German Society for the Friends of Radio History), Beim Tannenhof, D-89079 Ulm/D.10, Germany. Detailed monthly magazine, full of well-researched articles on specific sets and technologies.

HISTORICY PARDUBICKY RADIOCLUB (Pardubice Historical Radio Club), Ivan Marek, Za pasazi 1342, CR 53111 Pardubice, Czech Republic.
HISTORICY RADIOCLUB CESKOSLOVENSKY (HRCS), Dr F. Perina, Prikra 3495, CR 76000 Zlin, Czech Republic.
IRISH VINTAGE RADIO & SOUND SOCIETY, Henry Moore, 9 Auburn Close, Killiney, Co. Dublin, Ireland.
LONDON VINTAGE RADIO CLUB, 19 Honeysuckle Crescent, London, Ontario, N5Y 4P3, Canada.
NEDERLANDSE VERENIGING VOOR HISTORISCHES RADIO (Dutch Historic Radio Association), Paulus Pofferstraat 19, 6814 KT Arnhem, Netherlands.
NEW ZEALAND VINTAGE RADIO SOCIETY, 41A Eversleigh Street, Takapuna, Auckland 9, New Zealand.
NORSK RADIOHISTORISK FORENING (Norwegian Radio Historical Association – NRHF), Box 465 Sentrum, N-0105 Oslo 1, Norway.
NORTHEAST VINTAGE RADIO CLUB, Benella, Australia (62-3149, 61-5153 or 62-1454).
THE RADIOPHILE, 'Larkhill', Newport Road, Woodseaves, Stafford ST20 0NP. This magazine arranges regular swap-meets in Staffordshire and Lincolnshire. These are a ideal occasions to meet fellow enthusiasts and take part in the auction and perhaps enter a set for the *concours d'élégance*.
SOCIETY FOR THE PRESERVATION OF ANTIQUE RADIO KNOWLEDGE (SPARK), Harold Parshall, 2673 South Dixie Drive, Dayton, OH 45409, USA.
SOCIETY OF WIRELESS PIONEERS Inc., Paul N. Dane, 146 Coleen Street, Livermore, CA 94550, USA.
VINTAGE RADIO & PHONOGRAPH SOCIETY, Inc., P.O. Box 165345, Irving, TX 75016, USA. Dedicated to the historical preservation of early wireless, phonographs, telegraph, tubes and related items. The society has grown to a world-wide membership.
VINTAGE RADIO CIRCLE, Ringwood, Hants. Now disbanded.
VLAAMSE LIEFHEBBERS VAN DER HISTORISCHE RADIOTECHNIEK (Flemish Historic radio Enthusiasts – VLHR), Rudi Sillen, VLHR, Limberg 31, B 2230 Herselt, Belgium.
XTAL SET SOCIETY, Phil Anderson, 789 N. 1500 Road, Lawrence, KS 66049-9194, USA. For collectors of crystal sets.

VINTAGE TELEVISION SOCIETIES (ALL OF THESE GROUPS HOLD MEETINGS AND ISSUE PUBLICATIONS)

ALEXANDRA PALACE TELEVISION SOCIETY, Simon Vaughan, Archivist APTS, 30 Firsgrove Crescent, Brentwood CM14 5JL (0277-

226102). Membership is restricted to those who worked at Alexandra Palace during 1935–1952; the society is aiming to make as complete an archive as possible of photos, drawings, reminiscences and film concerning the AP studios and transmitters.

ALEXANDRA PALACE TELEVISION TRUST, David Binns, Alexandra Palace, Wood Green, London N22 4AY. The trust aims to open a museum of television history at the Palace.

BBC TEST CARD CLUB, Keith Hamer, 7 Epping Close, Derby DE3 4HR.

BRITISH VINTAGE WIRELESS SOCIETY, Gerald Wells, Vintage Wireless Museum, 23 Rosendale Road, London SE21 8DS. Many collectors belong to BVWS, who often have articles on vintage television in their magazine

BRITISH AMATEUR TELEVISION CLUB, Dave Lawton, G0ANO, Grenehurst, Pinewood Road, High Wycombe HP12 4DD. Again, many collectors also belong to BATC; the club has some supplies useful to people who restore old TV cameras.

405 LINE GROUP, Andrew Emmerson, 71 Falcutt Way, Northampton NN2 8PH (01604-844130, fax 01604-821647). Publishes *405 Alive*, the only magazine in the world devoted to the vintage television hobby in all its forms.

NARROW BANDWIDTH TV ASSOCIATION, Doug Pitt, 1 Burnwood Drive, Wollaton, Nottingham NG8 2DJ. The NBTVA newsletter covers Baird-era topics.

ROYAL TELEVISION SOCIETY, Holborn Hall, 100 Gray's Inn Road, London WC1X 8AL (0171-430 1000). The society caters for people professionally involved in television but many TV enthusiasts are also members.

SAVERS OF TELEVISION AND RADIO SHOWS (STARS), 96 Meadvale Road, Ealing, London W5 1NR.

TEST CARD CIRCLE, Curtons House, School Lane, Walpole St Peter, Wisbech PE14 7PA. Publishes a regular magazine full of articles on test card music and trade test films.

AUDIO EQUIPMENT AND RECORDED SOUND

AMERICAN HERITAGE ANTIQUE SOUND & MOTION PICTURE SOCIETY, (00 1 616-454 3467).

ASSOCIATION FOR RECORDED SOUND COLLECTIONS, P.O. Box 543, Annapolis, MD 21404-0543, USA (00 1 410-956 5600 ext 242, e-mail *peters@umd5.umd.edu*). The association is a non-profit organization founded in 1966 and dedicated to the preservation and study of recordings in all fields of music and speech. Its membership includes private collectors and representatives of some of the world's largest public record archives. Further information on ARSC may be obtained from the Executive Director, Peter Shambarger.

MILITARY RADIO, RADAR AND AVIONICS

CENTRE FOR THE HISTORY OF DEFENCE ELECTRONICS (CHiDE), Department of Conservation Sciences, Bournemouth University, Fern Barrow, Poole BH12 5BB (01212-595178). Impressive plans are afoot to create a research centre and museum here for the history of radar, sonar and electronic countermeasures. Major support is already pledged and the Centre is now looking for volunteers and members.

CROYDON AIRPORT SOCIETY, Geoff Hare, 12 Colburn Avenue, Caterham CR3 6HU (01883-342433). This group is fitting out the original control tower, complete with all radio equipment. Helpers welcome!

DUXFORD RADIO SOCIETY, Secretary Mrs B. I. Pope, 95 Northolt Avenue, Bishops Stortford CM23 5DS. Supports the interest in all military radio activities as well as the radio equipment displays at the Imperial War Museum, Duxford, near Cambridge.

HISTORICAL RADAR ARCHIVE, Squadron Leader Mike Dean, MBE RAF (Retired), Little Garth, High Street, Scampton, Lincoln LN1 2SD (01522-730338) or Ian Brown, 3 Kingsmuir Crescent, Peebles EH45 9AB. Research into radar activities of Word War II.

INTERNATIONAL ANGRY-NINE ASSOCIATION (IANA), Box 3170, 3502 GD Utrecht, Netherlands. For collectors and users of vintage US, Canadian and British military radio equipment, in particular the AN/GRC9 set.

MILITARY RADIO COLLECTORS GROUP, c/o Ed Zeranski, 4712 Coronado Avenue, San Diego, CA, USA (00 1 619-223 3242, e-mail *ezeran@cris.com*).

MILITARY WIRELESS AMATEUR RADIO SOCIETY (allied to the Military Vehicle Trust), John Taylor-Cram, 7 Hartplain Avenue, Cowplain, Waterlooville PO98 8RP (01705-250463). For collectors and users of vintage military radio equipment.

SURPLUS RADIO SOCIETY, Box 887, 3700 AW Zeist, Netherlands. For collectors and users of vintage military radio equipment.

WW II BATTLE RE-ENACTMENT ASSOCIATION, also known as WW II LIVING HISTORY, Membership Secretary C. Knight, Cheddar, Cattersall Lane, Godalming GU7 1LW. Leading organization for collectors, hobbyists and historians interested in World War II military equipment. Arranges public shows and mock battles.

WW2 AIR FORCES RADAR REUNION, Harry Jurd, 9 Chelmer Court, Basingstoke RG21 2DT (01256-25980). For all people who were involved with radar during the war.

TELEGRAPHY

CERCLE SAMUEL MORSE, Francis Feron F6AWN, BP 20, F-14480 Creully, France. For amateur radio CW operators.

FISTS, Box 47, Hadley, MI 48440, USA. For amateur radio CW operators; membership exceeds 2000.
MORSE TELEGRAPH CLUB Inc., 1101 Maplewood Drive, Normal, IL 61761, USA. For collectors and users of keys.
TELECOMMS HERITAGE GROUP, Box 561, Croydon CR9 7ZS (0181 406 3995). Offers four newsletters, two magazines and at least one swap-meet a year.
TELEGRAPH AND TELEPHONE STUDY CIRCLE, Box 62, York YO1 1YL; also Box 5627, Hamden, CT 06518, USA. The circle exists to bring together all those with an interest in the history of the telegraph and telephone, with particular reference to methods of prepayment (stamps, franks and phone cards) and ephemera (forms, publicity material etc.). Members are kept in touch with a 16–20 page quarterly journal *Across the Wire.*

TELEPHONES

ANTIQUE TELEPHONE COLLECTORS ASSOCIATION, Box 94, Abilene, KS 67410, USA. The USA's first telephone collectors' club, with a monthly bulletin devoted mainly to small ads. Headquarters is shared with the Museum of Independent Telephony.
AUSTRALASIAN TELEPHONE COLLECTORS SOCIETY, Box 566, Lane Cove, NSW 2066, Australia.
AUSTRALIAN HISTORIC TELEPHONE SOCIETY Inc., Box 194, Croydon, Vic. 3136, Australia.
NEW ZEALAND HISTORIC TELECOMMUNICATIONS COLLECTORS CLUB, 7 Stretton Street, Harewood, Christchurch 5, New Zealand.
POST OFFICE VEHICLE CLUB, 7 Bignal Rand Drive, Wells BA5 2EU (01749-675168). For everyone interested in Post Office and BT motor vehicles.
SVERIGES TELEHISTORISKA SAMLARFORENING (Swedish Telecomms History Association), Malungsvàgen 57, SE-192 71 Sollentuna, Sweden.
TELECOMMS HERITAGE GROUP, Box 561, Croydon CR9 7ZS (0181 406 3995). Offers four newsletters, two magazines and at least one swap-meet a year.
TELEGRAPH AND TELEPHONE STUDY CIRCLE, Box 62, York YO1 1YL; also Box 5627, Hamden, CT 06518, USA. The circle exists to bring together all those with an interest in the history of the telegraph and telephone, with particular reference to methods of prepayment (stamps, franks and phone cards) and ephemera (forms, publicity material etc.). Members are kept in touch with a 16–20 page quarterly journal *Across the Wire.*
TELEPHONE COLLECTORS INTERNATIONAL (TCI), 19 Cherry Drive North, Oswego, IL 60543, USA. America's second phone collectors' association.

TELEPHONE HISTORY INSTITUTE (THI), 4888 Davis Boulevard, Suite 145, Naples, FL 33942, USA.

RADIO AND TV NOSTALGIA

FRIENDS OF OLD-TIME RADIO, c/o Jay Hickerson, Box 4321, Hamden, CT 06514, USA (00 1 203 248-2887, fax 00 1 203 281-1322). Convention held in Newark, NJ, USA in October. If you would like a flyer about this and/or about Hickerson's book, *The Ultimate History of Network Radio Programming*, please send him your address.
NORTH AMERICAN RADIO ARCHIVES, NARA Membership Director Janis DeMoss, 134 Vincewood Dr., Nicolasville, KY 40356, USA.
OLDTIME RADIO COLLECTORS AND TRADERS SOCIETY, 725 Cardigan Court, Naperville, IL 60565-1202, USA.
OLD-TIME RADIO SHOW COLLECTORS ASSOCIATION (ORCA), Membership Secretary John Wolstenholme, 56 Melbourne Avenue, Dronfield Woodhouse, Sheffield S18 5YW. USA contact: Tom Monroe, 2055 Elmwood Ave, Lakewood, OH 44107, USA (00 1 216-226 8189). Runs a lending library of old radio shows and welcomes donations (send copies, not original recordings).
ROBERT FARNON SOCIETY, Stone Gables, Upton Lane, Seavington St Michael, Ilminster TA19 0PZ. The Farnon Society was founded in 1956, and its aims remain to publicize the work of Robert Farnon in particular and similar musicians involved in light and film music. The society is also the forum for people interested in library music, used (for example) as incidental and background music in the sound-tracks of films and radio and TV programmes. At least four magazines are published each year, and London meetings are held in April and November. Free sample magazine available from the Secretary, David Ades.
SAVERS OF TELEVISION AND RADIO SHOWS (STARS), 96 Meadvale Road, Ealing, London W5 1NR. Members lend each other copies of old programmes.
VINTAGE RADIO PROGRAMME COLLECTORS CIRCLE, Roger Bickerton, 3 Park Edge, Harrogate, HG2 8JU (01423-887452). Caters for collectors of the spoken word and other radio broadcasts.

PLASTICS

BAKELITE MUSEUM SOCIETY (01984-632133, 0181-852 3492 and 01374-126670). The original society for plastics collectors.
PLASTICS HISTORICAL SOCIETY, 19 Langdon Avenue, Bedgrove, Aylesbury HP21 9UL. Lively group, associated with the British Plastics Federation.

SCIENTIFIC INSTRUMENTS IN GENERAL

SCIENTIFIC INSTRUMENT SOCIETY, 31 High Street, Stanford-in-the-Vale, Faringdon SN7 8LH (01367-710223, fax 01367-718963).

PROFESSIONAL BODIES

AUDIO ENGINEERING SOCIETY, Inc., 60 East 42nd Street, Room 2520, New York, NY 10165-2520, USA (00 1 212-661 8528, fax 00 1 212-682 0477). Now in its fifth decade, the AES is the only professional society devoted exclusively to audio technology. Its membership of leading engineers, scientists and other authorities has increased dramatically throughout the world, greatly boosting the society's stature and that of its members in a truly symbiotic relationship.

INSTITUTE OF ELECTRICAL AND ELECTRONICS ENGINEERS, Inc., IEEE Operations Center, 445 Hoes Lane, PO Box 459, Piscataway, NJ 08855-0459, USA.

INSTITUTION OF BRITISH TELECOMMUNICATIONS ENGINEERS, 2–12 Gresham Street, London EC2V 7AG (0171-356 8008).

INSTITUTION OF ELECTRICAL ENGINEERS, Savoy Place, London WC2R 0BL (0171-240 1871, fax 0171-240 7735).

RADIO CLUB OF AMERICA Inc., Vivian A. Carr, Membership Chairman, 1620 Route 22, Union, NJ 07083, USA (00 1 908-687 3090). Fraternal organization for people in the radio and electronics industry and academia. It does not cater at all for the usual ham or collecting interests although many members are probably hams. Some excellent publications issued.

ROYAL TELEVISION SOCIETY, Holborn Hall, 100 Gray's Inn Road, London WC1X 8AL (0171-430 1000, fax 0171-430 0924).

SOCIETY OF BROADCAST ENGINEERS, Inc., 8445 Keystone Crossing, Suite 140, Indianapolis, IN 46240, USA (00 1 317-253 1640, fax 00 1 317-253 0418).

SOCIETY FOR THE HISTORY OF TECHNOLOGY, USA; information from The University of Chicago Press, Journals Division, Box 37005, Chicago, IL 60637, USA (00 1 773-753-3347, fax 00 1 773-753 0811). Formed in 1958 to encourage the study and teaching of the history of technology and its relations with society and culture. Membership is international, open to individuals and organizations, with some 3300 individual and institutional members around the world.

SOCIETY OF MOTION PICTURE & TELEVSION ENGINEERS, 595 W. Hartsdale Avenue, White Plains, NY 10607-1824, USA (00 1 914-761 1100, fax +1 914 761 3115).

SOCIETY OF WIRELESS PIONEERS, Box 86, Geyserville, CA 95441, USA. Dedicated to collecting, researching, recording and preserving the history of communications, particularly wireless and radio telegraphy. Although the foundation of the society's structure is the Professional Operating Member, SOWP recognizes that there are many whose interests in wireless communications are just as deep. SOWP also has a place for historians and collectors, who preserve our history and memorabilia.

TELEPHONE PIONEERS ASSOCIATION, Box 13888, 930 15th Street, Suite 1200, Denver, CO 80201-3888, USA (00 1 303-571 1200, 00 1 303-572 0520). With over 800 000 members, the Pioneers are active in virtually every corner of North America. The association has a long tradition in preserving old equipment.

PART 4

Extending Your Knowledge

CHAPTER 26

Recommended reading

It would be so handy if I could name just two or three books and say, 'There you are, just buy these and that's your lot!'. Life is not like that and you'll need to read several to really soak up the info. Some people consider buying books and reading a waste of money; they prefer to pick up their knowledge the practical way (I almost said the cheapskate way). Those people have my sympathy; obviously no-one ever introduced them to the delights of reading at the right age.

Well, I guarantee that if you read each of the books listed below you will be a much wiser person. There is a modest satisfaction that comes with actually knowing what you are talking about and the cheapest and easiest way of picking up knowledge is by reading. It's funny but this is one of those secrets that increasingly few people seem to find out and, yes, it is something I am passionately involved with. I believe that modern education methods (and the cut-backs in funding for schools) are setting people out in life with the wrong attitudes and a complete misunderstanding of education. Self-education is a marvellous thing but you need to be guided into it. Oh well, I've said my piece now.

Inevitably most books eventually go out of print, which means you will have to borrow copies from a library or try and find a second-hand copy. A list of selected booksellers and book search organizations is given in the 'Buyer's guide' chapter.

Tracing past copies of periodicals can be extremely difficult, although some sources are given in the 'Buyer's guide' chapter. As far as Great Britain is concerned, to trace a library with holdings of a particular magazine or newspaper the book to consult is the several volumes of the *British Union Catalogue of Periodicals* (*BUCOP* for short), which is kept in major libraries. You will find that there are some magazines which neither the original publisher nor the British Library nor any other library have, which can be extremely frustrating. In the main, however, you will be able to find what you are looking for, though quite possibly at the expense of considerable effort. It helps a great deal if you have the precise reference and page numbers; in this case you can quote these and order a photocopy without having to have the bound volume transferred to your local library.

There are three large libraries which are particularly recommended for tracing old technical books and magazines: these are the Science Museum Library, the Patent Office Library and the library of the Institution of Electrical Engineers, all in London. Telephone first to check opening hours; in many cases older items are not on the open shelf but held in store off-site, meaning it is necessary to order them in advance.

BRITISH LIBRARY, Euston Road, London NW1 2DB (0171-412 7000).

BRITISH LIBRARY DOCUMENT SUPPLY SERVICE, Boston Spa, Wetherby LS23 7BQ (01937-546000, fax 01937-546333).

BRITISH LIBRARY NEWSPAPER LIBRARY, Colindale Avenue, London NW9 5HE (0171-412 7353). Open 10.00–16.45 Monday to Saturday by reader's pass (bring proof of identity to apply for this).

INSTITUTION OF ELECTRICAL ENGINEERS, Savoy Place, London WC2R 0BL (0171-240 1871, library fax 0171-497 3557).

PATENT OFFICE LIBRARY (British Library Holborn Reading Room), 25 Southampton Buildings, London WC2A 1AW (0171-412 7494, fax 0171-412 7495, e-mail *sris-centre-desk@bl.uk*).

SCIENCE MUSEUM LIBRARY, Imperial College Road, London SW7 5NH (0171-938 8234).

What follows is a list of recommended book and magazine titles, divided up into different areas of interest.

BOOKS

CLASSIC AUDIO

AUDIO! AUDIO! by Jonathan Hill and John Howes. A comprehensive directory presenting 850 amplifiers from nearly 150 British manufacturers of the 1940s–1970s. £10 + £2.50 post & packing, from Sunrise Press, Spice House, 13 Belmont Road, Exeter EX1 2HF (01392-411565).

EVOLUTION OF THE AUDIO RECORDER (late 1940s–early 1970s), EC Designs, Inc., Box 33, Genesee Depot, WI 53127-0033, USA. More than 500 pages, claimed to be the definitive guide to wire and tape recorders.

SOUND RECORDINGS by Peter Copeland. British Museum Books, £7.60 post paid from CLPGS, Woodbine Cottage, Brigg Road, Caistor LN7 6RX. Superbly illustrated book telling the story of recording from Edison's first phonograph to the CD.
VINTAGE HI-FI SPOTTER'S GUIDE, Volumes 1 and 2. Both completely different, covering early tube hi-fi and sound reproduction equipment from the 1930s to 1965. Each has over 450 illustrations and an index. $17.95 each or $32 for both. Charles Kittleson, Box 691, Belmont, CA 94002, USA (00 1 415 595 4164).

JUKE-BOXES

JUKEBOX ART by Pearce. Evolution of the juke-box in 128 pages and 250 photos.
JUKEBOXES – OBSCURE, MYSTERIOUS AND INNOVATIVE by Adams. First volume of a series, 182 pages. Both titles $29.95, available by mail order (credit cards accepted) from ARC, Box 2, Carlisle, MA 01741, USA (00 1 508-371 0512, fax 00 1 508-371 7129).

PLASTICS

BAKELITE, AN ILLUSTRATED GUIDE TO COLLECTIBLE BAKELITE OBJECTS by Patrick Cook and Catherine Slessor, The Apple Press, 1992, hardback £12.95. Text is not without errors but the pictures are beautiful.
EARLY PLASTICS by Sylvia Katz. Brief but comprehensive and well illustrated paperback in the Shire Albums series; modestly priced. Shire Publications, Cromwell House, Church Street, Princes Risborough HP17 9AJ.

PRO AUDIO

AMPS! THE OTHER HALF OF ROCK & ROLL by Ritchie Fliegler. Evolution and operation of Fender, Vox, Ampeg and Marshall with photos and trouble-shooting guide. $24.95 available by mail order (credit cards accepted) from ARC, Box 2, Carlisle, MA 01741, USA (00 1 508-371 0512, fax 00 1 508-371 7129).
THE HISTORY OF MARSHALL by Michael Doyle. For musicians, technicians and collectors, 254 pages including 62 in colour. $32.95 available by mail order (credit cards accepted) from ARC, Box 2, Carlisle, MA 01741, USA (00 1 508-371 0512, fax 00 1 508-371 7129).

RADIO AND TELEVISION – REPAIRING OLD SETS

First off is Chas Miller's book, the *PRACTICAL HANDBOOK OF VALVE RADIO REPAIR*. Newnes Technical Books, hardback, £25, new edition in preparation. This is a delightful book, covering far more than its title might imply and comes highly recommended. It confines its attention to the electrical side of repair, so you still have to look elsewhere for information on other restoration topics. The other book worth looking out for is the *PRACTICAL WIRELESS SERVICING MANUAL* by F.J. Camm (George Newnes Ltd, various editions from 1938 onwards through to the 1960s). This is also a very comprehensive book and, although long out of print, it is easy to find in second-hand shops and at radio swap-meets.

As far as television is concerned, two books now out of print must be mentioned: the *TELEVISION SERVICING HANDBOOK* by Gordon J. King (Odhams Books, hardback) and *TV TECHNICIAN'S BENCH MANUAL* by G.R. Wilding (Fountain Press, hardback) are both indispensable and should be obtainable in second-hand bookshops and through your public library. I also happen to like *TV SERVICING MADE EASY* by Wayne Lemons (Sams, paperback), which is an out-of-print American book but still relevant in most respects.

RADIO AND TELEVISION – HOW TO RESTORE THEM

That's what you bought this book for but I am happy to acknowledge there are others. They are all foreign but can be found on the shelves of several of the specialist radio bookshops and mail order stockists (see 'Buyer's guide' chapter).

OLD TIME RADIOS RESTORATION AND REPAIR by Joseph Carr. Tab Books (USA), paperback. Price £14.95 in the UK. This is a thick book (256 pages) and full of good information on the theory and practice of repairs. The book says nothing about restoration though, at least not as I define restoration.

DISCOVERING VINTAGE RADIO by Peter Lankshear. *Electronics Australia* magazine format. Price $4.95 in Australia, varies in UK. This is a budget-priced reprint of a series of articles in the magazine *Electronics Australia* and has excellent coverage.

ANTIQUE RADIOS RESTORATION AND PRICE GUIDE by David and Betty Johnson. Wallace-Homestead, paperback. Price $14.95 in USA, varies in UK. More of a survey than a detailed book (a bit like this one!) but sound advice on things like choosing tools, repairing castings and staining wood.

RADIO AND TELEVISION – IDENTIFICATION AND COLLECTOR GUIDES

COMMUNICATIONS RECEIVERS by Raymond S. Moore. Comprehensive guide to 700 valve communications receivers from 1932 to 1981. $19.95, from ARC.

CRYSTAL CLEAR, Volumes 1 and 2 by Maurice L. Sievers. Descriptive history and photo guide of crystal sets. $29.95 per volume, from ARC (see below).

HISTORIC TELEVISIONS & VIDEO RECORDERS by Michael Bennett-Levy. Large illustrated paperback, £12.95 post-paid from MBL Publications, Monkton House, Old Craighall, Musselburgh EH21 8SF. The first book to look at televisions from the collector's viewpoint. A few mistakes but these are due to be corrected in Michael's next book.

MACHINE AGE TO JET AGE by Mark V. Stein. This book has 1400 American radios illustrated in 250 pages. $24.95 paperback + $7.50 postage from Radiomania, 2109 Carterdale Road, Baltimore, MD 21209, USA.

OLD RADIO SETS by Jonathan Hill. Shire Publications, paperback, price £2.25. This is a much cheaper book but you'd have to have a hard heart not to want a copy. As well as being eminently affordable, the book is pocketable and illustrated on every page. Expertly written, this book forms an admirable introduction to the subject.

OLD TELEVISION by Andrew Emmerson. Shire Publications, paperback, price £2.95. Modesty forbids me for pronouncing judgement on my own book, except to say that it makes a good companion to Jonathan's book (above).

RADIO ART by Robert Hawes. Green Wood Publishing, hardback, £12.95. A good survey of the more decorative radios. No longer available from the publisher but remainder stocks can be found in some shops and from specialist radio outlets such as *Antique Radio Classified* – see publications in the 'Buyer's guide' chapter or end-note to Collector guides section below.

RADIO MANUFACTURERS OF THE 1920s (three volumes) by Alan Douglas. The definitive book on American broadcast receivers of the 1920s. $29.95 per volume, from ARC.

RADIO! RADIO! by Jonathan Hill (second edition). Sunrise Press, hardback, price £25. Comprehensive, illustrated checklist of most radio sets made or sold in Britain from the 1920s to the 1960s. This is an ideal picture guide to identifying radio sets, with quite a bit of background material as well. Pricey perhaps at first sight, but then when you consider the cost of the detailed research and the vast number of photographs you realize what a good bargain it is. Knowledge is never gained at zero cost. Sunrise Press, Spice House, 13 Belmont Road, Exeter EX1 2HF (01392-411565).

RADIOS REDUX by Philip Collins. Showcases the sleek lines and luscious colours of over 80 sets of the 1930s and 1940s. $17.95, from ARC.

RADIOS: THE GOLDEN AGE by Philip Collins. The 'prequel' to *RADIOS REDUX*. $17.95, from ARC.
THE SETMAKERS by Keith Geddes and Gordon Bussey. Published by BREMA, the British Radio and Electronic Equipment Manufacturers' Association. Hardback, sold by booksellers and branches of Maplin Electronics (see telephone directory). Nearly 500 pages of glorious nostalgia. Although this book nominally celebrates the firms who produced British radio and television sets, it is also full of photos of sets plus reproductions or period advertisements and brochures, many in colour. A superb bargain.
SONY TRANSISTOR RADIOS: THE COMPLETE PHOTO GUIDE TO COLLECTIBLE MODELS by Eric Wrobbel; *COLLECTIBLE TRANSISTOR RADIOS FROM TOSHIBA AND TRANCEL* by Eric Wrobbel. Both published by Eric Wrobbel, 20802 Exhibit Court, Woodland Hills, CA 91367, USA (00 1 818-884 2282).
THE GOLDEN AGE OF TELEVISION by Philip Collins, $15.95, from ARC. Another sumptuous picture book of (American) sets.
THE RADIO COLLECTOR'S DIRECTORY & PRICE GUIDE 1921–1965 by Robert E. Grinder. More than 20 000 American models from 1921 to 1965 listed, with much background information. $26.95, from ARC.
TUBE LORE by Ludwell Sibley. Nearly 200 pages of detailed information on valves. $19.95, from ARC.

Most American radio books are available by mail order (credit cards accepted) from ARC, Box 2, Carlisle, MA 01741, USA (00 1 508 371 0512, fax 00 1 508 371 7129) or from Antique Electronic Supply, Box 27468, Tempe, AZ 85283 (00 1 602-820 5411, fax 00 1 800-706 6789).

RADIO AND TELEVISION – OTHER MUST-HAVE BOOKS

I would suggest that you need a valve data book, either one of the old *Wireless World* or Babani publications – these are easy to pick up second-hand at radio swap-meets – or if you want a nice, clean, new book check out the reprint volumes published by Radio Bygones, 9 Wetherby Close, Broadstone BH18 8JB (tel/fax 01202-658474).

Another valuable book is the current year's *SOUND & VISION YEAR-BOOK*, a modest but extremely comprehensive paperback published by the Sunrise Press, Spice House, 13 Belmont Road, Exeter EX1 2HF (01392-411565). The book costs £3.50 post paid; it is revised annually and contains all the latest information on dealers, specialist organizations and publications, museums and so on.

A handy book to have is *THE BBC: 70 YEARS OF BROADCASTING* by John Cain (BBC Books). This can be obtained from the BBC bookshops in London or though booksellers in general, and is a well illustrated and easy-to-read treatment of this subject.

RADIO AND TELEVISION – WORTH LOOKING OUT FOR

The following books are out-of-print but turn up regularly in second-hand bookshops and can also be ordered through your public library.
THE BOOK OF PRACTICAL RADIO by John Scott-Taggart (Amalgamated Press, 1934). Comprehensive book by the most prolific author of the inter-war years.
WIRELESS ENCYCLOPEDIA edited by Sir Oliver Lodge (Harmsworth, 1923). Three volume 'bible' of early radio.

RESTORATION TECHNIQUES IN GENERAL

RESTORATION & PRESERVATION OF SCIENTIFIC APPARATUS by Guy Biraud (self-published, 1987). A unique book written throughout in two languages (English and French) covering scientific instruments, radios, phonographs, toys and cameras. Available from Guy Biraud at the Musée Européen de la Communication, Château de Pignerolle, 49124 St Barthélémy d'Anjou, France (00 33 2-4193 3838, fax 00 33-2 4193 8934).

TELEPHONES AND TELEGRAPHS

OLD TELEPHONES by Andrew Emmerson and *PHONE BOXES* by Neil Johannessen. Brief but comprehensive and well illustrated paperbacks in the Shire Albums series, modestly priced at £2.25. Shire Publications, Cromwell House, Church Street, Princes Risborough HP17 9AJ.
OLD TIME TELEPHONES! TECHNOLOGY, RESTORATION AND REPAIR by Ralph O. Meyer. Published by Tab Books (division of McGraw-Hill) at £16.95. ISBN 0-07-041818-7. Paperback, 290 pages, many illustrations and diagrams. You'll probably have to order this book through W.H. Smith or a good bookseller but I promise you it's a good book. It's not often that I find nothing but praise for a book but this one is hard to beat. The title says it all and the book covers both classic and current telephones, providing schematics and hard-to-find technical data.

Oh yes, it *is* an American book so it covers American telephones rather than British ones, but the information is totally valid to British readers. Many pages describe magneto wall phones and candlestick instruments which are identical or virtually identical to those used over here. Even the American 200-type and 500-type phones are conceptually identical to our

200- and 700-types. Coverage is from 1878 to the Princess phone era (late 1960s/early 1970s), and minor manufacturers such as Kellogg and Stromberg Carlson are not ignored.

The photos have been taken specially for the book and are nice and clear. There are plenty of interesting and unusual phones, such as the ten-button touch-tone phones (from the time when the * or # buttons were not provided) as well as bell-sets arranged for harmonic ringing.

Circuit diagrams are provided and the author has gone to great lengths to explain inductor, booster and anti-sidetone circuits. Basic telephone technology and audio theory are not ignored either. The briefest section is on restoration techniques but the rest of the text more than makes up for this.

This book is a treasure, yet it's not expensive. If you call yourself a telephone collector you'll regret not having a copy.

TELEPHONES AND TELEGRAPHS – OTHER USEFUL BOOKS

Older textbooks on telephony and telegraphy are a valuable source of circuit diagrams and information in general. Two books, both called *TELEPHONY* and published by Pitman, are worth having. Both books are in two volumes; the pre-war edition was by Herbert and Procter, whereas the post-war revision was by Atkinson. Poole's *TELEPHONE HANDBOOK* (also published by Pitman) is also useful. The standard work on telegraphy is Herbert's *TELEGRAPHY* (Pitman).

MAGAZINES

This list covers magazines devoted specifically to our hobbies. Collectors wishing to explore pastures new and the wider field of 'general' collector publications should also read the chapter 'Collectors' luck'.

ALWAYS JUKIN', 221 Yessler Way, Seattle, WA 98104, USA (00 1 206-233 9460, fax 00 1 206-233 9871). Monthly magazine for juke-box collectors, with subscribers in 25 countries. Mastercard and Visa.

ANTIQUE RADIO, Mose' Edizioni, Via Bosco 4, 31010 Maser (TV), Italy (00 39 423-950385, fax 00 39 423-529049, e-mail *mose@tv.shineline.it*). Beautifully printed magazine; text in Italian with English summary.

ANTIQUE RADIO CLASSIFIED, Box 2, Carlisle, MA 01741, USA (00 1 508-371 0512, fax 00 1 508-371 7129). Visa/MC. Large-circulation articles plus ads monthly with worldwide readership.

BVWS JOURNAL. Magazine of the British Vintage Wireless Society (see 'Clubs and societies' chapter).

CARD TIMES, Magpie Publications, 70 Winifred Lane, Aughton, Ormskirk L39 5DL. For collectors of cigarette cards, tea cards and all other kinds of cards except picture postcards. Regular features on the stars of yesteryear, diary details of fairs and auctions, many dealers' and collectors' ads.

CHUCK SCHADEN'S NOSTALGIA DIGEST, The Nostalgia Digest, Box 421, Morton Grove IL 60053, USA. Regular old-time radio magazine and tape sales.

COMMON GROUND, Box 26, Blyth NE24 3YP. Described as a hands-on magazine for the hi-fi enthusiast. Subscription only.

EIGHT TRACK MIND, 8-TM Publications, Box 90, East Detroit, MI 48021-0090, USA. Quarterly magazine passionately devoted to collecting and preserving eight-track machines and cartridges. Magazine appears sporadically in UK but a subscription is safer.

ELECTRIC RADIO, 14643 Country Road G, Cortez, CO 81321-9575, USA (tel/fax 00 1 970-564 9185). Nicely printed 56-page monthly magazine for collectors and restorers of amateur and military radio receivers and transmitters. Annual subscription (international airmail) 70.

405 ALIVE, 'Larkhill', Newport Road, Woodseaves, Stafford ST20 0NP (tel/fax 01785-2846960. World's only magazine devoted to collecting and restoring old television sets and programmes, by subscription.

GLASS AUDIO, Audio Amateur Publications, Box 576, 305 Union Street, Peterborough NH 03458-0576, USA (00 1 603-924 9464, fax 00 1 603 924 9467). Quality magazine for the valve audio connoisseur. By postal subscription, sometimes found in a few specialist hi-fi outlets here.

HI-FI NEWS & RECORD REVIEW, Link House, Dingwall Avenue, Croydon CR9 2TA (0181-686 2599, fax 0181-760 0973). Bookstand magazine concentrating on modern hi-fi but with regular classic hi-fi supplements combining vintage equipment reviews and restoration articles.

HI-FI WORLD, 64 Castellain Road, London W9 1EX (0171-289 3533, fax 0171-289 5620). The only monthly bookstand magazine to take a sane attitude towards valve audio. Highly recommended. Has regular features on audio equipment from the past.

HIGH PERFORMANCE REVIEW, Box 346, Woodbury, CT 06798, USA (00 1 203-266 0084, fax 00 1 203-263 4730). Sold in Britain at Tower Records, Piccadilly Circus, London for £4.59 per issue. Glossy quarterly magazine concentrating on contemporary high-end hi-fi equipment, including vacuum tube items.

THE HISTORIC RECORD, John R. Wrigley, 185 The Wheel, Ecclesfield, Sheffield S30 3ZA (tel/fax 0114-246 0275). Quarterly magazine of interest to collectors and archivists of historic sound.

HOLLOW STATE NEWSLETTER, Ralph Sanserino, Box 1831, Perris, CA 92572-1831, USA. Another American newsletter for enthusiasts of valve audio equipment. Highly recommended.

JUKEBOX COLLECTOR, 2545 SE 60th Ct., Suite 105, Des Moines, IA 50317, USA (00 1 515 265 8324). Sample $3, annual subscription $30.

JUKEBOX JOURNAL, Box 545, Brighton, Sussex BN1 4HU (fax 01273-677922).
JUKEBOX MAGAZINE, Box 1156, Peterborough, Ontario, K9J 7H4, Canada (00 1 705-778 2628, fax 00 1 705-778 5291). Canada's juke-box collector publication.
MIX, 6400 Hollis Street #12, Emeryville, CA 94608, USA (00 1 510-653 3307, fax 00 1 510-653 5142). Also sold at Tower Records, London. Pro audio industry magazine with some historical articles and classified ads for vintage gear.
OLDE TYME RADIO SERVICES DIRECTORY, listing over 200 American suppliers. $4 plus postage from Andrew Mooradian, 5 Priscilla Lane, Winchester, MA 01890, USA.
PRIMYL VINYL, PVX, Box 67109, Chestnut Hill, MA 02167, USA. Newsletter for audiophile record collectors – send IRC for information.
RADIO BYGONES, 9 Wetherby Close, Broadstone BH18 8JB (tel/fax 01202-658474). Bimonthly magazine on old radio equipment and history, by subscription.
THE RADIO COLLECTOR, Box 1306, Evanston, IL 60204-1306, USA (00 1 708-869 5016, fax 00 1 708-869 5054). Bimonthly newsletter devoted to vintage radio history, hardware, restoration and theory. Annual subscription (international airmail) $35.
THE RADIOPHILE, 'Larkhill', Newport Road, Woodseaves, Stafford ST20 0NP (tel/fax 01785-2846960). Detailed and very readable magazine concentrating mainly on the restoration of old radio sets, by subscription.
THE RECORD MACHINE, 152 Ravensbury Road, Earlsfield, London SW18 4RU. Bimonthly subscription magazine, £6.25 a year.
SOUND PRACTICES, Box 19302, Alexandria, VA 22320, USA. 'The quarterly journal for the intrepid audio experimenter'. Independently recommended as a good magazine for DIY tube amplifier enthusiasts. Published quarterly, overseas subscription $26 yearly.
STUDIO SOUND, 245 Blackfriars Road, London SE1 9UR (0171-620 3636, fax 0171-401 8036). Leading pro audio magazine in the UK, occasional ads and articles relating to vintage gear.
TELEVISION CHRONICLES, 10061 Riverside Drive, #171, Dept. M, Toluca Lake, CA 91602, USA. Quarterly magazine for TV programme collectors. Each series covered is represented by a profile of the show's origins, production and evolution, followed by a complete episode guide. Also supplementary features, such as interviews, spotlights on individual performers, and book reviews. Four quarterly issues are $20 ($35 in Canada) by cheque or money order .
THRILLING DAYS OF YESTERDAY, Box 36106, Denver, CO 80236, USA (e-mail *JRayb71827@AOL.com*). For all fans of those days of the golden age of radio, this newsletter covers all the dramas, comedies, westerns, soap operas, variety shows etc. of those wonderful bygone days. It is

published six times a year with an annual subscription price of $15 ($20 foreign). Past issues are available.

TRANSISTOR NETWORK, RR1, Box 36, Bradford, NH 03221, USA (00 1 603-938 5051). Monthly newsletter for transistor radio collectors.

ULTRA HIGH FIDELITY, Box 158, Cheshunt EN7 6UH (tel/fax 01992-620905). Quarterly journal devoted to serious and vintage audio.

VACUUM TUBE VALLEY NEWS, 1095 E. Duane # 106, Sunnyvale, CA 94086, USA (tel/fax 00 1 408-733 6146). Educated and opinionated articles four times a year on pro sound, guitar amps, vintage hi-fi, classic radio and more.

WALLAGE REPRINTS, 48 Albert Road, Ashford TN24 8NU. A series of eighteen different radio books of the 1920s and 1930s reprinted at very moderate prices. Send SAE for book list and order form.

ON-LINE SOURCES

Antique Radio Classified: *http://www.antiqueradio.com.*
Antique Radios Online: *http://www.antiqueradios.com.*
Bill's Antique Radio Emporium: *http://www.flash.net/~billhar.*

CHAPTER 27

Museums and other places to visit

Apart from the other, obvious, reasons for having a day out at a radio, television or telecomms museum, these are good places to see nicely restored old apparatus (and some which is not so nicely restored or not even restored at all!). See if you can pick up any tips on how they go about the work. This list is a fairly broad selection, including some specialist displays (as indicated). You have to stop somewhere but we have covered much of Europe and North America.

Bear in mind that some telecomms museums also cover radio and vice versa. Generally they are not listed twice, so please check all listings.

The list is not exhaustive; not all of the museums listed specialize exclusively in 'our' kind of interest. Some are more professional than others whilst some are frankly a bit amateurish but still worth visiting. Owing to constant change, entries may become out of date, so it is always worth telephoning in advance. The lists were compiled with the help of the Edmonton Telephone Historical Information Centre Foundation (Canada), Rétro-Phonia (Besançon) and Wumpus Radio (Berlin).

RADIO AND TELEVISION INTEREST MUSEUMS IN BRITAIN

AIRCRAFT RADIO MUSEUM (John Coggins), 6 Market Corner, Baginton, Coventry CV8 3AP (01203-302668). Visitors welcome but strictly by appointment.

Figure 27.1 A number of privately run radio, television and telephone museums can be found in Britain and elsewhere. Early transistor radios, television sets of the early 1950s (one with a magnifying lens for its screen) and period literature all feature on this view.

AMBERLEY CHALK PITS MUSEUM, Amberley, Arundel BN18 9LT (01798-831370, fax 01798-831831). Opposite Amberley station, easy road access. Opening hours vary according to time of year, daily in high season. Comprehensive industrial archaeology museum, with a nice selection of old radio and television sets in a reconstructed radio and television showroom of the 1950s, also a reconstruction of WW II aircraft cabin with radio and radar apparatus. Further information from Ron Weller (01903-267839).
AUTOWORLD at the Patrick Collection, 180 Lifford Lane, Kings Norton, Birmingham B30 3NT (0121-459 9111). Presented as the Midlands' premier automobile museum, this venue also contains a pre-war television set in working order, TV sets of the 1950s and a music shop of the 1960s as part of period settings to bring to life the different eras of motoring. Opening times vary according to season so ring first to check.
BAMPTON COMMUNICATIONS MUSEUM. Now closed but may reopen nearby in future.
BAWDSEY RADAR MUSEUM, Bawdsey Manor, Woodbridge, Suffolk. Occasional open days. Information from Jim Reynolds (0181-948 3436 or 01394-411633).

BBC EXPERIENCE, Broadcasting House, London W1A 1AA. 'Theme park' exhibition devoted to the past, present and future of the BBC, with historical collection on display. Open seven days a week, from 9.30 a.m. to 5.30 p.m. Bookings line on 0870-603 0304 (local rates apply); tickets cost £5.75 for adults, £4 for children.
BOGNOR REGIS WIRELESS MUSEUM (Curator R.E. Simpson) located at the Local History Museum, Hotham Park Lodge. Open Wednesday, Friday and Sunday 13.00–17.00 from May Day bank holiday until mid-September.
BUCKLEY'S YESTERDAY'S WORLD, Battle, Hastings, Sussex (01424-775378). Wireless shop display.
CARPETBAGGER AVIATION MUSEUM, Sunny Vale Farm Nursery, off Lamport Road, Harrington, Northampton NN6 9PF (0160128-608). Privately run museum in restored operations room of World War II airfield with some radio equipment. Open Sundays in season (ring for opening times).
CENTRE FOR THE HISTORY OF DEFENCE ELECTRONICS, Department of Conservation Sciences, Bournemouth University, Fern Barrow, Poole BH12 5BB. Impressive plans are afoot to create a research centre and museum here for the history of radar, sonar and electronic countermeasures. Major support is already pledged and the Centre is now looking for volunteers and members.
CITY OF NORWICH AVIATION MUSEM, Old Norwich Road, Horsham St Faith, Norwich NR10 3JE. Display of early electronic warfare equipment.
DESIGN MUSEUM, Butler's Wharf, Shad Thames, London SE1 2YD (0171-403 6933). A small number of classic design sets are on show.
FENLAND AVIATION MUSEUM, Bamber's Garden Centre, Wisbech, Cambs. This is on the B198 road (old A47), half a mile from the A47 Wisbech Bypass. Open Weekdays and Bank Holidays, March to September, 09.30–17.00. Museum contains aeronautical artefacts, including airborne radio equipment, much of which has been recovered from crash sites. Further information from Mr D. Denchfield, 119 Cavalry Drive, March, Cambs. PE15 9DP.
GODFREY MANNING AIRCRAFT MUSEUM, Edgware, Middx. (0181-958 5113). Collection of aircraft equipment and instruments, mainly contemporary but some vintage as well. Visitors welcome by appointment.
HACK GREEN SECRET BUNKER, Box 127, Nantwich, Cheshire CW5 8AQ (01270-629219, fax 01270-629218). Originally built as a radar station during World War II, the bunker was rebuilt as a top-secret Regional Seat of Government during the 1970s, a function which it retained until 1993. Preserved in excellent order, the three-level structure contains what were government offices, dormitories, canteen and a BBC studio, and now fulfils a secure commercial purpose as well as housing an extensive museum of

civil defence and Cold War memorabilia as well as some military radio, radar, telephone and teleprinter equipment. Open daily from 10 a.m. to 5 p.m. until the end of October then at weekends and public holidays (except Christmas day and Boxing day) during November–February. From March it reverts to seven-day opening.

HMS *BELFAST*, moored near Tower Bridge, London. The museum is aboard a WW II cruiser. The radio room is fitted with sets of the period as well as other radio and radar-related items.

MILITARY COMMUNICATIONS MUSEUM (Curator Bob Francis), 163 Sherwood Park Avenue, Blackfen, Sidcup DA15 9JG. Write to the curator for further details.

MUCKLEBURGH COLLECTION, Weybourne Military Camp, Norfolk (on the A149 coast road between Sheringham and Blakeney). Large section covering military radio gear.

MUSEUM OF THE MOVING IMAGE (MOMI), South Bank, London SE1 (0171928 3535 switchboard or 0171-401 2636 recorded information line). Exhibits include old TV cameras, fifties living room with period set playing old programmes.

NATIONAL MUSEUM OF PHOTOGRAPHY, FILM AND TELEVISION, Prince's View, Bradford BD5 0TR (01274-727488). Open Tuesday to Sunday 10.30–18.00. Good displays of old TV cameras and receivers. Well-endowed library of old programmes, on open access.

NATIONAL VINTAGE WIRELESS & TELEVISION MUSEUM, The High Lighthouse, Harwich, Essex (01206-322606). Possibly not as elaborate as the title might suggest.

NATIONAL WIRELESS MUSEUM, Arreton Manor and Puckpool, Isle of Wight (two sites). Enquiries to Douglas Byrne, Lynwood, 52 West Hill Road, Ryde PO33 1LN (01983-567665). Many radios and one old TV set on display.

NAVAL COMMUNICATIONS & RADAR MUSEUM. Visits by appointment only. Write to Lt Cdr W.E. Legg RN (Retired), Museum Curator, HMS *Collingwood*, Fareham PO14 1AS.

ORKNEY WIRELESS MUSEUM. Enquiries to Peter MacDonald, 9 Quoybanks Place, Kirkwall KW15 1JQ. Mainly a radio collection, with replica Baird televisor.

PRITTLEWELL PRIORY MUSEUM, Southend, Essex. Two rooms of radios, gramophones and televisions, mainly of Ekco manufacture.

RADIO MEMORIES at Aerial Services, 203 Tankerton Road, Whitstable, Kent. Vintage wireless, gramophones and television.

RADIO MUSEUM (P. and J. Baker), 16 The Green, Ravensworth, Yorks. DL11 7ES (01325-718614).

ROYAL CORPS OF SIGNALS MUSEUM, Blandford Camp, Dorset DT11 8RH (01258-482248). Admission free, open 10.00–17.00 Monday to Friday, 10.00–16.00 weekends, June to September. The museum is clearly signposted from the B3082 Blandford to Wimborne road and contains a comprehensive display of military line and radio communications from the Crimean War to the present day.

SCIENCE MUSEUM, Exhibition Road, South Kensington, London SW7 2DD (0171-938 8000). Broadcast communications gallery with many interesting exhibits (plus many other items in store.)

TANGMERE MILITARY AVIATION MUSEUM, Tangmere Aerodrome, Chichester PO20 6ES (01243-775223). Display of communications electronics includes electronic countermeasures used to disrupt the German Knickebein bombing system in WW II. Open daily 10.00–17.30 March to October, 10.00–16.30 February and November, closed January and December.

TROPIQUARIA, Washford Cross, Watchet, Somerset TA23 0JX (01984-640688). Look out for tall radio masts on the A39 road between Williton and Minehead. 'Wireless in the West', is a small museum exhibition to celebrate sixty years of West and Welsh regional broadcasting. Tropiquaria animal paradise is on the site of the BBC's Washford transmitting station. Open daily 10.00–18.00 throughout the holiday season.

Joseph URBAN, 44 High Street, Anstruther, Fife (01333-310471). Old-style radio shop with built-in museum open 10.00–17.00 weekdays Easter to October.

VINTAGE WIRELESS MUSEUM (Curator Gerald Wells), 23 Rosendale Road, West Dulwich, London SE21 8DS (0181-670 3667). Delightful museum with working pre-war television sets and ex-BBC standards converters. Often featured on television, but not always open – visitors must telephone first for an appointment.

VINTAGE WIRELESS MUSEUM (Curator and owner Ray Leworthy), Old Brewery, 53 High Street, Lindfield, Sussex RH16 2HN (0444-484552). Open 10.00-16.00 Tuesdays, Thursdays and Fridays, 10.00–13.00 first and third Saturday in the month. Sundays by appointment.

YORK CASTLE MUSEUM, The Eye of York, York YO1 1RY (01904-653611). Open daily from 09.30. Exhibits include 1950s living room and a Bush TV22 television on which you can watch 'Hancock's Half Hour'.

YORKSHIRE AIR MUSEUM, RAF Elvington, near York. For military radio enthusiasts. Exhibits include mock-up of wireless operator's position in Halifax bomber and communications equipment in control tower.

Some other military and aviation heritage centres have radio exhibits on show, as do the larger museums of science and industry, and also some local museums of rural life. An excellent guide to these is the *Steam Heritage Preserved Railways, Industrial, Transport, Ship, Aircraft & Military Museum & Rally Guide* published annually on 1st March by TEE Publishing, The Fosse, Radford Semele, Leamington Spa CV31 1XN (01926-614101, fax 01926-614293) and available direct or from most large newsagents.

RADIO AND TELEVISION MUSEUMS IN AUSTRIA

RADIO MUSEUM GRÖDIG, Hauptstrasse 3, 5082 Grödig (00 43 6246-72857). Open Wednesday afternoons 16.00–19.00 or by appointment.
RADIO MUSEUM (Funk- und Radio-Museum), Eisvogelgasse 5, Vienna.
TECHNICAL MUSEUM (Technisches Museum), Vienna.

TELEVISION MUSEUMS IN CANADA

MZTV MUSEUM OF TELEVISION, 151 John Street #401, Toronto, M5V Z1Z (00 1 416-599 7339). The MZTV Museum of Television exists in several dimensions – as a physical museum open by appointment and located in the CHUM City complex right in the heart of Toronto; as an exhibition 'Watching TV: Historic Televisions and Memorabilia from the MZTV Museum' co-produced by the MZTV Museum and the Canadian Museum of Civilisation; and as an e-museum on a WEB site. Part of the MZTV Museum's mission is to reach audiences by means of electronic technologies, including the World Wide Web.

RADIO AND TELEVISION MUSEUMS IN FRANCE

DOUVRES-LA DELIVERANDE, near Caen, Normandy, France. New museum devoted to radar and telecommunications during World War II, opened in 1994 inside former radar tracking station.
EUROPEAN COMMUNICATIONS MUSEUM (Musée Européen de la Communication), Château de Pignerolle, 49124 St Barthélémy d'Anjou (00 33 2-41 93 38 38). Close to the large town of Angers.
MILITARY RADIO DISPLAY 'Le Stéphanois', Jean Gonneaud, 3 rue de l'Eglise, 30170 Saint Hippolyte du Fort (00 33 4-66 77 25 70). Thirty years of military radio collecting, 200 sets of all nationalities and even a jeep-radio!
MUSÉE DU DÉBARQUEMENT, Arromanches, Normandy (where the D-Day invasion took place). A variety of wartime military radio equipment is on display in this museum.
POST & TELECOMMS HISTORY MUSEUM (Musée de l'histoire des PTT), 68340 Riquewihr (00 33 3-89 47 93 80).
RADIO ARDENNES WIRELESS MUSEUM (Musée de la TSF – Radio Ardennes), 184 Avenue Charles de Gaulle, 08000 Charleville-Mézières (00 33 3-24 56 12 41).
RADIO FRANCE MUSEUM (Musée de Radio France), 116 Avenue du Président Kennedy, 75116 Paris Cedex (00 33 1-42 30 15 16). Guided tours (only) at 10.30, 11.30, 14.30, 15.30 and 16.30. Closed Sundays and public holidays.

RADIO MUSEUM (Musée de la Radio), Château de Creully, 14480 Creully (00 33 2-31 80 18 65).
RADIO MUSEUM (Musée de la Radio), 43500 St Victor sur Arlanc (00 33 4-71 03 34 25, fax 71 03 36 77). Open all year round during the afternoon from 14.40 to 19.00.
RADIO MUSEUM (Musée de la Radio TSM), ZA des Grossines, 17320 Marennes (00 33 5-46 85 37 60).
RADIO & PHONOGRAPH MUSEUM (Musée de la Radio et du Phonographe), route du Château de Val, 15270 Lanobre (00 33 4-71 40 32 89).
WIRELESS MUSEUM (Musée de la TSF), TH. B., 136 rue de Charnay, 79400 Nanteuil (00 33 5-49 05 55 93). Three hundred sets on show.
France also has some private collections open by appointment:
MUSEE COCSET, 35 rue Charles de Gaulle, 51420 Cernay Les Reims (00 33 3-26 07 30 58).
MUSEE LACROUTS, 33720 Podensac. Visits by appointment, preferably weekends after lunch. (00 3 5-56 27 01 04 evenings).
MUSEE 'LES SANGLOTS LONGS', Réguiny 56500 (facing lake). The product of two people passionately interested in the events of World War II and radio. Open from 1st March to 31st October (00 33 2-97 38 61 11).

RADIO AND TELEVISION MUSEUMS IN GERMANY

AMATEUR RADIO MUSEUM (Amateurfunkmuseum), Am Mittenfeld 3, 85658 Egmating (00 49 8095-24630).
BAVARIAN RADIO MUSEUM (Rundfunk Museum Schloss Brunn), Am Schloss 2, 91448 Brunn (00 49 9104-3290 or 2482).
BREMEN RADIO MUSEUM (Bremer Rundfunkmuseum), Findorffstr. 85, Bremen (00 49 421-357406).
DRESDEN TECHNICAL COLLECTIONS (Technische Sammlungen Dresden), Junghansstr. 1, Dresden (00 49 351-346 2293).
ELECTRICAL MUSEUM (Elektro-Museum), Hohe Strasse 24, 99094 Erfurt-Möbisburg.
ELECTRUM (Museum of Electricity), Klinikweg 23, Hamburg (00 49 40-6396 3641).
GERMAN NATIONAL MUSEUM (Deutsches Museum), Museumsinsel 1, 80538 München (00 49 89-217 9310).
GERMAN POSTAL MUSEUM (Deutsches Postmuseum), Schaumainkai 53, Frankfurt am Main.
GERMAN RADIO ARCHIVE (DRA, Deutsches Rundfunkarchiv), Bertramstrasse 8, 60320 Frankfurt (00 49 69-156870, fax 00 49 69-156872).
GERMAN RADIO MUSEUM (Deutsches Rundfunkmuseum e.V.), Hammarskoeldplatz 1, Berlin (00 49 30-302 8186).

Figure 27.2 Period room settings, a replica shop front and much more can be seen at the 'On The Air' museum in the historic centre of Chester. Unlike most museums, you can also buy period radios, televisions and telephones here – the ideal take-away souvenir!

GERMAN TELEVISION MUSEUM (Museum für deutsche Femsehgeschichte), Hindemith-Strasse 1–5, 55127 Mainz.

HEIMAT MUSEUM, Triberg, Schwarzwald (00 49 7722-4434). Some radios (Saba).

HEIMAT MUSEUM, Waldbronn bei Karlsruhe. Around two hundred radio sets.

INTERNATIONAL RADIO MUSEUM (Internationales Radio Museum) (Hans Necker), 5928 Bad Laasphe, Haus der Jugend, Bahnhofstrasse 33 (00 49 27 52-9798).

HERITAGE MUSEUM (Volkskunde-Museum), Prenzlauer Tor, 17268 Templin, near Berlin (00 49 3987-2725).

MUSEUM OF POST AND TELECOMMUNICATION (Museum für Post und Telekommunikation), Leipziger Str./Ecke Mauerstr., 10117 Berlin. Closed until 1999 for rebuilding. Temporary location An der Urania 15, 10787 Berlin (0049 30-7501 6890). Open Monday to Thursday 09.00–17.00, Saturday and Sunday 10.00–17.00.

MUSEUM OF POST AND TELECOMMUNICATION (Museum für Post und Telekommunikation), Stephansplatz 5, 20354 Hamburg (00 49 40-3503 7701). Open Tuesday, Wednesday and Friday 10.00–15.00, Thursday 10.00–18.00.

NOSTALGIA MUSEUM (Nostalgiemuseum), Zur Alten Donau 4, 93086 Wörth-Hofdorf (00 49 9482-90086). Open weekends and holidays 14.00–17.00 or by appointment.
RADIO & ELECTRICAL MUSEUM (Radio/Elektromuseum), Gymnasialstr. 72, 66557 Illingen (00 49 6825-402-0, fax 00 49 6825-402-20).
RADIO HISTORY MUSEUM (Museum für Radio- und Funkgeschichte), Bad Bentheim (00 49 5922-7312).
RADIO MUSEUM (Rundfunk Museum Fürth), Schlosshof 23, 90768 Fürth (00 49 911-752986).
SIEMENS MUSEUM, Prannerstr. 10, München (00 49 89-234 3210).
TECHNICAL MUSEUM (Landesmuseum für Technik, Bibliothek und Arbeit), Museumsstrasse 1, 68165 Mannheim.
URBAN & INDUSTRIAL MUSEUM (Stadt und Industrie Museum Wetzlar), Lottestr. 8-10, 35573 Wetzlar (00 49 6441-99221). Philips radio sets.
TRANSPORT & TECHNOLOGY MUSEUM (Museum für Verkehr und Technik), Trebbiner Str. 9, 10963 Berlin, (00 49 30-254 840).
WEHRHISTORISCHE MUSEUM, Röthenbach near Nürnberg. Military museum with comprehensive collection of military communication equipment.

RADIO AND TELEVISION MUSEUMS IN JAPAN

BROADCAST MUSEUM, Atagoyama. When opened in 1956 at the birthplace of Japan's broadcasting, this was the world's only museum dedicated to broadcasting. The museum possesses approximately 20 000 materials on broadcasting and its long history, and 6000 books related to broadcasting, some of which are exhibited to the public.

RADIO AND TELEVISION MUSEUMS IN THE NETHERLANDS

GERMAN COMMUNICATIONS TECHNOLOGY CENTRE (Zentrum für deutsche Nachrichten- und verwandte Technologie), Pater Perestraat 29, 1111 KR Diemen (00 31 20-699 6262).
OMROEPMUSEUM (Broadcasting Museum), Oude Amersfoortseweg 121–131, 1212 AA Hilversum (00 31 35-885888). Open Tuesday to Friday 10.00–17.00, Saturday and Sunday 12.00–17.00 (closed on Monday).

RADIO AND TELEVISION MUSEUMS IN POLAND

POST & TELECOMMUNICATIONS MUSEUM, skr. pocztowa 2030, 50-954 Wroclaw 1. Open weekdays (except Tuesday) 10.00–15.00 and Sunday

11.00–14.00. Wroclaw is the former German city of Breslau and the museum is in the pre-war telecomms headquarters. It is understood that a similar collection in Warsaw was lost during the war, so this is now the best telecomms museum in Poland.

RADIO AND TELEVISION MUSEUMS IN SWITZERLAND

AUDIORAMA (Musée suisse de l'audiovisuel), ancien Grand Hôtel, Avenue de Chillon, CH-1820 Territet/Montreux (00 41 21-963 22 33, fax 963 02 94).
MUSEUM FÜR KOMMUNIKATION, Helvetiastr. 16, Bern (00 41 31-357 5555, fax 00 41 31-357 5599).
RADIO MUSEUM (Musée Jurassien de la Radio), G. et N. Schnoebelen, 2952 Cornol (Jura), (00 41 32-462 27 74, fax 462 20 61).

RADIO AND TELEVISION MUSEUMS IN THE USA

ANTIQUE WIRELESS ASSOCIATION, Village Green, Routes 5 and 20 Bloomfield, NY, south of Rochester, NY, USA (00 1 716-657 6260, if no answer dial 716-657 7489). Open May to October, Sunday 14.00–17.00; additionally, June to August, Saturday 14.00–16.00 and Wednesday 19.00–21.00 (closed holidays). See an old shipboard wireless transmitter and an early Western Union telegraph office with original instruments.
HISTORICAL ELECTRONICS MUSEUM, Box 1693, MS 4610 Baltimore, MD 21203, USA (00 1 410-765 2345).
MUSEUM OF BROADCAST COMMUNICATIONS, 78 E. Washington (corner of Washington and Michigan Ave), Chicago (00 1 312-629 6000). Exhibits showcase the evolution of radio and television broadcasting. The wonderful exhibits contain much memorabilia as well as a number of old broadcast radios and a bunch of early TV sets and cameras; also a shop and viewing theatre. Free admission, open Monday to Saturday. There is a computer catalogue of their archive holdings and for a small fee you can view the show of your choice in their special screening room.
MUSEUM OF TELEVISION & RADIO. The museum has two sites: 25 West 52 Street, New York City, USA (00 1 212-621 6800 for daily information on scheduled activities, 621 6600 for all other information) and 465 North Beverly Drive, Beverly Hills, CA, USA (00 1 310-786 1025 for daily information on scheduled activities). The main activity of the two museums is to present exhibitions and classic programmes from their collection, as well as enabling visitors to come in and hear or watch more than 75 000 programmes of their choice. The museums also offer live and taped radio broadcasts throughout the year, where stations from across the country

come there to broadcast their shows. The museum shops are the place to purchase books, tapes, memorabilia, and also Museum publications on television and radio.
NEW ENGLAND MUSEUM OF WIRELESS AND STEAM, 697 Tillinghast Road, East Greenwich, RI 02818-1424, USA (00 1 401-884 1710). Open by appointment. 'In the early days of radio, Massie of Providence raced neck and neck with Marconi. The museum preserves the original Massie coastal wireless station which is the oldest surviving, originally equipped, wireless station anywhere.'
SMITHSONIAN INSTITUTE National Museum of American History (00 1 202-357 1840, fax 00 1 202-357 4256).

Obviously this is only a small fraction of all the relevant museums in North America, although it does contain many of the key sites that overseas visitors find most interesting. Further museums with radio collections are listed in the book *Antique Radios Restoration and Price Guide* by David and Betty Johnson (details in 'Recommended reading' chapter).

TELEPHONE MUSEUMS AND RELATED PLACES TO VISIT IN BRITAIN

THE ALMONRY HERITAGE CENTRE, Abbey Gate, Evesham, Worcs. (01386-446944). Open 10.00–17.00 except Sunday 14.00–17.00. Excellent local museum with surprisingly large collection of old telephones.
AMBERLEY CHALK PITS MUSEUM, Amberley, Arundel BN18 9LT (01798-831370, fax 01798-831831). Opposite Amberley station, easy road access. Opening hours vary according to time of year, daily in high season. Comprehensive industrial archeology museum, with a nice section on telecomms, including a restored rural telephone exchange. Further information from Ron Weller (01903-267839).
AVONCROFT MUSEUM OF BUILDINGS, Stoke Heath, Bromsgrove, Worcs. B60 4JR (01527-31886/31363). Buildings spanning seven centuries have been re-erected here, from windmills to prefabs. British Telecom has assisted by supplying examples of all its designs of telephone kiosk. Open from early March to late November from 11.00 onwards. Daily in summer, weekends and bank holidays at other times. Closed some Mondays though (best to ring first). Admission charge.
BAMPTON MUSEUM. This delightful museum, in Bampton, Devon, has now closed. The instigator hopes to re-establish it on another site nearby.
BATTLE OF BRITAIN CONTROL ROOM, RAF Uxbridge, Uxbridge, Middx. (01895-237144). Underground operations room with all the plotting tables and other paraphernalia, preserved exactly as it was on the morning of 15th September 1940. Individual and group visits, by appointment only. Contact Warrant Officer Chris Wren.

CABINET WAR ROOMS, Clive Steps, King Charles Street, London, SW1 (0171-930 6961). Churchill's underground wartime headquarters, recreated to look as if they have been left as they were at the time of the blitz. All telephone equipment on show. Open daily 10.00–17.15. Admission charge.

DESIGN MUSEUM, Butler's Wharf, Shad Thames, London SE1 2YD (0171-403 6933). A small number of classic design phones are on show. The shop sells a booklet on telephone design.

ESSEX SECRET BUNKER, Crown Buildings, Shrublands Road, Furze Hill, Mistley, Essex CO11 1HS (01206-392271). Open daily 10.30–16.30. Former Cold War period bunker, later the Essex county war headquarters. This underground structure was built in 1951 and has been packed with 80 tons of authentic Cold War period equipment, with the original plotting board, telecomms equipment and other effects on display.

FLEUR-DE-LIS HERITAGE CENTRE, Stone Street, Faversham, Kent. Here, in one of the finest heritage centres in the country, they are working on a UAX13 (rural automatic exchange) and a kiosk. The exchange came from Littlebourne, near Canterbury, and is of a type widely used in east Kent and elsewhere.

GOONHILLY EARTH STATION, Cornwall. Part of the former Taunton Museum collection is on display at the visitor centre here.

GUERNSEY TELEPHONE MUSEUM, Hermes, La Planque, Cobo Road, Castel (01481-711221). Free admission Tuesday and Wednesday evenings (19.00–21.00), April to September. 'Telephones, switchboards and other equipment as used in the Bailiwick of Guernsey since the first telephone exchanges were opened on 28th July 1898.'

HELLFIRE CORNER at Dover Castle, Kent. Tour of once secret wartime installations buried deep inside the cliffs includes viewing of the communications room, with large quantities of old switchboards, amplifiers and other equipment. Dover Castle is in the care of English Heritage and a single admission fee is charged for admission to all the exhibits. Tours of the secret underground tunnels leave approximately every fifteen minutes in summer and every forty-five minutes in winter; the tour takes fifty minutes. The last tour starts at 17.00 in summer and at 15.00 in winter. The castle opens daily at 10.00 and closes at 18.00 (16.00 October to March).

KELVEDON HATCH R4 BUNKER, Kelvedon Hatch, Essex (01277-364883). Open daily 10.00–16.00 with tours on the hour. Formerly top secret, this huge underground structure is an excellent example of an R4 Rotor (radar defence) bunker built in the early 1950s as a Sector Operations Centre for the Metropolitan Sector of Fighter Command. It later became a Sector Operations Centre for the UK Warning and Monitoring Organisation. Remaining top secret until recently, it is now easy to find, with a brand-new entrance from the main Brentwood–Ongar road (north-west of Brentwood, near Kelvedon Hatch). You are welcome

to ring for directions. Here too there is a large collection of telephone and teleprinter equipment, and volunteer helpers are sought – come and have a look around, then sign the book to volunteer.
KIDDERMINSTER RAILWAY MUSEUM, reached via the Severn Valley Railway Station. Primarily railway relics but with some nice single needle telegraphs and railway telephones.
KINGSTON-UPON-HULL, KINGSTON COMMUNICATIONS, formerly the Hull Corporation Telephone Department. Has some items that are viewable on application to Telephone House, Carr Lane.
MANCHESTER, GREATER MANCHESTER MUSEUM OF SCIENCE & TECHNOLOGY, Liverpool Road Station, Liverpool Road, Castlefield, Manchester M3 4JP (0161-832 2244). Open 10.30 a.m.–5 p.m. daily.
MUSEUM OF COMMUNICATION, Saltire Court, Castle Terrace, Edinburgh. Open Tuesday to Saturday 10.00–17.00, also in summer Sunday 14.00–17.00. Enquiries on 0131-473 3939 or to C.H. Matthews, Curator, Museum of Communication, 22 Kinglass Avenue, Bo'ness, West Lothian EH51 9QA (01506-824507).
MUSEUM OF SUBMARINE TELEGRAPHY, Porthcurno, Cornwall. Open April to October, 11.00–15.00 on Wednesday and Friday only.
NARROW GAUGE RAILWAY MUSEUM, Wharf station, Tywyn, Gwynedd LL36 9EY. Two single-needle telegraphs.
NATIONAL RAILWAY MUSEUM, Leeman Road, York YO2 4XJ (01904-621261). Mainly trains but some telephone and telegraph exhibits also.
NORWICH TELECOM MUSEUM. Closed but planned to reopen under volunteer management.
OXFORD TELECOM MUSEUM, BT Museum, 35 Speedwell Street, Oxford OX1 1RH (01865-246601). Ring for opening times. Not a huge museum but a very comprehensive collection.
ROYAL CORPS OF SIGNALS MUSEUM, Blandford Camp, Dorset DT11 8RH (01258-482248). Admission free, open 10.00–17.00 Monday to Friday, 10.00–16.00 weekends June to September. The museum is clearly signposted from the B3082 Blandford to Wimborne road and contains a comprehensive display of military line and radio communications from the Crimean War to the present day.
ROYAL MUSEUMS OF SCOTLAND, Chambers Street, Edinburgh EH1 1JF (0131-225 7534). Good collection of early telegraphs. Also has a Sinclair telephone switch and a semi-automatic exchange which predates Strowger.
SCIENCE MUSEUM, Exhibition Road, South Kensington, London SW7 2DD (0171-938 8000). Telecommunications gallery with many interesting exhibits (plus many other items in store.)
SCOTLAND'S SECRET BUNKER, Troywood, St Andrews, Fife KY16

8QH (01333-310301, fax 01333-312040). This is one of the underground nuclear war command centres equipped in the 1960s, known to devotees as Anstruther. It is open from Easter until the end of October, 10.00–18.00. Parties and coaches welcome. The communications centre, emergency broadcasting studios and computer room are all preserved with authentic artefacts.

STORY OF TELECOMMUNICATIONS, formerly the Telecom Technology Showcase, London. BT's premier museum; sadly closed in August 1997.

TIME-BALL TOWER, Victoria Parade, Deal, Kent (01304-360897). A four-floor museum of time and telegraphy. The time-ball drops at 1 p.m. daily. Tuesday to Saturday 12 noon–4.45 p.m., Easter to September.

WOLLATON PARK INDUSTRIAL MUSEUM, Nottingham (0115-928 4602). Contains the former Ericsson museum collection of telephones and switching equipment, presented by the Plessey Company. Many more items in store than on show.

WESTERN APPROACHES COMMAND CENTRE, Derby House, 1 Rumford Street, Liverpool, L2 3SZ (0151-227 2008). This underground headquarters was the command centre for the Battle of the Atlantic during World War II. Visitors today can see the RAF plotting room, the main operations room, the cypher room, teleprinter room and the telephone exchange. Open all year round but confirm by ringing the number shown.

Many other museums have telephone exhibits on show, particularly the larger museums of science and industry, some local museums of rural life and several preserved railways, tramway and transport museums. An excellent guide to these is the *Steam Heritage Preserved Railways, Industrial, Transport, Ship, Aircraft and Military Museum and Rally Guide* published annually on 1st March by TEE Publishing, The Fossem Radford Semele, Leamington Spa CV31 1XN (01926-614101, fax 01926-614293) and available direct or from most large newsagents.

TELEPHONE MUSEUMS AND RELATED PLACES TO VISIT IN AUSTRIA

POST- UND TELEGRAPHENMUSEUM. In the Technical Museum, Mariahilfstrasse 212, Vienna.

TELEPHONE MUSEUMS AND RELATED PLACES TO VISIT IN BELGIUM

MUSÉE DES POSTES ET TELECOMMUNICATIONS, Grand Sablon 40, Brussels.

TELEPHONE MUSEUMS AND RELATED PLACES TO VISIT IN CANADA

BELL CANADA MUSEUM, 6055 Monkland Avenue, Montreal, H4G 1H3 (00 1 514-870 4696). Bell Canada's historical collection of equipment, opened on request. Archive centre, also open by appointment, is at 1055 Beaver Hall Hill, Montreal, H2Z 1S4 (00 1 514-870 7088, fax 00 1 514-870 2537).
BELL HOMESTEAD, 94 Tutela Heights Road, Brantford, Ontario, N3T 1N3 (00 1 519-756 6220, fax 00 1 519-759 5975). Historic residence of the Bell family, open 10.00–18.00 summer, closed Monday.
CANADIAN FORCES COMMUNICATIONS & ELECTRONICS MUSEUM, Vimy Post Office, Kingston, Ontario, K7K 5Z0 (00 1 613-541 5395, fax 00 1 613-541 4489). Large display of military communications equipment.
EDMONTON TELEPHONE HISTORICAL INFORMATION CENTRE FOUNDATION, Box 4962, Edmonton, Alberta, T6E 5GB, Canada (00 1 403 441 2077, fax 00 1 403 433 4068). Located at 10437 - 83 Avenue, Edmonton. Publishes a very handy and comprehensive *World Directory of Telephone Museums*.
NATIONAL MUSEUM OF SCIENCE & TECHNOLOGY, Box 9724, Ottawa Terminal, K1G 5A3 (00 1 613-991 3083, fax 00 1 613-3636). Hands-on exhibit from Strowger to optical fibre.

TELEPHONE MUSEUMS AND RELATED PLACES TO VISIT IN CROATIA

PTT-MUSEJ, Jurisiceva 13, Zagreb (274811).

TELEPHONE MUSEUMS AND RELATED PLACES TO VISIT IN THE CZECH REPUBLIC

POSTOVNI MUZEUM, Holeckova 10, Prague.

TELEPHONE MUSEUMS AND RELATED PLACES TO VISIT IN DENMARK

TELEFONMUSEET, Svanemollevej 112A, 2900 Hellerup near Copenhagen (00 45-3399 4050, fax 00 45-3162 3770). Open Sunday 14.00–16.00. A large collection of telephones and switchboards on display, including just about every variant possible of Ericsson magneto table telephones.

TELEPHONE MUSEUMS AND RELATED PLACES TO VISIT IN EIRE

DUBLIN. The training section of Telecom Eireann intends to set up a museum here shortly. Currently the items are in store at Kimmage Manor.
MAYNOOTH COLLEGE, Maynooth, County Kildare. This is the home of the electrical collection of the Reverend Nicholas Callan, who invented the induction coil. Among items on display are a single needle and other telegraph apparatus. Free admission, by appointment.
VALENTIA ISLAND MUSEUM, Valentia, County Kerry. Pays special attention to the European terminal of the 1858 Atlantic cable.

TELEPHONE MUSEUMS AND RELATED PLACES TO VISIT IN FINLAND

TELEGALLERIA, Elimaenkatu 9A, Helsinki (00 358-0704 2050, fax 00 358-0704 2045).

TELEPHONE MUSEUMS AND RELATED PLACES TO VISIT IN FRANCE

MUSÉE REGIONALE 'LES TÉLÉCOMMUNICATIONS EN FLANDRE', 12 avenue Foch, Marcq-en-Baroeul/Nord (00 33 3-2072 3028). Open Wednesday, Saturday and Sunday 10.00–12.00 and 14.00–16.00. Large collection of telegraphs, manual telephone switchboards and auto equipment. Model of Chappe telegraph, captured German field teletype.
MUSÉE DES ARTS ET METIERS, Paris (00 33 1-4027 2372, fax 00 33 1-4027 2662). Metro station: Arts et Métiers. Very disappointing for a national science museum; the whole collection is stuck in a time-warp circa 1961 with no new additions or evidence of renovation since then. Small collection of telephones and telegraphs.
MUSÉE DE LA POSTE, 34 Boulevard de Vaugirard, Paris.
MUSÉE DES PTT, Riquewihr/Alsace,
MUSÉE DES TÉLÉCOMMUNICATIONS, Pleumeur-Boudou. This earth station in Britanny is the French equivalent of Britain's Goonhilly Down and houses a museum spanning 150 years of telecommunication history. English-speaking telephone service (00 33 2-9646 6380).
TELECOM MUSEUM, 45 rue de Soupetard, 31049 Toulouse (00 33 5-6148 4284). Described as 'un petit musée des telecommunications. Le responsable est M. Michel.'

TELEPHONE MUSEUMS AND RELATED PLACES TO VISIT IN GERMANY

GERMAN NATIONAL MUSEUM (Deutsches Museum), Museumsinsel 1, 80538 München (00 49 89-217 9274, fax 00 40 89-217 9324).
GERMAN POSTAL MUSEUM (Deutsches Postmuseum), Schaumainkai 53, Frankfurt am Main (00 49 69-606 0230, fax 00 49 69-606 0123).
MUSEUM OF POST AND TELECOMMUNICATION (Museum für Post und Telekommunikation), Leipziger Str./Ecke Mauerstr., 10117 Berlin. Closed until 1999 for rebuilding. Temporary location An der Urania 15, 10787 Berlin (0049 30-7501 6890). Open Monday to Thursday 09.00–17.00, Saturday and Sunday 10.00–17.00.
MUSEUM OF POST AND TELECOMMUNICATION (Museum für Post und Telekommunikation), Stephansplatz 5, 20354 Hamburg (00 49 40-3503 7701). Open Tuesday, Wednesday and Friday 10.00–15.00, Thursday 10.00–18.00.
PHILIPP-REIS-SAMMLUNG, Hugenottenstrasse 93, Friedrichsdorf/Taunus. The Germans consider Reis the true inventor of the telephone.
POST AND TELECOMMUNICATIONS MUSEUM (Post- und Fernmeldemuseum), Freidrichstrasse 13, Stuttgart (00 49 711-2000 3344, fax 00 49 711-2000 3456).
POSTMUSEUM, Zeppelinstrasse 24, Hannover.
POSTMUSEUM, Koblenz.
TRANSPORT & TECHNOLOGY MUSEUM (Museum für Verkehr und Technik), Trebbiner Str. 9, 10963 Berlin, (00 49 30-254 840).
VERKEHRSMUSEUM, Postabteilung, Lessingstrasse 6, Nürnberg.

TELEPHONE MUSEUMS AND RELATED PLACES TO VISIT IN GREECE

POSTAL MUSEUM, 5 Rue du Stade Pagrati, Athens.

TELEPHONE MUSEUMS AND RELATED PLACES TO VISIT IN HUNGARY

TELECOMMS MUSEUM (Telefónia Múzeum), Budapest 1014, Úri Utca 49, on Castle Hill (201 8188). Fun place for children, where they can make a hundred-year old exchange spring to life. Open Tuesday to Sunday, 10.00–16.00, closed Mondays.
TRANSPORT & COMMUNICATIONS MUSEUM (Közlekedési Múzeum) with old railway telegraphs and telephones, 1146 Budapest, Városligeti Street 11 (342 0565, fax 344 0322). Open Tuesday to Friday

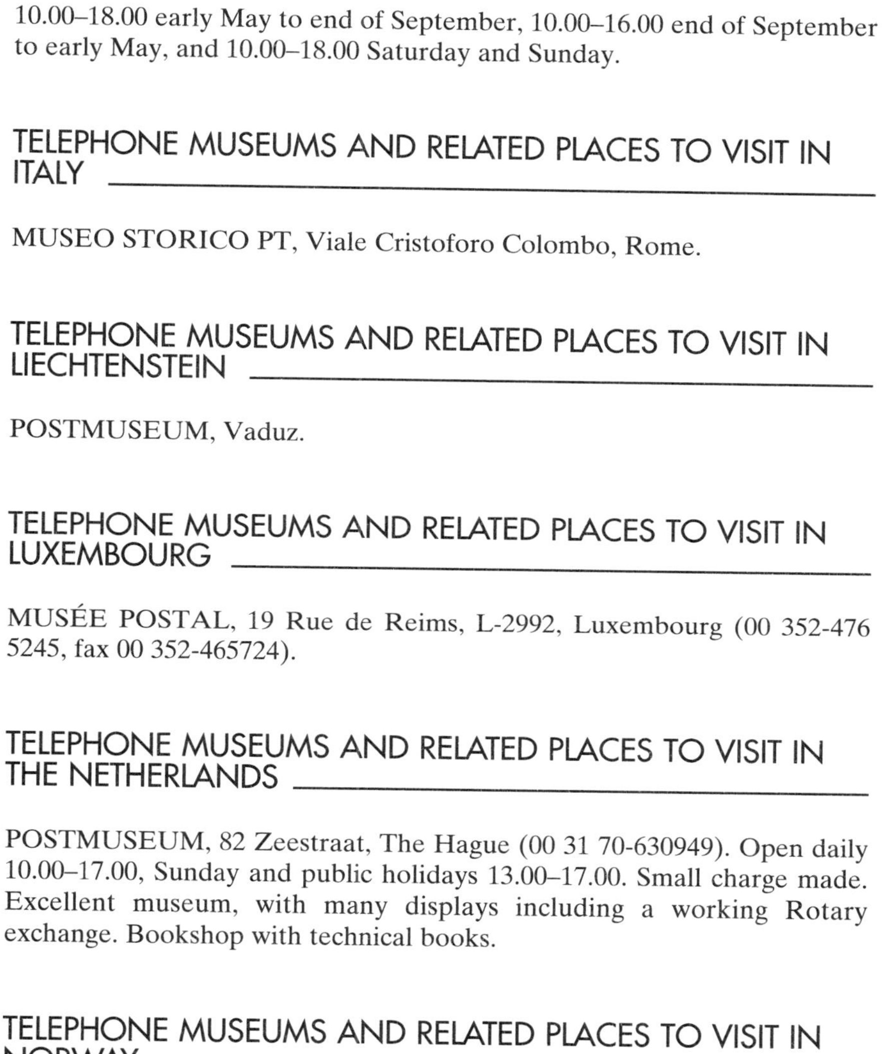

10.00–18.00 early May to end of September, 10.00–16.00 end of September to early May, and 10.00–18.00 Saturday and Sunday.

TELEPHONE MUSEUMS AND RELATED PLACES TO VISIT IN ITALY

MUSEO STORICO PT, Viale Cristoforo Colombo, Rome.

TELEPHONE MUSEUMS AND RELATED PLACES TO VISIT IN LIECHTENSTEIN

POSTMUSEUM, Vaduz.

TELEPHONE MUSEUMS AND RELATED PLACES TO VISIT IN LUXEMBOURG

MUSÉE POSTAL, 19 Rue de Reims, L-2992, Luxembourg (00 352-476 5245, fax 00 352-465724).

TELEPHONE MUSEUMS AND RELATED PLACES TO VISIT IN THE NETHERLANDS

POSTMUSEUM, 82 Zeestraat, The Hague (00 31 70-630949). Open daily 10.00–17.00, Sunday and public holidays 13.00–17.00. Small charge made. Excellent museum, with many displays including a working Rotary exchange. Bookshop with technical books.

TELEPHONE MUSEUMS AND RELATED PLACES TO VISIT IN NORWAY

POSTMUSEET, Tollbugata 17, Oslo (00 47 2-408059).

TELEPHONE MUSEUMS AND RELATED PLACES TO VISIT IN POLAND

POST & TELECOMMUNICATIONS MUSEUM, skr. pocztowa 2030, 50-954 Wroclaw 1. Open weekdays (except Tuesday) 10.00–15.00, Sunday

11.00–14.00. Wroclaw is the former German city of Breslau and the museum is in the pre-war telecomms headquarters. The museum has Morse, Hughes and Baudot telegraphs and old telephone apparatus. It is understood that a similar collection in Warsaw was lost during the war, so this is now the best telecomms museum in Poland.

TELEPHONE MUSEUMS AND RELATED PLACES TO VISIT IN PORTUGAL

MUSEO DOS CTT, Rua de D. Estefania 175, Lisbon.

TELEPHONE MUSEUMS AND RELATED PLACES TO VISIT IN RUSSIA

POLYTECHNICAL MUSEUM, Moscow.
POPOV CENTRAL MUSEUM OF COMMUNICATIONS, St Petersburg.
POSTAL & TELEGRAPH MUSEUM. Inside Kazan railway terminus, Komosolskaya Plotschad, Moscow.

TELEPHONE MUSEUMS AND RELATED PLACES TO VISIT IN SAN MARINO

MUSEO POSTALE, Borgo Maggiore, San Marino.

TELEPHONE MUSEUMS AND RELATED PLACES TO VISIT IN SPAIN

MUSEO POSTAL Y TELECOMUNICACION, Pasaje de Montalban, Madrid (00 34 1-521 4260).

TELEPHONE MUSEUMS AND RELATED PLACES TO VISIT IN SWEDEN

POSTMUSEUM, Lilla Nygatan 6, Stockholm.
SCIENCE MUSEUM, Stockholm. Large section on telecommunications, subsidized by Ericsson.
TELEMUSEUM, Museivagen 7, S-115 27 Stockholm (00 46 8-663 1085, fax 00 46 8-663 2604).

TELEPHONE MUSEUMS AND RELATED PLACES TO VISIT IN SWITZERLAND

MUSEUM FÜR KOMMUNIKATION, Helvetiastr. 16, Bern (00 41 31-357 5555, fax 00 41 31-357 5599).
SWISS TRANSPORT MUSEUM, Lidostrasse 5, CH-6006 Luzern (00 41 41-314444, fax 00 41 41-316168). Open 10.00–16.00 daily. Two halls devoted to post and telecomms, including working exhibits of all types of automatic telephone selectors used in Switzerland. The museum is mainly orientated to non-technical visitors and thus fascinating for children (huge model railway!). Hefty admission charge.

TELEPHONE MUSEUMS AND RELATED PLACES TO VISIT IN THE USA

BELL'S WORKSHOP, 125 High Street, Boston, MA 02107 (00 1 617-743 1167, fax 00 1 617-743 6057). Telephone museum containing a replica of Bell's original workshop. Another related museum at 185 Franklin Street.
GTE COMMUNICATIONS MUSEUM, 303 East Berry Street, Fort Wayne, IN. Open 10.00–16.00 Monday to Friday. Hundreds of items, many rare phones plus a working Strowger exchange demonstrator.
HENRY FORD MUSEUM and GREENFIELD VILLAGE, 0900 Oakwood Boulevard, P.O. Box 1970, Dearborn, MI 48121-1970, USA (00 1 313-271-1620). Open daily 09.00–17.00, closed Thanksgiving and Christmas Day. Interiors of Greenfield Village buildings are closed January through to March.
HISTORIC SPEEDWELL, 333 Speedwell Avenue., Morristown, NJ 07960, USA (00 1 201-540-0211). Exhibits highlight the work of the Vail family and their contributions to steam-powered transportation and telegraphic communication; the factory served as a secret laboratory for Alfred Vail and Samuel F. B. Morse in 1837. Open Thursday and Friday 12.00–16.00, Saturday and Sunday 13.00–17.00 May to October. Closed 4th July.
ILLINOIS BELL MUSEUM, Chicago (00 1 312-964 8444).
MUSEUM OF INDEPENDENT TELEPHONY, Box 625, 412 South Campbell Street, Abilene, KS 67410 (00 1 913-263 1757).
PIONEER MUSEUM OF LINCOLN TELEPHONE & TELEGRAPH, Lincoln, NB (00 1 402-476 4321).
PIONEER MUSEUM OF NEW YORK, 57 East 14th Street, Huntington, NY 11746 (00 1 516-423 2275).
SMITHSONIAN INSTITUTE National Museum of American History, Division of Electricity, Washington DC, USA (00 1 202-357 1840, fax 00 1 202-357 4256).

TELEPHONE PIONEER COMMUNICATION MUSEUM, 140 New Montgomery Street, San Francisco, CA 94105 (00 1 415-542 0182, fax 00 1 415-4133). Extensive collection of telecomm exhibits. Archive Centre is at 1515 19th Avenue, San Francisco, CA 94122 (00 1 415-661 6469, fax 00 1 415-661 1077) and is the largest catalogued archival collection in North America.
US ARMY SIGNAL CORPS MUSEUM, Attn: ATZH-SM, Building 36305, Ft Gordon, GA 30905-5293, USA (00 1 706-791 3856). The exhibits and administration building is located at Avenue of the States and 37th Street. Open Tuesday to Friday 08.00–16.00, Saturday 12.00–17.00. Closed Sunday, Monday and all Federal Holidays.

This is only a fraction of the dozens of telephone-related museums in the USA. For more information buy the guide published by the Edmonton Telephone Historical Information Centre – entry under Canada.

TELEPHONE MUSEUMS AND RELATED PLACES TO VISIT IN YUGOSLAVIA

PTT-MUSEUM, Palmoticeva ul br. 2, Belgrade.

PLASTICS MUSEUMS

THE BAKELITE COLLECTION AT ORCHARD MILL MUSEUM, Williton, Somerset TA4 4NS (01984-632133). 'A unique display of vintage plastics and domestic items set within a seventeenth century watermill.' Shop, licensed café, bed and breakfast also on site.
DESIGN MUSEUM, Butler's Wharf, Shad Thames, London SE1 2YD (0171-403 6933). Paradise for style freaks with lots to see, including some bakelite items. Good bookshop and food.

CHAPTER 28

The Internet – a global network of kindred spirits

It's difficult to pick up a newspaper or turn on the television these days without hearing someone spouting on about the Internet, Infobahn, Information Superhighway or whatever. As a result, the subject is likely to arouse strong passions of interest, apathy or total antagonism, which is a great shame. That said, most people have lived their lives quite happily without these strange-sounding concepts and, understandably, the idea of sharing their living room with an inhuman machine sounds to them like insanity and purgatory combined.

Most of us can remember back to the days before compact discs, microwave ovens and so on . . . and probably decided we could manage well enough without these new-fangled gadgets when they first came out. Gradually we have come to recognize the merits of these innovations and doubtless the idea of the Internet will become less threatening to people at large.

After more than twenty-five years now, the Internet is scarcely new but access to it has only recently become feasible (and affordable) for most people. All you need is a personal computer, a modem, a telephone line and suitable software. This book is not the place to tell you how to hook up to the 'Net' but, in case you thought it had no connection with vintage radio, telephones and the like, this is the time to correct that impression.

It is simplest to treat the Web as a huge information resource from which data can be retrieved and to which data can be added by any authorized user. Usenet is a global conferencing system separated by topic where users post messages, queries and answers on thousands of different subjects of interest.

To the user there are many attractions. The Internet acts as a single network, without censorship or international borders, and once the slightly strange (but standardized) commands have been learned, you can navigate the system with no more difficulty than using the telephone. Charges are low in relative terms and, apart from telephone dial-up charges (normally local rate) and a modest fee to the Internet service provider, there is nothing more to pay.

One of the most useful elements of the Internet is Usenet, a world-wide talking shop where people share their interests with others. For just about every hobby interest imaginable there is a 'newsgroup' to which you can subscribe (at no cost). Then, every time you log on to the Net, your computer will download the newest messages from your chosen newsgroups. You can read through advertisements, questions raised by readers, answers posted to the newsgroup by other readers and so on. It's like a giant discussion group or conference; alternatively you can treat it as your own on-line magazine.

There are plenty of books about the Internet (and a number of methods of getting on-line); you can find these at your local bookstore or newsagent. What they won't tell you is where you can find the specialist newsgroups that cover our interests; but I can. The situation is constantly changing but here are some you may like to try. Many of these newsgroups publish an FAQ bulletin, which will save you asking obvious questions; these are often extremely detailed documents and well worth downloading. An FAQ, by the way, is a frequently-asked question.

Where the Internet comes into its own for our kind of hobbies is as a world-wide social club where you can ask for help, discuss things that interest you or scan the sales and wants ads. To give just one example, I needed a manual for a twenty-five-year-old Sony product. Sony's service centre could not help so I posted a request in the *sci.engr.television.broadcast* group. Within three days some kindly person in Washington State informed me he could supply a photocopy for the cost of postage alone and a fortnight later I had my manual. Through the Internet I have tracked down pre-war radio recordings and a complete Marconi Mk II television camera chain (which made the Museum of Photography, Film and Television happy because they believed none existed).

But perhaps best of all is the free exchange of help and advice. No matter what your problem is, the chances are that someone has the answer. There's a tremendous amount of good advice and cheerful good will to be found and it will all come flooding into your computer every day without any extra charge if you want it. There may be a lot of dross on the Internet but it certainly has its uses. Also, please do not take the descriptions given here as exclusive; many sites have valuable links which will lead you to many more fascinating sites.

RESOURCES ON THE INTERNET (CHOSEN FOR COMPREHENSIVENESS)

RADIO (USENET)

- *rec.antiques.radio+phono*: This group was chartered to meet the needs and interests of collectors of antique (generally, pre-1960s) home-entertaiment radios, phonographs and related materials. Advertising the occasional buy/sell/swap item is within the group's charter, as is the occasional and timely announcement of events of interest to collectors.
- *rec. radio.swap*: Mainly ham radio but some vintage material.
- *sci.electronics.repair*: From post-war valve radios to stuff made yesterday.
- *rec.radio.shortwave*: For short wave listeners.
- *rec.radio.info*: Miscellaneous.

RADIO (WEB PAGES)

- *http://www.antiqueradio.com/*: Antique Radio Classified Web.
- *http://www.ggw.org/awa*: Antique Wireless Assocation.
- *http://rnw.nl/en/pub/antique.html#top*: Antique radio links.
- *http://cpunet/classicradio/radio_links.html*: Classic radios.
- *http://www.ndirect.co.uk/~e.tedeschi*: Hove Radio Museum.
- *http://www.accessone.com/%7Ephiln*: Phil's Old Radios.
- *http://www.inx.de/%7Ewumpus/radio.htm*: Radio Wumpus.
- *http://members.aol.com/retrophoni/bienvenue.htm*: Retro-Phonia.
- *http://www.paulandmark.u-net.com/valve*: Superb valve radio restoration guide.
- *http://www.neca.com/~radiodoc*: Antique radio restoration and repair.
- http://www.cs.ruu.nl/people/gerard/radios: *Radio restoration, dating, valve data.*

MILITARY RADIO (WEB PAGES)

There are some pages that list the descriptions of thousands of military radios (American of course!). There are even pictures and/or schematics for some of the sets.

- *http://linux.cec.army.mil/Bille/m1.htm#A016*
- *http://linux.cec.army.mil/hamradio/equipment/mi.html*
- *http://telalink.net/~badger/millist/mi.html*
- *http://linux.cec.army.mil/*

- *http://www.dse.nl/ws19/index.html*
- *http://www.iaehv.nl/users/pb0aia/srs/index.html*: Home of the Surplus Radio Society.
- *http://chide.museum.org.uk/reme.technology/reme.technology.index.html/*: REME, Royal Electrical and Mechanical Engineers Museum of Technology, with many British sets.

BOATANCHORS LIST

An active mailing list devoted to the discussion of pre-1970s vintage communications equipment. Discussion covers amateur radio receivers, transmitters, microphones, Morse code keys, accessories, certain military radio equipment and various other related subjects. Wanted and items for sale are posted. A frequently asked questions (FAQ) file and daily digest format are available.

To subscribe, send an e-mail message to: *LISTPROC@thePorch.com*. In the message put this:

subscribe boatanchors <your full name & callsign>. Leave the 'Subject:' blank.

To receive the daily digest, send an e-mail to: *LISTPROC@thePorch.com*. In the message put this: *set boatanchors mail digest*. Leave the 'Subject:' blank.

To receive the 'BOATANCHORS-LIST FAQ', send an e-mail message to: *LISTPROC@theporch.com*. In the message put this: *get boatanchors faq*. Leave the 'Subject:' blank. You will receive this via e-mail in a short while.

TELEVISION (USENET)

- *sci.engr.television.broadcast*
- *uk.tech.broadcast*

And see Radio above

TELEVISION (WEB PAGES)

- *http://www.users.dircon.co.uk/~bandc/a.wiseman/625*: Andrew Wiseman's Television Room: old idents, public service announcement films and much more.
- *http://www.islandnet.com/~ianc/dm/dm.html*: Dead Media Project.
- *http://ourworld.compuserve.com/homepages/steveroberts/restorat.com*: Doctor Who Restoration Team, proper technical stuff.

- *http:www.cram.nl/cram/i&i/940310.htm*: Dutch television history.
- *http://soli.inav.net/~jebraun/old_tv.htm*: Early television in the USA and Germany.
- *http://avimall.com/entertain*: Entertainment archives: episode guides to 500-plus shows, test cards, the lot.
- *http://songs.com/noma/philo/index.html*: Farnsworth Chronicles: Philo Farnsworth, television's unsung inventor.
- *http//:www.petford.demon.co.uk/kaleidos/405aliv.htm*: 405 Alive.
- *http://www.compulink.co.uk/~donore/405.htm*: 405 Lines in Ireland.
- *http://www.traum-von-sehen.de*: German television history.
- *http://www.antiqueradios.com/germanyv.shtml*: German television history.
- *http://www.users.zetnet.co.uk/itw/index.html*: Independent Television History Web site.
- *http//:www.petford.demon.co.uk/kaleidos/*: Kaleidoscope: classic British television.
- *http://www.neog.com/mbc/index.html*: Museum of Broadcast Communication, USA.
- *http://www.mtr.org:80/*: Museum of Television & Radio, USA.
- *http://www.citytv.com/mztv/gallery.html*: MZTV virtual gallery of classic TVs.
- *http://www.nmsi.ac.uk/nmpft*: National Museum of Photography, Film & Television of Great Britain.
- *http://homepages.enterprise.net/gdixon*: New techniques for old television.
- *http://src.doc.ic.ac.uk/public/media/tv/collections/tardis/index.html*: Archive old TV programmes.
- *http://www.meldrum.co.uk*: Teletext then and now, test card gallery.
- *http://ftp.wi.net/~rkurer/*: Toon Tracker: an excellent collection of references, pictures and sound clips of early TV animation.
- *http://www.geocities.com/hollywood/5147*: TV idents of the UK.
- *http://www.garm.com/tvsmikes.htm*: TV receivers for sale.
- *http://www.harryposter.com*: TV receivers for sale.
- *http://www.tvtrecords.com/tvbytes/tvthemes.html*: TV Themes.
- *http://src.doc.ic.ac.uk/public/media/tv/collections/tardis/trading/index.html*: UK Video Tape Trading Homepage.
- *http://members.aol.com/mcleandon/tv_index.htm*: World's oldest video recordings.

OLD-TIME RADIO PROGRAMMES

There is lots on the Net! For starters, check out this Web site:

:• *http://www.crl.com/~lgenco/otr.html*

This is the mother of all OTR (old-time radio) repositories, with links to just about everywhere and everything plus an OTR FAQ. You can get this FAQ from:

- *oak.oakland.edu/pub/hamradio/arrl/infoserver/faqs/oldtimer.txt*

To subscribe to the OTR electronic newsletter, send an e-mail to:

- *old.time.radio-request@airwaves.com*. Subject: SUBSCRIBE.

To unsubscribe, send an e-mail to: *old.time.radio-request@airwaves.com*. Subject: UNSUBSCRIBE. (The body of the message is ignored in both cases.)

CLASSIC AUDIO (USENET)

- *rec.antiques.radio+phono*
- *rec.audio.high-end*
- *rec.audio.marketplace*

CLASSIC AUDIO (WEB PAGES)

- *http://members.aol.com/edisonshop/edison3.jpg:* Antique talking machines information, parts, repairs, even hear the voice of Edison himself!
- *http://www-odp.tamu.edu/~schulte:* Dedicated to the masterpieces of all antique phonographs and the golden age of juke-boxes.
- *http://www.78rpm.com:* Vintage record and phonograph publications and equipment auction.

CLASSIC MICROPHONES (WEB PAGES)

- *http:/k-bay106.com/mics.htm*: Microphone 'museum'.
- *http://soli.inav.net/~jebraun/mikes.htm*: Microphone 'museum'.
- *http://davinci.csun/edu/~kwebb*: Microphone 'museum'.
- *http://soli.inav.net/~jebraun/radio_links.htm*: Links to many other sites.
- *http://www.garm.com/tvsmikes.htm*: Mikes for sale.
- *http://www.neumann.com/history/welcome.htm*: Microphone 'museum'.

TELECOMMS (USENET)

- *alt.dcom.telecom.ip*: Occasional historical discussions.

- *comp.dcom telecom.tech*: As above.
- *uk.telecom*: As above.

TELECOMMS (WWW PAGES)

- *http://www.morsum.demon.uk/*: Morsum Magnificat, devoted to Morse communication.
- *http://www.thg.org.uk*: Telecommunications Heritage Group, UK.
- *http://www.btinternet.com/~fyneview/light.straw.html*: Enthusiast site.
- *http://www.seg.co.uk/telecomm/index.htm*: Strowger pages.
- *http://www.telephone-pioneers.org/*: Telephone pioneers.
- *http:/www.omen.com.au/~larry/tgram.html*: Telegram collecting.
- *http://www.chss.montclair.edu/~pererat/telegraph.html*: Vintage telegraph keys.
- *http://www.cybercomm.com/~chuck/atca.html*: Antique Telephone Collectors Association, USA.
- *http://www.cybercomm.com/~chuck/phones.html*: Old telephones in North America, with many, many links.
- *http://dcpu1.cs.york.ac.uk:6666/fisher/telecom*: Tony Fisher.

COLLECTING IN GENERAL

- *http://www.antiques.tm.fr*: Large number of useful links.

PART 5

Reference Department

CHAPTER 29

Safety first

Yes, it looks more like safety *last* putting this section at the back of the book but I reckoned you might be tempted to skip this section if it was at the beginning.

NO APOLOGIES!

You may be a professional, experienced engineer or simply seeking to gain satisfaction from a creative and enjoyable hobby. We stress the dangers not to insult anyone's intelligence but to ensure that you continue to enjoy your pastime – in safety! (Special thanks to Malcolm Burrell for his help with this section.)

Mending old radios and TVs should be a pleasant pastime and relaxation, but don't relax too much. There is grief lurking inside every old set and whilst Gerry Wells at the Vintage Wireless Museum in London will cheerfully tell you of the first man to be electrocuted by a TV set, death is no laughing matter. So this chapter is intentionally laden with doom.

Watch my lips – safety isn't an optional extra with old electrical equipment, it is an absolute must. You may not know anyone who has been killed while fooling with old sets but I can take you straight to people who do. We don't want dead readers.

Really, this is an attitude affair. We have many people who mistakenly feel that 'old technology' is somehow more user-friendly, in some strange way automatically good – merely because it is *old*. Don't be fooled!

Approach old equipment with an open and alert mind and realize that a hot chassis, or a resistor line cord, or asbestos insulation, or selenium rectifiers require much more thought and consideration for safety. Remember, too, that the old-timers who talk so glibly about safety received an on-the-job training, having worked man and boy in the business. We enthusiasts are mainly part-timers or newcomers to the hobby and haven't had the benefit of that training. And just because it's a hobby, that doesn't exempt us from safety drill. Live chassis are indiscriminate in whom they kill and even if you are a thoughtful, careful kind of person, that doesn't mean the last person who handled the set was. Never drop your guard – you may know why wax was melted over grub screws but another clown may not have done. He lost the proper grub screws and used normal bolts . . . and the next person who grasped the on-off/volume knob of the radio wondered why it always gave a tingle! Does this sound tedious? It shouldn't do, because safety is never boring.

Safety!

Vintage radio and television receivers use 'live chassis' techniques, in which the chassis is connected directly to one side of the incoming mains supply. This means they can be lethal to carry out repair or servicing work on, unless the appropriate safety measures are in place.

These include feeding the supply to the set through a proper double-wound mains isolating transformer having an adequate insulation rating, thereby reducing the possibility of severe electric shock or death. A residual current device protector, on sale at all do-it-yourself shops, is a further precaution.

If you are not satisfied that you fully understand the risks involved in this sort of work, *do not proceed any further.* Instead seek advice and assistance from a competent technician or engineer.

CONGENIAL ENVIRONMENT

If you are setting up a work area for your hobby, make it as practical and pleasant as possible. Plenty of light, natural and artificial, is a must, as is an opening window for ventilation (to get rid of solder fumes, chemical cleaners and so on). An anglepoise lamp is helpful, preferably with a blue-tinted 'daylight' bulb for better colour recognition. The lamps with a built-in magnifying lens are also very useful. A rubber mat with criss-cross ribs is ideal for the bench top; small components will drop through the holes and not

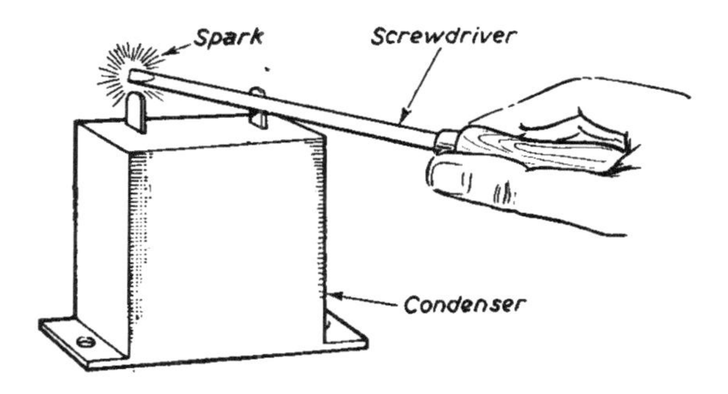

Figure 29.1 This is a method of testing the ability of a high-capacity capacitor to take a charge. Capacitors should always be discharged to chassis before starting repair work.

roll away or scratch cabinets. Failing this, a plastic kitchen-sink mat makes a good substitute. A rubber mat to stand on is also a good policy if you are likely to handle high voltages.

A carpeted floor is also helpful; experts say the benefits of using a short, thick-nap carpet on the floor and on part of the bench-top are many, including:

- dropped small parts tend to stay in the area they are dropped, rather than bouncing across the room;
- dropped valves have a greater potential for surviving;
- it provides better electrical insulation and your feet stay warmer.

KITCHEN CHEMISTRY

Hank Van Cleef has some strong words to say on what he calls 'kitchen chemistry' and they deserve a wider audience. The kitchen makes a wonderful workshop for many of us. A kitchen is a place for food preparation, not chemistry experiments or industrial processes. Be very careful to keep solvents where they cannot contaminate foods or anything used for food preparation. Store chemicals separate from food items, and away from the inquiring hands of small children. Also keep in mind that many of the preparations sold in grocery stores for kitchen cleaning purposes are, in reality, very strong chemicals and may have very little information on the chemical content or processes.

He continues that almost any solvent or process has safety considerations to consider . . . and almost any solvent or process will damage something in a device. You may want to use it over here on this metal part, but if you get it on that plastic part or electronic component, it may destroy it.

Additionally, there are the issues of fire hazards, fumes, violent reactions with other chemicals and safe storage to consider. Some domestic oven cleaners are more violent than many an industrial cleaning process, with labels giving a litany of safety precautions for their use.

Hank concludes you should know your products and know your processes. Test powerful cleaning products first; many of the spray-can cleansers that are so easy to use take off markings and finishes along with the dirt. You have been warned!

DON'T PLUG IT IN!

Whenever you acquire a new treasure there's always a terrific temptation to try it out. With mains-driven equipment that means plugging it in and seeing if it works. Well don't, not until you have made some quick checks.

Is the mains cable (line cord) complete? It may have frayed or have brittle insulation, allowing two wires to touch (or giving you a shock). An input filter capacitor (if fitted) may have broken down and could give a dead short. Electrolytic capacitors in the power supply may not be able to stand the surge of full voltage after many years of lying dormant. Plugging in elderly appliances without making first-line checks is asking for trouble!

ELECTRICITY DOESN'T WARN YOU FIRST

One side of the mains supply is 'live' with respect to earth whilst the 'neutral' side approximates to earth potential. A large number of radio and TV receivers were produced using the 'universal' or 'AC/DC' technique which permitted the omission of large, expensive mains transformers. When switched on these sets employed a chassis common to one side of the mains supply. Incorrect polarity of the mains input connection usually had little effect with AC mains operation but could result in the chassis being LIVE! Some more recent receivers have used a technique where the chassis is always at half the mains potential regardless of the polarity of the mains plug connection!

Another thing about live-chassis sets – live spindles. We've touched on this already but it's worth making the point once more. The shafts of switches and potentiometers fixed to the chassis may well be at chassis potential and thus live. The bakelite or wood cabinet is insulated but these shafts are not, and if someone lost the proper grub screw and replaced a knob using a cheesehead screw, the next person to grip that knob may get a dose of 250 volts. Originally these grub screws were sealed and embedded in wax but you cannot rely on subsequent tinkerers having the same high standards.

Oh, you're wasting your time with wireless – the thing now is television!

Figure 29.2 Television was a ready inspiration for humour in the 1930s.

Even in more orthodox apparatus, standards of insulation were not always as high as they are now. Soldered connections to HT and mains wiring should always have rubber or plastic sleeving, but in times gone by this was often omitted (or it may since have perished). Beware too of kinked and frayed braiding on cloth-covered mains cords, particularly when the cord has a dropper conductor.

ISOLATE FOR SAFETY

Anybody who employs staff to repair domestic electrical goods is obliged to provide isolating transformers for their use. They are not expensive, especially if you buy overstocks at swap-meets, and if they are good enough for employees, they must be equally good for equipping the home enthusiast's test bench. Only a fool would set a lower price upon life.

Think: how often do you work on your own? If you had an accident and had to shout for help, would anyone hear you? If not, your personal safety must be worth an isolating transformer. This is particularly vital with the adjustment and repair of TV receivers; these must be connected to the mains supply via an isolation transformer. In field servicing this item is frequently considered inconvenient and engineers frequently service live equipment which is not isolated from the supply.

The best advice is to keep one hand behind your back at all times you are working on powered equipment. If you need two hands, they say, you should find another way. It may be hard to comply but you should try. Also learn where the anode and screen pins are on the more common valve types (e.g. pins 3 and 4 on many octals) and be careful around them. Keep your fingers off any terminal inside the equipment. Use good test leads and you won't ever need to touch anything inside (except possibly a control grid connection to see if it causes a hum – not the best technique but frequently used).

NEVER ASSUME ANYTHING!

So you've correctly connected the brown or red wire to the right-hand pin of your plug and the blue or black wire to the left-hand pin. So what? Did a DIY enthusiast fit your mains supply socket correctly? Or has the connection of the mains cable to the TV been incorrectly wired at some distant point in time? Or perhaps the receiver mains switch is faulty or has had one pole bridged? Any of these items could result in a LIVE chassis!

Don't even rely on the judgement of others. I remember I was once preparing some flex using nature's own wire strippers (i.e. my teeth, not a clever thing to do) and asked my helper, 'Is the mains off?'.

'Yes, of course it is', came the reply. But needless to say, it wasn't, as I rapidly found out. The taste of mains and saliva mixed is distinctly unpleasant, I can tell you.

FUSES CAN FOOL YOU

A 13-amp fuse in a 13-amp plug offers very little protection, whatever you may think. As Bill Launer, WB0CLD says, one of the most difficult concepts to grasp is that the basic purpose of a fuse (or circuit-breaker) is to protect the wire or mains cable, not the equipment. In other words, fuses do not provide total equipment protection, only short circuit protection; if some degree of equipment protection is desired they will be selected to over-protect the electric wire.

A typical 'fast-blow' fuse will carry 135 per cent of rated current for up to 3600 seconds before opening. It takes 300 per cent of rated current to cause instantaneous opening. A 'slow-blow' fuse will carry 135 per cent of rated current for up to 3600 seconds before opening, and takes 300 per cent of rated current for 6 seconds (minimum) to cause instantaneous opening.

Since we normally put the fuses on the primary side of a power transformer (especially in HV supplies), and the output filters have large capacitors which can act as nearly infinite current sources, you should *never* expect a fuse to protect you from shock (or worse!). Whilst the dangers of high voltage are well known, low voltage–high current power supplies can also be hazardous. You won't be shocked by 12 or 28 volts DC but you still need to be cautious and not wear rings or watches when you might come in contact with them – the resulting burns can be extremely nasty, as any telephone technician who has brushed his watch against the 50-volt busbar will tell you!

It hardly needs saying that people who replace fuses with ones of higher value or with nails (they laughingly call these 'supply stiffeners'), or even bypass fuse-holders and wire around them, are not clever, just plain stupid.

NEON SCREWDRIVER

The neon screwdriver is a cheap, simple and invaluable tool. Keeping one hand in your pocket, hold the screwdriver in the other and, lightly touching the cap with one finger, bring the blade into contact with the chassis. If the neon glows, the chassis is LIVE. You can check the neon by similarly touching the blade to a portion of the set which should be live, e.g. a tag on the mains dropper resistor (but it will not usually become lit if the chassis is LIVE instead).

Do remember that various receiver faults can render the chassis LIVE even if correct polarity is observed in the mains connection. Three common occurrences could be a bridged live pole on the mains switch, an open circuit neutral pole on the mains switch or even an open circuit neutral connection in the mains cable. On sets fitted with fuses in each pole of the mains input, failure of one could render the chassis LIVE.

WORK AREA

If you are working with electricity and your work area has a concrete floor, a rubber mat is essential, particularly during damp weather! Where possible try to arrange a neat working area away from water or central heating pipes. For safety try to arrange that this area is separate from the area occupied by your family. This is emphasized because, inadvertently rushing to answer a telephone, you might just leave a TV chassis connected to a supply and curious little fingers know nothing of the dangers of electricity – or, for that matter, the lethal vacuum encased within every picture tube!

VISUAL INSPECTION!

Before contemplating connecting any unknown receiver to the mains supply, spend a little time inspecting it for signs of missing or loose parts, blown fuses, overheating or even fire damage. Use a meter to check obvious points to ensure no short circuit exists (e.g. across the mains input). If you then decide to apply power keep clear but be observant since an elderly electrolytic might explode! This can be avoided if you can apply power gradually through a variac. Auto-transformers are handy for supplying reduced power to sets being repaired but they are not a substitute for a proper isolation transformer!

CATHODE RAY TUBES – STORAGE AND HANDLING

Many younger enthusiasts may not be aware of the dangers of mishandling tubes, in particular the old round types found in early TVs. When handling these tubes eye protection should be worn and tubes must not be left lying around, they must be stored in boxes. The glass is surprisingly fragile and can implode without any provocation or warning. Bits of glass flying around at high speed can be deadly. The notes following are inspired by Malcolm Burrell again.

Picture tubes are perhaps one of the most hazardous items in any TV receiver. This is because most are of glass construction and contain a very high vacuum. If you measured the total area of glass in any picture tube then estimated the pressure of air upon it at 14.7 pounds per square inch, you would discover that the total pressure upon the device could amount to several tons! Fracturing the glass suddenly would result in an extremely rapid implosion such that fragments of glass, metal and toxic chemicals would be scattered over a wide area, probably causing injury to anyone in close proximity. In modern workshops it is now a rule that protective goggles are worn when handling picture tubes.

The weakest point in most picture tubes is where the thin glass neck containing the electron gun is joined to the bowl. It is therefore essential that you refrain from handling the tube by its neck alone. Once a tube is removed from the receiver hold it vertically with the neck uppermost and one hand beneath the screen with the other steadying the device by the neck. With larger devices it is sometimes easier to grip the periphery of the screen with both hands.

Until the advent of reinforced picture tubes, most were mounted in the cabinet or on the TV chassis by some form of metal band clamped around the face. Never support the weight of the tube by this band since it has been known for the tube to slide out! Some of the larger tubes are extremely heavy. It may, therefore, be easier to enlist assistance.

Before starting to remove a tube, first discharge the final anode connection to the chassis metalwork and preferably connect a shorting lead to this connection whilst you are working. It might be convenient to keep a spare piece of EHT cable with a crocodile clip at one end and a final anode connector at the other.

Exercise care when removing picture tubes from elderly equipment. You may find that the deflection coils have become stuck to the neck. It is extremely dangerous to use a screwdriver to prise them away. Gently heating with a hair-dryer or soaking in methylated spirit is safer.

Disposal of picture tubes also requires care. Unless rendered safe they should never be placed in dustbins or skips. Many engineers swipe the necks off tubes in a cavalier fashion using a broom handle but this is not recommended. A safer method is to make a hole in the side of a stout carton, preferably one designed to hold a picture tube. The tube is placed in the carton and the neck broken using a broom handle. The carton should then be clearly labelled that it contains chemicals and broken glass!

MAINS-DERIVED EHT

It is impossible to over-stress the dangers of mains-derived EHT systems found in pre-war and some early post-war sets. These systems are lethal, so treat them with respect. Unlike modern EHT systems, which are limited in the amount of current they can supply, these older systems using transformers straight off the mains can deliver enough current to kill an army. If you are not sure what you are doing, please ask a friendly colleague. For testing, use an EHT meter; they are still used by the TV servicing trade.

A few simple points will prolong your well-being. Keep one hand behind your back. This will prevent a shock across your body when testing. Don't laugh – a number of early TV engineers were electrocuted servicing these units. Note that the chain of 4.7 mega-ohm bleed resistors can go open circuit, leaving the capacitors fully charged after switch-off.

When starting work, switch off the set. Remove the mains plug. Discharge all EHT smoothing and, if fitted, reservoir capacitors by shorting them out with a 1 kilo-ohm resistor for 30 seconds and then by permanent links whilst working on the unit. Make up the links using plastic

Yes, I understand the music comes out of the box but where is the man that's talking?

Figure 29.3 Radio humour of the 1930s.

insulated rods, crocodile clips and probes. Do not come into contact with any part of the EHT circuit until you have carried out the above. Do not forget to remove all shorting links before applying power again. HVR2 and V16 valves are difficult to obtain!

So you have a faulty EHT power pack on your vintage television set. How do you start? Common faults are:

1. Breakdown of the high-voltage secondary causing arcing, smoke, very low EHT or, in the early stages, white interference spots on the picture with a general lack of brilliance.
2. Faulty EHT rectifier valve such as the Mullard HVR2, Mazda V22 or Marconi V16. Symptoms such as internal sparking and blue glow (soft valve) are common.
3. Failure of smoothing condenser, typically 0.1 μF at 7 kV. This can go short circuit and can damage the valve and, if the fuses don't fail, the transformer as well.

If you have a failed transformer and cannot wait to get that raster, you may wish to try a few of these suggestions.

Silicon diodes of the BY182 type will act as temporary test EHT rectifiers if used with a 47 kilo-ohm resistor at the anode to act as a surge limiter. Transformers removed from small neon signs produce around 2–4 kV and make good emergency replacements. Wire in with EHT cable and take care.

The focus voltage in most old colour TV sets is around 4.5 kV and can be hitched up to test. Note, the colour TV will no doubt have a live chassis. Run it from an isolating transformer for safety.

EHT triplers removed from old colour TVs and Thorn mono chassis will produce EHT when their input is connected to the line output valve anode (top-cap).

Mains EHT transformers can be rewound to complete your restoration but the cost is high. Not surprising when you consider that the high-voltage secondary may contain up to 7000 turns of hair-thin wire in layers. And that's to say nothing of the wax and pitch that have to be removed to gain access.

CHARGED CAPACITORS

Leaving charged-up electrolytics may be a legitimate prank in a shared workshop but it's a recipe for disaster for forgetful 'lone wolf' repairers. You can discharge capacitors to earth with a screwdriver or a wire but the preferred method is to use a resistor built into a grounding lead with a test lead end on it. A 10 kilo-ohm resistor works fine to reduce or eliminate the spark. Just be sure to hold it on for a second or three. And yes, it was a favourite trick in old-time television workshops to toss a charged electrolytic to a colleague with the word 'Catch!'. It's a nasty trick because one's natural reaction is to obey.

NOXIOUS CHEMICALS, FUMES AND COMMON SENSE

Electricity is not the only hazard that lurks in the workshop. It's funny (but only in an ironic kind of way) how many of the things we used to do are now considered highly dangerous. Remember the tin of paraffin or petrol in every workshop for cleaning grease and dirt off metal components . . . and good old 'carbon tet' (carbon tetrachloride) or 'trike', the solvents that would get the muck out of everything. Naughty, naughty. We have to use safer methods these days.

But some of the chemicals that are still on the market are potentially lethal. The consumer magazine *Which*? has highlighted that paint and varnish removers such as Strypit (made by Rustins Ltd) contain methylene dichloride (alias dichloromethane or DCM). This chemical is now listed as a Class Three carcinogen 'with the possible risk of irreversible side-effects'. As well as leading to nausea and drowsiness, it is a powerful skin irritant, meaning that you should wear rubber gloves and ensure good air ventilation when using these products. Better still, choose a non-toxic paint stripper.

If you must use a chemical paint stripper or any other volatile chemical always use it out of doors to avoid breathing the harmful vapours. You don't notice it at the time but these fumes can make you feel really bad for three days afterwards if you're not careful. Current industrial safety regulations even limit the amount of soldering you can do in an unventilated room, so play safe. If you're handling acid, solvents or other noxious liquids, wear rubber gloves. They are not expensive and save a lot of pain, eczema, sores, and so on.

As Martin Ackroyd has said, the choice of paint stripper is easy: sodium hydroxide (lye) merely causes burns that a good skin graft will soon put right; methylene chloride destroys internal organs that are less easily replaced.

Oh yes . . . it's awfully easy to ignore warnings about inhaling solder flux fumes or using powerful solvents such as xylene without wearing rubber gloves. Usually there are no immediate ill-effects and yet long-term exposure to low amounts is more of a problem than short-term exposure to somewhat larger concentrations. The thing is to use common sense. Work outdoors, wash your hands well afterwards, use a simple dust-mask if necessary, avoid certain materials completely etc. You need have no worry about working with many toxic substances if you know what to expect beforehand, and can plan your actions and safety measures accordingly. Ignorance is what is really dangerous.

MORE TRAPS FOR THE UNWARY

Arden Allen points out that most, if not all, of the combined clean and lubricate preparations sold in aerosol cans have become more dangerous lately. That's because the manufacturers are replacing the conveniently non-flammable chlorinated hydrocarbon solvents (that help destroy the ozone layer) with more environmentally friendly flammable solvents. Read the label and look for the word *flammable*. Flammable liquids burn when they are heated to the point where they evaporate fast enough to produce a combustible mixture of gaseous fuel and oxygen (the flash point). When you add the third ingredient of the fire formula, namely heat, you've got problems, so don't add the heat.

WD-40 is a typical example of these products and, being a petroleum derivative, it is flammable (although its flash point is reasonably high and you need a pretty hot spark to ignite it). All the same, if you are using WD-40 to lubricate power switches, switch off the device first. Real fires have been started in this way. You can also start fires when certain switch cleaners (used for cleaning valve pins) soak into fibre or phenolic valve sockets if the cleaner has not had time to dry (being wet, it is conductive). Arden Allen advises that you cannot remove the ionic contamination out of the insulation once it is in. Even after you dry things out the insulation, being porous, will absorb moisture from the atmosphere and eventually it will fail due to tracking and then arcing. You should clean sockets etc. only with cleaners designed for cleaning electrical equipment. If you are in doubt about the condition of the insulation, saturate it with a moisture displacer, such as WD-40 then, to avoid any risk of fire, allow the WD-40 (which contains flammable solvent) to soak in and then dry for several hours before doing anything like throwing a high voltage switch or unplugging tubes while the set is powered up.

SOME MORE NASTIES LURKING IN THE WORKSHOP . . .

Mercury, which is poisonous, is used as a getter to absorb any remaining oxygen molecules in valves; this is what causes the internal silvering on some valves. There are also some old rectifiers that used mercury in them; when the tube is cold, you can see little drops of mercury adhering to the inside of the glass. In any case, be careful not to break any tubes that have little blobs of silver liquid clinging to the glass.

The risk of **radioactivity** is sometimes brought up and it should certainly not be ignored. Some USA-made common voltage regulator tubes had radioactive isotopes added, apparently to achieve more stable operation, whilst a number of WW II vintage tubes (voltage regulators and radar T/R switches) contain radium 226 in quite large amounts. Another source of radioactivity is the luminous paint containing radium used for dial markings on Second World War radios, such as the British Army's 19 Set. American military sets considered radioactive include ATD, BC-620, BC-654, BC-659, BC-1335, R-392, T-195, TBX and TBY.

Although post-war productions used a different, non-harmful luminous material, the original zinc sulphide and radium paint is now considered hazardous to health and, according to one report, one type of ex-Royal Navy ship's compass had to be recalled from the surplus market for this reason. Other equipment said likely to harbour radium paint includes the 18 Set, some domestic radios from the early 1950s with built-in clock/timers and American Jeep speedometers, plus virtually all aircraft instruments through to the 1960s.

Radioactivity can even lurk in old hi-fi components. Sold at one time for turntable use was the Mercury Disc-Charger, a red plastic cylinder with a clip on its side, weighing half a gram, and a small metal square plated with about 1 microcurie of radium 226 mounted in a shallow recess on the end. It was very effective for neutralizing static on records, and was used by simply clipping it on to the front of the head-shell. There was also a blue plastic teardrop-shaped version sold by a different company, but it contained a little less radium.

A considered assessment of the risks arising from radioactive equipment is fairly reassuring, thankfully. The first issue is potential exposure to radioactive materials when using our radios; this can be termed the irradiation hazard. The second issue is the risk of accidental ingestion of radioactive materials; we can call this the contamination hazard.

In practice, the irradiation hazard from most radios is minimal; the radiation dose rate at normal operating distances will probably be barely discernible above background levels. Good ventilation will minimize the hazard of inhaling radon gas given off by decaying radium. The contamination issue is very different though. There's no cause for concern if the radium paint is inside a meter where you can't touch it, but if it is on a front panel control, then a real possibility exists that you could wipe small amounts of radium off on your fingers, then eat or smoke, ingesting the radium. The same could happen if a tube containing radioactive material is broken. The small amount of material may not represent a significant health risk but it's a risk that's well worth avoiding.

With radioactive markings the best advice is to seal them with a coat or two of clear lacquer, to prevent picking up dust accidentally. They are easy to spot once you know what to look for, according to Lenox Carruth. Originally (and often on more recent equipment) the radium paint was light yellow in colour and appears thicker than the typical silk-screened markings. Old stuff now looks brownish and no longer glows. If you clean equipment with radioactive markings, you should wear rubber gloves and dispose of all rags, paper towels and gloves afterwards, sealing them in a Ziploc polythene bag. Do not store radioactive equipment where it will be in close proximity to people for long periods.

Beryllium oxide, found in some power transistors, VHF 'ceramic' transmitting triodes and klystrons, is very nasty stuff (if you inhale just a few micrograms, you could be dead in a week or two). Luckily, if you avoid filing, sanding or smashing the stuff, you are quite safe in handling it – large chunks are safe, micro-sized particles are dangerous.

Probably the best (and most obvious) advice is: don't put anything in your mouth; avoid handling broken glass, chemicals or powders, using gloves when necessary; and work in a clean area with adequate light and ventilation.

OIL-FILLED TRANSFORMERS AND CAPACITORS

Many oil products are considered harmful if misused; some engine oils are carcinogenic and should not be allowed into contact with your skin. Other types, containing PCBs (polychlorinated biphenols), are deadly if allowed to overheat and vaporize. Older television sets contained many oil-filled products – high-voltage capacitors, transformers and EHT triplers found in projection sets – and these may contain PCBs. If these do not leak or overheat you have nothing to worry about but you should be aware that environmentalists are extremely concerned about PCBs. Their use is banned in most situations nowadays and there are strict rules governing the disposal of products containing PCBs. To avoid the risk of skin cancer and other things too horrible to mention, wear rubber gloves when handling oil-filled devices. You have been warned.

SELENIUM RECTIFIERS

Selenium dioxide is the major compound produced when a selenium rectifier is overheated. It can cause severe burns to the mucous membranes and severe respiratory tract, skin and eye irritation. It can also promote allergic reactions with the skin. Fortunately it is not considered a carcinogen.

Another fortunate thing is the awful smell which gives it away. If you smell something really rotten, like decaying onions and garlic, coming from your equipment, it is best to leave the area immediately, opening some windows on the way out. Allow the selenium dioxide vapours to dissipate for several days before you go back – you will not want to go back very soon anyway!

EXPLOSIVE CHARGES

This, fortunately, is a low risk but still a real one. Military radio and radar equipment made during World War II, in particular by the Americans and Germans, was sometimes fitted with explosive charges. These were intended to destroy the equipment in the case of accident (impact) or capture by the enemy (tampering). In recent years at least one collector has seen his garden shed go up with a bang (spontaneous detonation) and another, to his horror, had an innocent-looking 'component' identified as a charge. Frequently these detonators look extremely similar in shape and size to electrolytic capacitors and issue 117 (1998) of *Funkgeschichte*, the German Historical Wireless Society's magazine, has photographs which illustrate this similarity well. It also quotes a newspaper report of 1949 in which a domestic radio technician testing war-surplus radio components for reuse lost his sight when one of these 'capacitors' blew up in his face.

CHAPTER 30

Data and more

This is the data that you always thought you could find in other books (but it turns out that you couldn't). Notes are at the end of the chapter.

PREFERRED VALUES FOR RESISTORS AND CAPACITORS

Series	*Basic values*								
E6	10				15				22
E12	10		12		15		18		22
E24	10	11	12	13	15	16	18	20	22

Series	*Basic values*											
E6				33				47				68
E12		27		33		39		47		56		68
E24	24	27	30	33	36	39	43	47	51	56	62	68

COMPONENT COLOUR MARKINGS

The colour codes used for fixed-value capacitors and resistors are generally as follows. This is, however, a simplification, as will be seen.

Colour	*Figure*	*Number of noughts*
Black	0	None
Brown	1	0
Red	2	00
Orange	3	000
Yellow	4	0000
Green	5	00000
Blue	6	000000
Violet	7	
Grey	8	
White	9	

On components marked in the 'body-tip-spot' scheme, the order of reading these colours is body, tip, spot. So a resistor with a red body, black tip and orange spot will have a value of 20 000 ohms (20 kilo-ohms). If there is no spot on the body, this indicates that the 'invisible spot' is the same colour as the body.

FOUR-BAND RESISTORS (FOURTH BAND = TOLERANCE)

First band		*Second band*		*Third band*	*Fourth band*
Black	0	Black	0	Multiply by 1	
Brown	1	Brown	1	Multiply by 10	
Red	2	Red	2	Multiply by 100	± 2%
Orange	3	Orange	3	Multiply by 1000	
Yellow	4	Yellow	4	Multiply by 10 000	
Green	5	Green	5	Multiply by 100 000	
Blue	6	Blue	6	Multiply by 1000 000	
Violet	7	Violet	7		
Grey	8	Grey	8		
White	9	White	9		
Gold				Divide by 10	± 5%
Silver				Divide by 100	± 10%

FIVE-BAND RESISTORS (FIFTH BAND = TOLERANCE)

As above but bands 1–3 are numeric, band 4 is a multiplication/division factor and band 5 is tolerance, using the same colour codes as above and additionally brown = ± 1%.

SIX-BAND RESISTORS (FIFTH BAND = TOLERANCE, SIXTH BAND = TC)

As above but a broad sixth band indicates the temperature coefficient in ppm/°C, where brown = 100, red = 50, yellow = 25, orange = 15, blue = 10, violet = 5, white = 1.

'READING' RESISTORS MARKED IN BANDS

For resistors marked in bands the decoding of the colours varies according to the number of bands present. The bands are displaced towards one end of the resistor and are read starting with the band closest to that end.

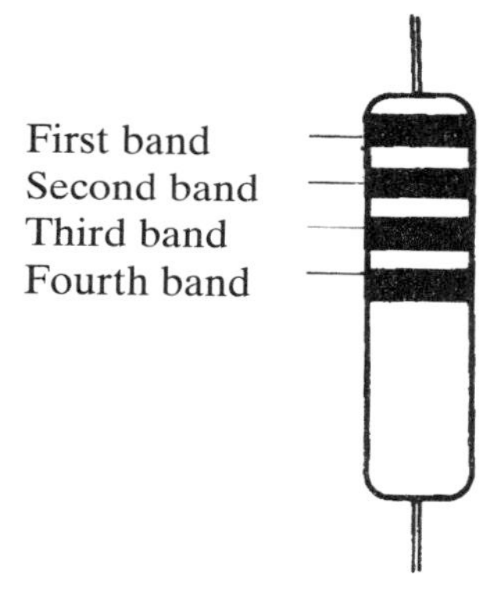

LETTER CODE FOR RESISTORS AND CAPACITORS (WORLD WAR II)

Pn = 0.n
Qn = 0.0n
Rn = 0.00n
Xn = 0.000n
Yn = 0.0000n

.... where n is the numeric value. For example, a 0.002 microfarad capacitor would be marked R2, whilst a 30 kilo-ohm resistor (= 0.3 mega-ohms) would carry Q3.

RESISTOR COLOUR CODES (ATWATER KENT)

300	Maroon/blue
500	Blue/red and green
1100	Black/blue or Black/red/blue
3300	Green/red
5000	Blue/yellow
6000	Purple
7500	Yellow
10 K	Maroon
15 K	Grey/yellow
20 K	Black/red
30 K	Grey
40 K	White
50 K	Blue
65 K	Black
100 K	Red/blue
250 K	Red/yellow
500 K	Purple/black

CAPACITOR COLOUR CODES FOR POLYESTER CAPACITORS

1st band		*2nd band*	
Black	0	Black	0
Brown	1	Brown	1
Red	2	Red	2
Orange	3	Orange	3
Yellow	4	Yellow	4
Green	5	Green	5
Blue	6	Blue	6
Violet	7	Violet	7
Grey	8	Grey	8
White	9	White	9

Third band (multiplier)		*Fourth band (tolerance)*		*Fifth band (working voltage)*	
Orange	× 0.001 μF	White	± 10%	Red	250 V DC
Yellow	× 0.01 μF	Black	± 20%	Yellow	400 V DC
Green	× 0.1 μF				

MOULDED 'DOT' CAPACITORS

This information is taken from *Radiotron Designers Handbook*. There were at least four different 'dot' systems for moulded mica capacitors:

1. The old RMA six-dot system. Dots 1, 2, 3 are capacitance; 4 is voltage; 5 is tolerance; 6 is a multiplier. Dot 1 is any colour other than white or black.
2. The RTMA (RCM) six-dot system. Dot 1 white means RTMA code, dot 1 black means JAN code. Dots 2, 3 are capacitance. Dot 4 is 'class' (RTMA) or 'characteristic' (JAN). Dot 5 is tolerance. Dot 6 is multiplier. Dot numbering runs left to right on both rows. Orientation is white/black (dot 1) in upper left corner or multiplier (black/brown/red/orange/yellow) in lower right corner. To summarize, for a six-dot cap: if dot 1 is *not* white or black, dots 1, 2, 3 = capacitance; 4=voltage; 6=multiplier. If dot 1 *is* white or black, dots 2, 3=capacitance; 6=multiplier.
3. The three-dot system. Dots 1, 2=capacitance; dot 3=multiplier. Orientation is 'name' above the dots or 'arrows' pointing right.
4. The four-dot system. Looks like a three-dot with a smaller dot in upper left corner. The small dot indicates tolerance. Dot meaning and orientation as for three-dot.

RTMA table:

Colour	*Value*	*Multiplier*	*Tolerance*
Black	0	1	20%
Brown	1	10	
Red	2	100	2%
Orange	3	1000	3%
Yellow	4	10000	
Green	5		5%
Blue	6		
Violet	7		
Grey	8		
White	9		
Gold		0.1	
Silver		0.01	10%

COMPONENT TOLERANCES

Capacitor voltage ratings (USA)

Colour	Rating	Colour	Rating
Brown	100 V	Violet	700 V
Red	200 V	Grey	800 V
Orange	300 V	White	900 V
Yellow	400 V	Gold	1000 V
Green	500 V	Silver	2000 V
Blue	600 V	None	500 V

Capacitor and resistor tolerances (USA)

Colour	Tolerance	Colour	Tolerance
Brown	1%	Violet	7%
Red	2%	Grey	8%
Orange	3%	White	9%
Yellow	4%	Gold	5%
Green	5%	Silver	10%
Blue	6%	None	20%

FIXED INDUCTORS

	Below 10.0 μH			*10.0 μH and above*		
Black	n/a	0	0	n/a	0	0.0
Brown	1	1	1	1	1	0
Red	2	2	2	2	2	00
Orange	3	3	3	3	3	000
Yellow	4	4	4	4	4	0000
Green	5	5	5	5	5	n/a
Blue	6	6	6	6	6	n/a
Violet	7	7	7	7	7	n/a
Grey	8	8	8	8	8	n/a
White	9	9	9	9	9	n/a
Gold			n/a	n/a	n/a	n/a

Tolerances: None = ± 20%
Silver = ± 10%
Gold = ± 5%

Examples:

Less than 10.0 μH	Brown-orange-grey	= 138 μH
	Brown-gold-red	= 1.2 μH
	Gold-red-green	= 0.25 μH
10.0 μH and above	Brown-black-yellow	= 100 000 μH
	Brown-black-black	= 10.0 μH
	Brown-black-brown	= 100 μH

FIXED CAPACITORS – BRITISH COLOUR CODE

Colour	*Value*
Red	Highest capacity +
Yellow	Second highest capacity +
Green	Third highest capacity +
Blue	Fourth highest capacity +
Violet	Fifth highest capacity +
Black	Principal negative connection
Brown	Second negative connection
Grey	Third negative connection
White	Centre connection of voltage doubler capacitors

Where only two leads are used, positive is red and negative is black. Where two capacities are of the same value, the one with the higher voltage rating has the higher colour in the table.

FUSES – BRITISH FUSE COLOUR CODE

Colour	*Value*	*Colour*	*Value*
Black	60 mA	Dark blue	1 A
Grey	100 mA	Light blue	1.5 A
Red	150 mA	Purple	2 A
Brown	250 mA	White	3 A
Yellow	500 mA	Black & white	5 A
Green	750 mA		

Fuse markings: AGC or F (German *Flink*) = fast blow, quick acting
FF = ultra-fast action
MDL or T (German *Traege*) = slow blow or antisurge

LOUDSPEAKER WIRING COLOUR CODES

British code for loudspeaker field coils

Brown High potential
Black with brown tracer Other connection

British moving coil loudspeakers

Green (outer end)	Output transformer, primary ends of winding
Brown (inner end)	Output transformer, primary ends of winding
Red	Output transformer, primary centre tap

Maroon	Output transformer, secondary end, inside
White	Output transformer, secondary end, outside
Yellow	Field winding, outside end
Black	Field winding, inside end

RMA colour code for loudspeaker field coils

Black and red striped	Start
Yellow and red striped	Finish

Note: The centre-tap (if any) will be grey and red striped.

RMA colour code for loudspeaker voice coils

Black	Start
Green	Finish

MAINS CORDS

By this we mean flexible appliance wiring or line cords, not the fixed cable or wiring inside the walls. All colour code combinations are given in the order L, N, E: in other words line (live), neutral and earth (ground).

The current general standard is Brown, Blue, Yellow/Green striped and this is mandatory in most European countries and on items intended for sale in those countries. The colours were chosen not for their association (otherwise brown would be earth!) but for reasons set out below.

Elsewhere in the world and also in Europe in previous times, these colour codings vary widely. The following combinations can be encountered. Again the order is L, N, E and the list is not intended to be exhaustive.

Belgium:	Red, yellow or blue, Grey, Black
Germany:	Usually Grey, Black, Red
Great Britain:	Red, Black, Green
Netherlands:	Any colour but grey or red, Red, Grey
Russia:	Red, Grey, Black
Switzerland:	Red, Grey, Yellow or yellow/red
USA, Canada:	Black, White, Green (if you remember Black Death you won't go far wrong!).

Under the American system, the black wire is known as the hot or ungrounded wire and goes to a brass terminal. The white wire (grounded in a two-wire system) has a plated terminal and is connected to the broader of the two flat pins of the plug.

The foregoing should make it clear why a unified colour coding of wires was necessary. Green is by no means the obvious colour for earth either:

before standard colours were adopted, earth was red in Austria, Finland, Germany, Norway and Sweden, black in Belgium and Russia, grey in the Netherlands and Poland, yellow in Switzerland and green in Britain and North America. Three cheers for standardization!

The choice of colours for the world-wide system was not arbitrary either: earth (ground) had to be distinguished and making it striped was an inspired choice. The other two colours had to be clearly distinguishable, even by people who suffer colour-blindness, and blue and brown were judged the most clearly different shades by experts in this field.

COMMONLY ENCOUNTERED SUPPLY VOLTAGES

AC		DC	
100 V	(Kansai region of Japan)	12–13.8 V	(car battery)
110 V	(Europe)	24–28 V	(lorry or boat battery)
117 V	(N. America)	32 V	(rural electric power plants in USA; some 32 V radios were made)
220 V	(Europe)		
230 V	(new standard for Europe)		
240 V	(Great Britain*)		

*Up to 1946, 240 V, 230 V and some lower voltages were permitted; from then until the European harmonization, the voltage was fixed at 240 V.

SUPPLY FREQUENCIES ENCOUNTERED

25 Hz	A few locations in Britain served by industrial rather than commercial supplies until the early 1950s, parts of the USA, also Ontario, Canada (Ontario Hydro changed to 60 Hz during the 1950s); many radios and other appliances were made for 25 Hz supplies).
50 Hz	Europe, Africa, much of Asia, Australasia, parts of Japan (Kansai region, 100 V), formerly some districts in the USA.
60 Hz	North America, parts of Japan (Kanto region, 220 V).
400 Hz	Equipment used aboard aircraft.

RADIO BATTERY COLOUR CODES – BRITISH COLOUR CODES FOR BATTERY CORDS

Highest positive voltage	Red	Red
Second positive voltage	Maroon and red	Yellow
Third positive voltage	Maroon	Green

Negative HT	Black with red tracer	Black
Positive LT voltage	Yellow	Pink
Negative LT voltage	Black with yellow tracer	Black
Positive bias voltage	Green	
Maximum negative bias	Black with green tracer	Brown
Second negative bias	Black and green	Grey
Third negative bias	White	

In the second (right-hand) scheme, any additional point, such as the fourth greatest GB negative, or fifth greatest HT positive, or positive bias, is violet and any centre tap is white.

RADIO RECEIVER WIRING

American colour code for receiver wiring (suggested, not mandatory)

Black	Grounds, grounded elements and returns
Brown	Heaters of filaments off ground
Red	Power supply, B+
Orange	Screen grids
Yellow	Cathodes
Green	Control grids
Blue	Plates (anodes)
Violet	Not recommended
Grey	Power line (AC)
White	Above or below ground returns, AVC etc.

Grey and violet are often omitted because of the difficulty in distinguishing them under certain artificial lights.

GEC (Britain) colour code for receiver wiring

White	High-potential connections to the aerial circuits, first section of the bandpass circuit and non-earth side of speech-coil
Green	Grid connections and high-potential ends of signal circuits
Blue	Screening grid circuits
Pink	Cathode connections
Orange	Connections to valve anodes
Black	Earth wiring
Slate	HT negative when not connected to earth
Red	HT smoothed
Red/white	HT unsmoothed, where smoothing choke is in positive leg
Green/white	Grid decoupling and AVC circuits

Black/red	LT positive in battery sets
Black/red, or black/white	Heaters

TRANSFORMER WIRING COLOUR CODES

NOTE: Ageing or overheating may cause colour changes to occur in wiring. Also, a different colour code may have been used by the manufacturer.

British colour code for power transformers

Primary 10 V tapping – black and green.
Primary 210 V tapping – black and yellow
Primary 230 V tapping – black and red
Primary 250 V tapping – black and brown
Primary zero tapping – black
Secondary HT ends – red
Secondary HT centre tap – red and yellow
Secondary rectifier heater ends – green
Secondary rectifier centre tap – green and yellow
Secondary valve heater ends – brown
Secondary valve heater centre taps – brown and yellow
Secondary additional LT winding ends – blue
Secondary additional LT centre tap – blue and yellow
Earthing lead – bare wire

RMA (American) colour code for power transformers

Primary leads – black
Primary leads if tapped: common – black; tap – black and yellow striped; finish – black and red striped
High-voltage plate winding – red; centre tap – red and yellow striped
Rectifier filament winding – yellow; centre tap – yellow and blue striped
Filament winding no. 1 – green; centre tap – green and yellow striped
Filament winding no. 2 – brown; centre tap – brown and yellow striped
Filament winding no. 3 – grey; centre tap – grey and yellow striped

Note how these colours are almost (but not quite) the same as British.

RMA colour code for audio-frequency transformers

Blue	Plate (finish) lead of the primary
Red	B+ lead (this applies whether the primary is plain or centre-tapped)
Brown	Plate (start) lead on centre-tapped primaries

Blue	Same (if polarity is not important)
Green and black	For centre-tapped push–pull operation
Green	Grid (finish) lead to secondary
Black	Grid return (this applies whether the secondary is plain or centre-tapped)

These markings apply also to line-to-grid and tube-to-line transformers.

RMA colour code for intermediate-frequency transformers

Blue	Plate (primary) lead
Red	B+ (primary) lead
Green	Grid (secondary) or diode lead
Black	Grid or diode return or AVC (secondary) lead

WANDER PLUGS – BRITISH COLOUR CODE FOR WANDER PLUGS

Red	Highest positive HT
Yellow	Second highest positive HT
Green	Third highest positive HT
Blue	Fourth highest positive HT
Pink	LT positive
Black	Negative (LT–, HT–, GB+)
Brown	Highest GB negative
Grey	Second highest GB negative
White	Third highest GB negative

Any additional battery lead is violet and any centre tap is white.

NB:
B+ is the highest voltage of the main DC power supply, generally supplying anodes.
Bifilar (or **bif**) is an anti-inductive winding, where a wire is, so to speak, coiled back on itself to cancel out radiation.
Filament is the same as heater.
GB is grid bias.
Ground in American parlance equals earth or chassis, according to context.
Plate (USA) equals anode (UK).

TELEPHONE MARKINGS

Markings found on and inside telephones normally identify terminals for connection or else (on British-made telephones) the manufacturer.

Terminals can be marked in a number of ways, which may be confusing when interpreting old circuit diagrams. The 'line' wires connecting telephones are taken to the terminals marked A and B, La and Lb, L1 and L2, or L and E. (L stands for line and E for earth). The battery connections for local battery telephones are marked C and Z, MC and MZ, or more obviously + and –. Z stands for zinc and C for carbon, since a dry battery has a carbon rod (the positive terminal) and an outer case of zinc (the negative terminal). The microphone (M) voltage is 3 V: anything higher will soon burn out the transmitter.

Battery call phones may have terminals marked RC and RZ for the ringing voltage (normally 6 V). Long-distance battery call telephones require a separate bell supply of 4.5 V, across BC and BZ. Sometimes all positive poles are taken to a common connection marked C. E denotes the earth connection, which implies that the telephone incorporates a lightning arrester or is intended for party-line working in which the bell is connected between one line and earth. ER stands for extra receiver and two terminals marked EB are for connecting a series extension bell after the strap connecting the EB terminals has been removed.

BPO MANUFACTURER CODES

Manufacturer codes were allocated by the BPO and are found both on instruments made for the BPO and on others. These are the most commonly found codes and, when followed by two digits, the number indicates the year of manufacture.

AK: Peel Conner (later GEC)
C: GEC (now GPT)
E: British Ericsson (later Plessey, then GPT)
FBR: Refurbished at Post Office factory, Birmingham.
FHR: Refurbished at Post Office factory, Holloway (London)
FWR: Refurbished at Post Office factory, Wales (Cwmcarn)
H: Automatic Telephone and Electric (later Plessey, then GPT)
PL: Plessey (became GPT)
PX: Phoenix
S: Siemens Brothers (later AEI, then GEC, now GPT)
TE: Telephone Manufacturing Company
W: Western Electric (and later STC)
Z First letter of code on items made for the British Army

All BPO telephone sets (and piece parts) carried a pattern number, in which table telephones, portable telephones and piece parts such as handsets carried even numbers whilst wall telephones had odd numbers. Special and experimental telephones had numbers prefixed SA (special apparatus), whilst designs inherited from the former National Telephone Company were prefixed NT. The pattern number is often followed by a mark number, generally starting with mark 234, which would be equivalent to Mk 1 in any other organization's series.

Note that type numbers of telephones beginning AP or Z are not BPO codes. AP stands for Admiralty Pattern, denoting a Royal Navy design, and YA is a British Army prefix. Type numbers beginning H are generally ATE products (the company originated at the place of Helsby), N denotes British Ericsson products, S are Siemens Brothers or AEI numbers, whilst a number beginning SA on BPO instruments denotes special apparatus, generally a prerelease prototype or an item made specially for a particular customer.

UNITS OF MEASUREMENT

One inch	= 0.0254 m
One foot	= 0.3048 m
One yard	= 0.9144 m
One square inch	= 6.451 square cm
One ounce	= 28.35 g
One pound	= 453.592 g
One ton	= 1017 kg (1.017 metric tonnes)
One cubic inch	= 16.387 cubic cm
Degrees Fahrenheit	= 1.8 degrees Centigrade/Celsius + 32
Degrees Kelvin	= add 273 to number of degrees Centigrade/Celsius

CHAPTER 31

Materials and finishes

You didn't expect to find a chapter on materials technology in a book like this – but unlike school or college, this kind of study is fun or at least enlightening. That's because once you start conserving and refurbishing old radios, televisions and other electrical apparatus you do almost need to become a materials technologist! You may choose to skip this chapter but it will probably give some helpful guidance, possibly even answer a few questions that have long puzzled you. If nothing else, you'll certainly be able to impress others with your newly found in-depth knowledge!

This section deals purely with materials and finishes; conservation and preservation techniques are treated earlier in this book.

METAL SURFACES

You will encounter a number of metal finishes in the restoration of old equipment and it's worth knowing and understanding them. External metal parts or components are usually plated or coated to provide protection, to improve appearance, or both. The most common finishes are:

Stove enamelling. A very hard paint-like finish achieved through a baking process. It is very hard-wearing.

Powder coating. The modern equivalent of stove enamelling and gives a textured finish which hides minor imperfections. Epoxy powder is baked on to the metal.

Enamel, known as porcelain in the USA. A thin film of white or coloured glass is melted over the sheet metal. The finish is very fine, almost like real porcelain, but it is easily damaged.

Nickel plating. Deposited direct on brass or over copper on steel. Reasonable appearance and protection. Rather soft and does not wear well with severe handling.

Chromium plating. Best deposited over a base film of nickel with consequent improvement in protection. Very bright and hard film which retains its appearance under hard-wearing conditions.

Cadmium plating. Greyish appearance when fresh. Usually applied to steel, which it protects very effectively. A chemical treatment ('passivation') is often applied, resulting in the familiar yellowish appearance. This provides improved protection. Suitable for the most arduous conditions. Loose deposits of cadmium sulphate resulting from previous cadmium plating are said to be toxic.

Zinc plating. A less expensive substitute for cadmium. Brighter, silvery appearance. Less effective protection, but adequate for the majority of commercial uses. May also be passivated to advantage. External contacts and small components may be plated to improve solderability.

Hot-tinning and electro-solder coating. These processes provide a substantial tin/lead coating which may be readily soldered even with some surface deterioration. These finishes may present a dark appearance on new components. This does not affect their solderability, which is much better than for the bright-looking pure tin electro-coating processes.

Gold flashing. A very thin, rather soft coating sometimes applied to protect silver plating from tarnishing. Internal contacts may be plated to improve performance without the expense of pure metal, or to prevent tarnish in store.

Silver plating. Used on light-duty wiping contacts.

Gold plating (hard). To cover the above to lower contact resistance and increase working life at low voltage and current levels.

PLASTICS

There are many, many plastic and allied materials, all having different characteristics.

THERMOPLASTICS

Thermoplastics are materials which melt when hot and set again when cold. They provide a wide range of inexpensive parts produced by high-speed injection moulding techniques.

ABS. A relatively modern plastic used for strong structural mouldings, around since the 1960s for telephones and radio cabinets. Looks like polystyrene but tougher and feels heavier.

Acetals (delrin). Relatively expensive, used to replace many structural metal parts.

Acrylics (perspex sheet and diakon moulding powders). Crystal clear or coloured material ideal for moulding and extruding. Polymethyl methacrylate (perspex, lucite) is used for all manner of clear dial covers, whilst diakon was used to mould the cases of telephones and radios (often mistakenly called coloured bakelite).

Nylon. Wear-resistant structural parts, connector parts etc. Often glass-filled for greater rigidity.

Polycarbonate (makrolon). Structural parts with good finish. Very high impact strength. Also used for some capacitors.

Polythene (polyethylene, alkathene). General insulation in component assemblies, cables etc.

Polypropylene. Similar to polyethylene but stronger and tougher.

Polystyrene (bextrene, distrene, trolitul). Inexpensive structural mouldings of good appearance. Clear and coloured, it came into widespread use in the 1960s for casings, as well as models and toys. Also used since the 1930s for small components.

Polysulphone. Generally transparent, for general purpose structural parts.

Polyvinyl chloride (PVC, welvic). Used in rigid form for general structural mouldings of good appearance.

PTFE (teflon, fluon). Outstanding insulator for highest performance requirements, also self-lubricating on account of its 'slippery' surface.

Terylene (melinex). Used in film form in 'polyester' capacitors and cable insulations. Member of the polyester or alkyd family of plastics.

Vinylidene chloride polymers (saran). Extrusions and tubing of extreme durability.

THERMOSETTING PLASTICS

Thermosetting plastics are those which cure under heat and pressure in a mould. They cannot be remelted but will decompose at high temperatures. They are mostly hard, rigid materials.

Phenolics such as phenol formaldehyde (bakelite, mouldrite, paxolin, tufnol). Bakelite and its look-alikes are generally found as black or brown general-purpose structural mouldings produced in dies under pressure. A variant, known as catalin, was made in a number of solid and transparent colours and was cast in moulds and then vulcanized. Radios moulded in catalin are extremely highly prized but are not without their problems; the material is extremely brittle and darkens or changes colour with age. A liquid catalin resin was used to produce radio dials and indicators.

Urea formaldehyde (beetle). White or light colours, general-purpose structural mouldings. The surface appearance is similar to bakelite and is often wrongly called bakelite.

Melamine formaldehyde (melamine, melmex, beetle). As for urea formaldehyde.

OTHER PLASTIC-TYPE MATERIALS

Cellophane. A viscose film material once used for decorative wrapping purposes but without other uses in 'our' field.

Celluloid. This name is applied to two variants of this early plastic, which is normally (but not always) clear and transparent, turning yellow with age. Cellulose nitrate (xylonite) was the original celluloid and is inflammable. Cellulose acetate (bexoid, cellomold, celastoid, erinofort) was developed later and is non-flammable. Many older mouldings shrink and crack as a result of polymer structure degradation and loss of plasticizer; the cause is residual traces of the acid mixture used to make the plastic from purified cotton or wood. The damage cannot be reversed and valuable pieces should be kept in a low-humidity environment.

Ebonite. Used for front panels, knobs, terminal blocks. Made of hard rubber and sulphur, originally black in appearance but fades to pale brown.

Epoxy resins. Widely used as surface coatings and in paints to give durability. Also used in adhesives to give great strength and for potting or embedding electronic components. When mixed with inert fillers, can be cast into shape to form dies, moulds, patterns and jig fixtures.

Formica. A trade name which, according to its makers, should not be used indiscriminately for all laminates of this kind. It is made by impregnating paper with a phenolic resin, overlaid by a melamine impregnated surface.

Glass fibre. When reinforced with resin, this forms a tough and resilient material and can be moulded to make apparatus cases. The surface finish is not of the highest quality and it tends not to be used for domestic equipment (but it was found on industrial television cameras, test gear and the portable radios made by GEC for the GPO radio investigation service). The fibres are often visible near the surface of the material. Bondaglass and Fibreglass are both trade names for the combined product.

Gutta-percha. The solidified juice of a tree and a non-conductor of electricity, used for covering telegraph wires and other electrical components during the nineteenth century. It is a very tough substance, without the elasticity of India rubber. It is softened by gentle heat and can also be welded in this manner; it takes sharp impressions and can be moulded very beautifully. Light and air make it brittle, but in the dark or underwater it will retain its original properties indefinitely.

Ivorine. A synthetic ivory made of cellulose nitrate. Usually off-white but often fades to clear and very prone to warping. Used mainly for engraved labels, with the engraving filled with black paint. Superseded by traffolyte. Also see celluloid above.

Paxolin. Similar to SRBP but with a fabric, rather than paper, base.

Presspahn. A kind of pressed paper or fibre-board, used for back panels of radios and TVs.

SRBP (single resin bonded paper). Used mainly for printed circuit boards and, in tubular form, for insulated spindles. Often (wrongly) called paxolin.

Traffolyte. A sandwich material of plastics of contrasting colours, so that engraving letters and numbers reveals a different colour. Commonly used for engraved signs, name-plates, panels etc. Also known as trefoil.

Tufnol. Very tough and dense laminated phenol material made of resin-bonded paper or fibre, used for shaft couplers and gears. Easy to drill and tap, very resistant to wear and tear. Also very expensive now. Brown in appearance.

Vulcanite. Used for front panels, knobs, terminal blocks. Made of hard rubber and sulphur, dull red in appearance but do not confuse with fibre.

Xylonite. See celluloid.

WHY PLASTIC MATERIALS DETERIORATE

Many of the plastics used in the past were not stable over long periods. Ultraviolet radiation from strong sunlight and heat above human body temperature accelerate distortion and discolouring, for which there generally is no repair other than replacement of the affected part. The best advice is that items in good condition should be cleaned up and positioned where they will not be subjected to strong sunlight or heat.

ABS. You might think that modern plastics would be immune from deterioration problems . . . but you'd be wrong. ABS, polystyrene and some allied plastics are susceptible to attack from the additives contained in PVC cables, which is why the latter are always wrapped in separate polythene bags when you open a new video recorder or whatever. A PVC mains lead wrapped closely around a device with an ABS case will 'melt' or 'burn' its way into the plastic and, over a few years, will leave deep gouges. It's all to do with the plasticizer in the PVC leaching out. People who lay sheets of expanded polystyrene as loft insulation have found they have problems where this has been in contact with ring main cabling, with the latter's sheath being embrittled in extreme cases.

Another problem with ABS is discolouring, which is discussed in Chapter 20, 'Conservation techniques'.

CELLULOID

Have you ever had a 'bakelite' phone (such as the BPO Tele. 232) where the cradle forks or stag's horns were disappearing in a damp, drippy mess? I have! The problem is well known to some collectors in Britain. The bakelite handset cradles of our 200-type telephones were recognized as being particularly fragile and, after the war, some bright spark decided to have these made of cellulose acetate instead. This plastic was after all more resilient. However, forty-plus years on many of these are deliquescent, absorbing moisture from the atmosphere and self-destructing at the same time. The following notes from Barry Ornitz explain the mechanism in more detail . . .

These are cellulosic plastics – either cellulose nitrate, cellulose acetate, or possibly mixed esters like acetate-butyrate. In the presence of moisture and oxygen, these plastics can decompose, releasing the acid from which they are made. The acid acts as a catalyst for the further decomposition – so, once started, it is difficult to stop and it occurs faster as it progresses. We are probably all familiar with the destruction of old celluloid movie film from the early years and the efforts to save these old films. The released acid is extremely corrosive to paint and metal alike.

At the first sign of yellowing or acrid fumes (possibly vinegar-like), wash the plastic immediately in warm water and soak it in a solution of baking soda (sodium bicarbonate) for a few hours. This will neutralize the acid, but it *may* remove any lettering on the plastic too so you are on your own here.

Whilst oxygen and humidity are necessary for the destruction process to take place, tightly boxing up your old treasures is not the solution either. It is better for any acid fumes to dissipate than be trapped with the plastic. Museums have discovered this fact the hard way.

Ordinary old-fashioned chemist's litmus-paper makes a quick and simple test. Moisten the blue paper and lay it on the plastic for a few hours. If it turns red or pink, you have a problem, and the washing and neutralization should be done soon.

This deterioration should not be seen as a manufacturing defect, by the way. The washing and neutralization steps are important ones for short-term stability, but after forty years, even the best cellulose acetate will decompose. Cellulose nitrate is even worse. There is not much you can do once the decomposition starts so early prevention (if that is ever completely possible) is the best.

Most of these cellulosic plastics contain a plasticizer that can slowly evaporate ('leach out') too. This leads to shrinking and extreme brittleness. You may have noted old radio dials and engraved instrument labels made out of celluloid; frequently these craze or turn brown, warp into crazy shapes and become extremely brittle. Attempts at correction may well end in tears too. Toy and model collectors suffer these problems even more; it's not uncommon to see model railway carriages which have turned banana-shaped or swollen plastic wheels which fall to pieces when touched. If considering making replacement parts or finding substitutes, Mylar or acrylic makes a much longer-lasting replacement plastic.

RECORDING TAPE

Audio and videotapes also suffer from the dreaded loss of plasticizer problem or something akin to it and, in consequence, lose their suppleness and flexibility, causing poor contact with the record/playback head as well as a binding action and squealing noise. An alternative explanation names hydrolysation as the culprit, stating that the binder compound used to stick the magnetic particles to the plastic backing has absorbed water from the air (possibly replacing the lost plasticizer). The water molecules actually make the tape expand a bit, so it doesn't fit the machined tape-guides properly any more. They can also interfere with the lubrication impregnated into the tape; and it is theorized that they can even interfere with the polished smoothness of the tape surface.

Why this should affect some tapes and not others depends on the formulation of the plastic backing and binder. Tapes of the 1950s and 60s are unaffected but these were made using whale oil (it is asserted), and the problems occurred when manufacturers started to look for a synthetic substitute. In the mid-1970s, 3M (Scotch) and Ampex, both major tape manufacturers, started experimenting with their formulas. They thought they were introducing major improvements, but instead created a tape much more prone to hydrolysation than anything had ever been. Because the problem did not show up for years, the formulas did not get corrected until some time in the mid-1980s. Theoretically any tape could get hydrolysed over a long period of time, especially if stored in a high-humidity situation, but in practice most squeaky tapes were made (roughly speaking) between 1975 and 1985.

One 'answer' is given by Richard Fish on the Internet. He cites a method of baking the tapes in a convection oven for eight hours at 130°F. It is entirely possible to bake a tape twice if the first time doesn't do the trick. You get about a three-week 'window', says Tom, before the tape starts to reabsorb water. So the best deal is to bake the tape and immediately make a copy. But if you forget to do it and it rehydrolyses, you can bake it again. There are other fixes for squealing tapes, involving water baths and lubricating films.

COMMON ADHESIVES

Cyanoacrylates (superglue). These adhesives are surface activated. They bond best in extremely thin coats and they bond to glass well. Being a low viscosity liquid, they can seep into an existing base by capillary action. Unfortunately they will not handle elevated temperatures and the joint is quite brittle. Thus if you expect thermal expansion to be a problem, you might consider something else.

Hot-melt adhesives (glue sticks etc.). Once again, these cannot take elevated temperatures. Being rubbery, they do allow a little give in the joint. For CRT socket attachment, where the base is often relatively cool because the electron gun is several inches forward of the base, these might work well. However, they all are high viscosity so expect little penetration by capillary action.

Epoxies (Araldite, Devcon etc.). Epoxy cements are available in two common forms: quick curing and slow curing. The five minute epoxies produce quite brittle joints and should probably not be used where thermal expansion is likely. The twenty-four hour epoxies are far more flexible when cured. Be careful in the proportions used and in the mixing.

Silicones (GE RTVs, Dow Bathtub Caulk, Permatex). Silicones usually do quite well at elevated temperatures. For home use, you will likely find only the one-part materials. These release acetic acid as part of their cure (polymerization) which can cause corrosion problems with certain reactive met-

als. They will also rarely cure thicker than 1/4 in sections. The two-part materials are like epoxies, i.e. a resin and a hardener (catalyst). They can do deep section cures and can be obtained in non-acid releasing forms.
For each of these adhesives, give joints twenty-four to forty-eight hours before applying stress. Also expect a little smell, both during mixing and application. Follow the safety precautions on the packages carefully and use the minimum of adhesive necessary to do the job.

CLEANING MATERIALS AND POLISHES

Combined cleaning and polishing liquids containing silicone or solvents can be hazardous to certain plastics. They can cause 'stress corrosion cracking' in some thermoplastic materials and it is suggested that these (and other) organic liquids can permeate the crystalline regions of plastics, allowing external stresses to do more damage than they normally would.

Armor-All (a form of silicone dissolved in a solvent – believed glycol) is fine on plasticized polyvinyl chloride plastics (PVC) but can cause problems with ABS plastics. Similarly GoJo is an emulsion of a detergent in water and organic solvents, the latter helping it to cut grease. Both of these products could cause deterioration of certain plastics and, if you have to use them, the safest plan is to test them thoroughly on some inconspicuous surface first. For clear dials and meter faces, you are safer using something much milder to clean them and, if strong solvents are essential, use these carefully and rinse off as quickly as possible.

LUBRICANTS AND CONTACT CLEANERS

Probably because manufacturers spend a lot of money convincing the public of the merits of their all-purpose products, people can easily end up believing all the claims made. Some of these preparations are not as suitable as they seem for 'our' hobby applications and, whilst the intention is not to criticize any particular product, users should at least be aware of their shortcomings.

WD-40 is a product to use with some care. The version sold in aerosol cans is made of around 50 per cent Stoddard solvent (a refined kerosene), about 15 per cent petroleum-base oil (actually a paraffinic solvent-refined mineral oil), 25 per cent LPG (isobutane and propane liquefied petroleum gas) and 10 per cent of a proprietary corrosion inhibitor. This makes it pretty flammable and calls for thoughtful use. Its original purpose was as a water-displacing compound; the Stoddard solvent gives it a low surface tension, which allows it to penetrate small cracks and crevices. After the kerosene has evaporated, the oil is left with a little corrosion inhibitor.

For car engines and industrial machinery, WD-40 has a number of applications but in domestic electrical apparatus, many of its properties are counterproductive. The oil in it has some limited lubricating properties, but it does nothing to clean oxidized switch contacts and acts as an insulator. The only way it can protect a surface from oxidation is by physically covering it. And, like all paraffinic oils, it reacts slowly with oxygen in the air to cross-link and polymerize, producing the well-known 'gummy mess'. As an insulator, it can lead to carbon arc tracks on switches and electrical contacts. As a cleanser it has some effect; the kerosene dissolves old grease, but it is not suitable for electrical contacts or carbon potentiometer tracks.

For cleaning and preserving electronic contacts there is nothing to beat two American products, De-Oxit D5 and the very similar Cramolin. The former is now available in Europe (see Probus Electronics Ltd under Electronic components in the 'Buyer's guide' chapter).

CLEANSERS

Distilled water is an excellent cleanser and dries very cleanly. It is safe on most things, but may dissolve inks used on dial markings, and should be used with care around electronic components. Tap water may leave stains.

Household ammonia (without additives) is fine for most metal parts but rinse it off well with clean water and allow things to dry thoroughly. It is a strong alkali and will dissolve shellac very quickly. A mild ammonia solution generally does a good job of dissolving dirt on painted and metal surfaces, although there is a growing concern over stress-corrosion cracking (SCC) of brass caused by ammonia. As soon as ammonia comes into contact with brass parts it attaches itself to the surface, so well that washing in solvents won't remove it, and causes microscopic corrosion of the surface of the metal. Subsequent mechanical stress on the component will cause it to fail and SCC has been found to be the cause of major failures in many restored clocks. Some metal polishes such as Brasso also contain ammonia, and accordingly they should be used with caution on 'our' kind of equipment.

Mineral spirits should be avoided. Every capacitor in the equipment will likely absorb the solvent and be destroyed. Even resistors, if they have microscopic cracks in their phenolic cases, are likely to be destroyed. Any plastic parts? Gone too. Vinyl or rubber insulated wire? Damaged or even destroyed.

Rust remover. There are several products offering a choice of liquid and jelly; brand names include Naval Jelly (USA) and Jenolite (UK), and nearly all of them contain phosphoric acid. They are extremely effective but must be washed off thoroughly afterwards.

Lighter fuel is a highly flammable light petroleum product containing naphtha and some pentane. It is ideal for removing adhesive label and tape residue, grease and dirt from most surfaces, dissolving most oil and grease without doing much damage to enamel paint (it can harm some lacquers). It is safe on most plastics but can leave a slightly dull surface.

Aerosol foaming cleanser has glycol ethers as its active ingredients. These are relatively mild solvents but will attack some surfaces.

Isopropyl alcohol, also sold as 'rubbing alcohol' is an alcohol/water mix, and is sold in various concentrations. It will attack marking inks and painted surfaces, but will sometimes cut adhesives and things like chewing gum that other products have difficulty with.

Cellulose thinners (automotive lacquer thinners) is extremely flammable. Excellent for cleaning petroleum and oily residues off metal parts to leave them absolutely clean. Dissolves many thermoplastics and will damage paint surfaces. Removes most marking inks very quickly as well.

Chlorinated hydrocarbons, including carbon tetrachloride, trichloroethane, and various other cleaning solutions, such as paraffin, are now considered undesirable (some are carcinogenic) and are difficult to obtain. Nowadays other solvents are sold which are just (or nearly) as effective.

Several of the chemicals listed above are noxious; they all require care and precaution in use and handling, and present serious safety hazards if not used properly. In many cases they are best used in a 'controlled' fashion, on the end of Q-tips or a small paintbrush. A soft toothbrush can reach into inaccessible places and is particularly valuable in cleaning up knurled metal parts and fluted knobs.

Finally, avoid the use of household dishwashers and dishwasher detergents. They are often put forward as an 'easy cleaning' method but are not to be recommended. Almost all dishwashers use high temperatures in their washing cycle, and the detergents used are a strong caustic solution. They may be great for washing dishes but not for delicate electronics!

(With thanks to professional chemist Barry Ornitz for his good advice on many of the substances discussed here.)

CHAPTER 32

Abbreviations and expressions used

These technical notifications include some that you will hear regularly but won't find in other books. Not all of these are used in this book but they are all likely to be encountered in technical literature or as markings on equipment. Abbreviations related exclusively to buying and selling among collectors are shown in the chapter 'Buying and selling'.

AC	Alternating current
AF	Audio frequency (often called low frequency in Continental Europe)
AGC	Automatic gain control
Aircheck	Recording of a radio programme with the music removed (apart from the opening and close of records), so that presenters can review their programmes without having to listen through the records
Airplane dial	Round dial on radio receivers (like the old Zenith consoles had). Most times they have a vernier tuning knob and span 300 degrees or so.
AM	(British) Air Ministry
Amp	Amplifier
ANB	Army-Navy-British (on headsets etc.)
AP (plus number)	Admiralty Pattern (type number, on all manner of British Navy equipment)
APC	Automatic picture control (to maintain contrast level)

AVC	Automatic volume control (in radio, effectively the same as AGC)
AVO	A particular UK make of multimeter (AVO stands for amps, volts, ohms) but often applied to any multimeter
AWG	American Wire Gauge – almost the same as the British SWG
B+	High-voltage supply, generally termed HT in Britain
BA	(Humorous) Boatanchor
BC	Broadcast
BF	*Basse fréquence* (French); see AF
BGM	Background music
Black Beauty	(Ironic) A type of shiny black capacitor, popular in the 1950s and 60s but now likely to cause severe problems in equipment
Boatanchor	(Humorous) Although there are different interpretations of this term, generally any radio that is 'hollow state' or tube-type is liable to be considered a 'boatanchor'. Units produced from WW II until the mid-1960s are probably the most popular examples of the genre. The origin of the term 'boatanchor' becomes self-evident after carrying a DX100 (a classic anchor) to your car and noticing how much longer your arms have become. Boatanchors may be transmitters or receivers . . . and no 'anchor' station is complete without an assortment of similar-aged station accessories like microphones, transmit/receive relays, vintage SWR bridges and so on
Bottle	(Humorous) Valve
BPO	British Post Office
Breveté	Patent (French word)
Bug, dead bug	(Humorous) Integrated circuit
BVA	British Radio Valve Manufacturers Association
BWO	Backward-wave oscillator
Cascade	In cascade = in series
Cermet	Ceramic-metal construction
Chassis potential	Potential of the chassis. This may be the same as earth if the chassis is earthed (grounded); alternatively on AC receivers it may be connected to the return of the mains supply and thus be live. Some TV receivers had chassis at half-mains potential

Cheater cord	TV sets in particular had plug-in power cords that were captive to the back of the set so that, when the owner removed the cover in order to pull the tubes and check them on the tester at the supermarket, he wouldn't get shocked. The cord that the repair shop used to plug into such a chassis was known as a cheater cord
Condenser	Older name for capacitor
CRT	Cathode ray tube (TV picture tube)
Crystal triode	Early name for transistor
CTV	Colour television
CW	Continuous wave, also applied to Morse code operation over radio links
DAT	Digital audio tape
DBGM	Deutsche Bundesgebrauchsmuster, post-war German for 'registered design'
DC	Direct current
DCC	Digital compact cassette
Deck	Colloquial term implying chassis, at chassis potential
DRGM	Deutsche Reichsgebrauchsmuster, pre-1945 German for 'registered design' (colloquially 'dirty rotten German make')
DSB	Double side-band
DSP	Digital signal processing
DVM	Digital voltmeter
DX	Long-distance
EHF	Extra-high frequencies, 30–300 GHz
EHT	Extra-high tension (measured in thousands of volts)
EIA	Electronics Industry Association (USA body)
EIAJ	Electronics Industry Association of Japan
Elco	Continental name for electrolytic condenser (capacitor)
EMF	Electromotive force (measured in volts)
FAQ	Frequently asked question
FET	Field-effect transistor
Firebottle	(Humorous) Valve
Flower power	(Humorous) Germanium transistor, so-called because of the common misprint in books arising from confusion during the hippy (flower power) era with the flower Geranium.
Fluid state	See hollow state

Freezing lotion	Aerosol can of freezing liquid (useful for identifying faulty capacitors and transistors)
GB	Grid bias or Gaumont-British
G-G	Grounded grid
Gimmick	A capacitor made by twisting two pieces of insulated wire together
GPO	General Post Office (Great Britain) or Government Printing Office (USA)
Griefkit	(Humorous) Rhyming name applied to a certain brand of do-it-yourself radio and electronics construction kits, popular in the USA and Britain from the 1950s to 1980s. The name is cruel because the quality of these kits has not been surpassed since, although not every constructor was up to the task
Ground	North American term for earth
HC	Hard copper (wire)
HF	High frequency, hoch Frequenz, haute fréquence (used in Continental Europe to mean what most English speakers call RF or radio frequency), otherwise the frequency band 3–30 MHz
Hollow state	(Humorous) Thermionic, the opposite of solid state
HT	High tension (measured in hundreds of volts) or handie-talkie (hand-held transceiver)
HV	High voltage (same meaning as HT)
Ident	Short for station identification, either visual (television) or audio (radio and television)
I.p.s.	Inches per second (tape recording speed)
ISEP	International standard equipment practice – a style of plug-in cards and racks popular in the 1960s and 70s
JAN	Joint Army–Navy (USA)
KG	Dutch for short wave
KW	German for short wave
LF	Low frequency (radio, 30–300 kHz) or, in Continental Europe this means what we English speakers call AF or audio frequency
LO	London (BBC/ITV telegraphic code) or local oscillator
LOPT or LOPTX	Line output transformer
LS	Loudspeaker
LT	Low tension
LW	Long wave

m	Metres
mF	Microfarad (logically it should stand for millifarad but nobody counts in millifarads)
MF	Medium frequency (radio, 300–3000 kHz) or multifrequency
Mil. spec.	Military specification (high quality standard)
mmF	Micromicrofarad (older name for picofarad)
Mod	Model, modification or modulation
MW	Medium wave
NF	Noise figure, a measure (in dB) of the noise generated by an amplifier compared to a perfect (noiseless) amplifier
NSN	(On components) NATO stock number
OB	Outside broadcast (known as a 'remote' in the USA)
OC	French for short wave
O/C or open	Open circuit
Octalox	Loktal (series of valves)
OM	French for medium wave
OT	Old-time, old-timer
OTR	Old-time radio
P as B	Programme as broadcast, i.e. a recording made 'live' at the time of broadcast, with nothing deleted
PA	Power amplifier or public address
PCB	Printed circuit board or polychlorinated biphenol, a hazardous substance
pF	Picofarad
PS	Power supply
PSU	Power supply unit
PTT	Press-to-talk (method of switching from receive to transmit) or Post, Telegraph and Telephone administration
PU	Power unit or (gramophone) pick-up
PWB	Printed wiring board (alternative expression for PCB)
Pwr sup	Power supply
Q	Abbreviation for capacitor or transistor on circuit diagram or alternatively, the figure of merit for a tuned circuit
Rat Shack	Colloquial name for a nation-wide chain of electronics hobby shops in the USA, often used semi-humorously
RCM	Radio Component Manufacturers

RCMF	Radio Component Manufacturers Federation (former UK trade body)
Rcvr	Receiver
Resistance	Older name for resistor
RLM	Reichsluftfahrtsministerium (former German air ministry)
RMA	Radio Manufacturers Association (former UK and US bodies, independent of each other)
Rock	Colloquial term for crystal
Röhre	German for valve
RS	Can stand for Radio Shack (Tandy) or RS (Radiospares) when used as a prefix to part numbers. Check the context
RTMA	Radio & Television Manufacturers Association (former US trade body)
RTR	Ready to run or reel-to-reel
RTTY	Radio teletype (teletypewriter traffic sent over a radio circuit rather than by wire)
RU	Rack unit; see U
Rx	Receive, receiver
SAE	Self-addressed envelope
SASE	Self-addressed stamped envelope
Sand-state	Solid-state or silicon (humorous)
S/C or short	Short circuit
S.c.c	Silk-covered cotton (wire)
Secteur	French for mains
SGDG	*Sans garantie du Gouvernement*, literally 'not guaranteed by the government', often found next to patent markings on French items
SHF	Super-high frequencies, 3–30 GHz
Shunt	'In shunt' is an older expression for 'in parallel'
Smoothing bomb	Electrolytic capacitor (colloquial)
Solid state	Device using semiconductor technology or components
Spaghetti	Systoflex (or similar) sleeving, for protecting bare wires (colloquial)
Spkr	Speaker (loudspeaker)
SSB	Single side-band
Suicide cord	Also known as a widow maker, this is a power cord with a plug on one end and either bare wires or alligator clips on the other. It was used in radio and TV workshops to apply power to a chassis that was out of the cabinet
Sustaining programme	Unsponsored programme provided without

	advertisements or idents by the network, which can then be rebroadcast by other stations without the need to obliterate commercials or another station's ident
SW	Short wave
SWG	Standard Wire Gauge
Three-wire fuse	Humorous term for transistor (also known as a solid-state electric fire)
Talkie-walkie	French for walkie-talkie
TI	Technical instruction
TR	Transmit–receive
Trafo	(German) Transformer
Tranny	(Colloquial) Transformer or transistor
Transcribed programme	Recorded programme as opposed to a programme transmitted live
Transcontinental	Series of valves introduced in Europe during the late 1930s. Most of these had side contacts and red metallized envelopes or steel cases
Transmission copy	The play-out copy of a programme which was recorded prior to being transmitted
TTY	Teletype (often used as a synonym for teleprinter)
Tube	USA: (electronic) valve; UK: if used as an abbreviation, generally means a CRT
Tuning wand	A non-metallic (usually Delrin) tool made to fit and adjust a screw/slug/core etc. of a tuned circuit component. Being both non-magnetic and non-conductive, the tool itself will not cause a shift in the tuned circuit, either by adding inductance (adding permeability) or by adding capacitance
TWA	Travelling wave amplifier
TWT	Travelling wave tube
Tx	Transmitter, transmission, transmission date
U	Unit of height (1.75 in) in 19 in equipment rack practice. Panels are measured in number of U height
uF	Microfarad (the u stands for the Greek letter mu)
UHF	Ultra-high frequency, 300 MHz–3 GHz (before WW II this term also embraced what we now call VHF)
UKG	Dutch and Polish for VHF
UKW	German for VHF

UL	Underwriters' Laboratory, a test house in the USA
U/S	Useless, unserviceable
USW	Ultra-short wave, older term for VHF
Vacuum state	See hollow state
Vacuum tube	Valve
Valve	France: rectifier valve. Lampe is the general term for a valve
Variac	Trade name for variable transformer but now used generically
VC	Variable capacitor
V	Video frequency or voice frequency (i.e. audio frequency) according to context
VGC	Very good condition
VHF	Very high frequency (30–300 MHz)
VLF	Very low frequency (3–30 kHz)
vol	Volume
VOM	Volt-ohm meter, multimeter
VR	Variable resistor
VSB	Vestigial side-band
VT	Valve, thermionic or valve, transistor (abbreviation used on circuit diagrams) or videotape
VTR	Videotape recorder
VTVM	Vacuum tube voltmeter, i.e. VVM
VVM	Valve voltmeter, what people used before high-impedance multimeters became widely available
Walkie-talkie	Portable hand-held radio transceiver
WARC	World Administrative Radio Conference
WG	Waveguide
Widow-maker	See suicide cord
x	General cypher used to shorten words (see examples below)
xcvr	Receiver
xfmr	Transformer
xistor, xstr	Transistor
xmit	Transmit
xmtr	Transmitter
xtal	Crystal
73	Best regards (telegraphic and amateur radio expression)

CHAPTER 33

Decoding valve, transistor and CRT numbers

This chapter explains the various series of component designators; detailed information on component characteristics can be found in the various data books and on-line sources of information described in the chapter 'Service data and other information'.

Note that at least three, entirely different, series of valve designations begin with the letters VT (US Signal Corps, British Royal Air Force and British Post Office). This can be a major source of confusion, although not if you know the source of the device in question.

COMMERCIAL VALVES

EUROPEAN PRO-ELECTRON VALVE SYSTEM, ALSO KNOWN IN BRITAIN AS THE MULLARD CODE

Two codes are used, one for transmitting and industrial valves, the other for receiving valves. Both consist of two or more letters followed by a serial number, for example ECC83, EL34. The code for receiving valves was introduced around 1934.

TRANSMITTING VALVES

First letter (general class of valve):

D Rectifying tube (including grid-controlled tubes)
M LF power amplifier or modulator triode

P RF power pentode
Q RF power tetrode
R Rectifier
T RF power triode

For tubes having dual systems, two of these letters are used, as in QQC04/15.

Second (or third for dual-system tubes) letter (type of cathode):
A Directly-heated tungsten filament
B Directly-heated thoriated tungsten filament
C Directly-heated oxide-coated filament
E Indirectly heated oxide-coated cathode
G Oxide-coated filament in mercury-vapour rectifier
V Indirectly heated oxide-coated cathode
X Directly-heated pure tungsten filament
Y Directly-heated thoriated tungsten filament
Z Directly-heated oxide-coated filament (except in mercury-vapour rectifiers)

Third (or fourth) letter:
G Mercury vapour filling
H Helix or other integral cooler
L Forced-water cooling
S Silica envelope
W Water cooling
X Xenon filling

The absence of a letter here indicates radiation cooling.

First number (anode voltage in kV) followed by hyphen:
05 0.5 kV
5 5 kV
12 12 kV and so on

This is for transmitting tubes. For rectifying tubes the first group of figures is the approximate DC output voltage in kV in a three-phase half-wave rectifying circuit.

Second number (output):
For valves up to 5 kW anode dissipation, the figures indicate maximum anode dissipation in watts.
For water-cooled valves above 5 kW dissipation, the figures indicate the maximum output in kilowatts.
For rectifiers, the figures indicate the maximum permissible rectified current per valve in milliamps or the approximate DC output power in watts or kilowatts per tube in a three-phase half-wave rectifying circuit.

Suffixed letters:

B	Cables
E	Medium seven-pin base
ED	Edison base
EG	Goliath base
G	Medium four-pin base
GB	Jumbo four-pin base
GS	Super Jumbo four-pin base
N	Medium five-pin base
P	P-base.

RECEIVING AND GENERAL PURPOSE VALVES

First letter (heater rating):

A	4 V AC
B	180 mA DC series connection
C	200 mA AC/DC series connection
D	0.5–1.4 V DC (normally directly heated, can be 0.625, 1.2 or 1.25 V)
E	6.3 V series or parallel supply (by far the most common prefix)
F	12.6 or 13 V
G	5 V, later defined as 'miscellaneous voltage, parallel supply'
H	150 mA series connection
K	2 V DC
L	450 mA series connection
O	Cold device, not heated, e.g. cold cathode tube, voltage stabilizer or semiconductor device
P	300 mA AC/DC series connection
U	100 mA AC/DC series connection
V	50 mA AC/DC series connection
X	600 mA series connection
Y	450 mA series connection

The second and, where applicable, third and fourth letters in multi-electrode valves indicate the types of device in the valve. They are normally listed in alphabetical order.

A	Signal diode (excluding rectifiers)
B	(=AA) Double diode with common cathode (excluding rectifiers)
C	Signal triode (excluding power output triodes)
D	Output or power triode (e.g. TV shunt stabilizer)
E	Signal tetrode (excluding power output triodes)
F	Signal or RF pentode (excluding power output pentodes)
H	Hexode or heptode (hexode structure)
K	Heptode or octode (octode structure)
L	Output tetrode, beam tetrode or pentode (including TV line output valves)

M	Magic-eye tuning indicator
N	Gas-filled triode or thyratron
P	Secondary emission tube
Q	Nonode or enneode
S	Scanning valve for television
T	Miscellaneous (including thyratrons)
X	Full-wave rectifier or double diode (gas-filled)
Y	Half-wave rectifier or single diode (vacuum)
Z	Full-wave rectifier or double diode (vacuum)

Digits indicate the base (first digit) and a code to distinguish valves that would otherwise have identical numbers (e.g. EL84 and EL85 are both output pentodes with a 6.3 V heater on a B9A base, but they are otherwise different). The last figure of tetrodes and pentodes (excluding power output tubes) indicates the type of characteristic; even figure denotes sharp cut-off characteristic, odd for variable-mu characteristic. With professional and industrial valves, serial numbers for prototypes always end in zero and those for variants in one of the figures 1–9.

1–10	Side contact
11–20	Footless 8-pin Telefunken
21–30	B8G (Loctal)
31–39 and 300 series	International Octal
40–49, 140–149	B8A (Rimlock)
50–59	B9G (Magnoval)
60	Subminiatures
70	Miscellaneous
80–89, 180 and 800 series	B9A (Noval)
90–99	B7G (Miniature seven-pin)
200 series	B10B
500 series	B9D

BRITAIN: GEC CODE (MARCONI/OSRAM AND EMI BRANDS ALSO)

Consists of letter (or letters) followed by digits (e.g. L63, KT88). The digits are simply to distinguish similar valves (like all triodes) and cannot be decoded. The letters have the following meaning.

A	Industrial valve (maybe almost anything – power triode etc.)
B	Double triode
D	Diode or double diode
GU	Gas-filled rectifier
GT	Gas triode (thyratron)

H	Signal triode (high impedance)
KT	Kinkless tetrode (beam tetrode)
L	Signal triode (low impedance)
M	Metallizing (when used later in code)
MU	Indirectly heated rectifier
N	Output pentode
P	Output triode
QP	Quiescent push–pull double pentode
S	Tetrode (screen grid valve)
U	Rectifier
VS	Variable-mu tetrode
W	Variable-mu pentode
X	Triode, hexode, heptode, octode, frequency changer
Y	Tuning indicator
Z	Sharp cut-off screened HF pentode

BRITAIN: EDISWAN AND MAZDA CODE

Consists of digits, letters, digits. Do not confuse with a US code.

First digits give heater voltage: 1 = 1.4 V, 6 = 6.3 V whilst 10, 20, 30 indicate 100 mA, 200 mA, 300 mA for series connection.

Letters give type of valve:

C	Frequency changer
D	Signal diode or double diode
F	Signal amplifier, tetrode/pentode
K	Thyratron (small gas triode or tetrode, according to another source)
L	Signal triode or double triode
M	Tuning indicator
P	Output (power) tetrode/pentode
U	Half-wave rectifier, without first figure
UU	Full-wave rectifier, without first figure

Mazda codes tend not to double-up letters – L is used for multiple triodes also.
Final digits distinguish between otherwise identical codes.

BRITAIN: STC BRIMAR (BRITISH-MADE NORTH AMERICAN RANGE) CODES

Three codes are used. If the valve is for export, the American RTMA code is adopted, sometimes with a prefix SV. Valves for the home market employ a code consisting of a figure or figures such as:

1	Half-wave rectifier
8	RF pentode
9	Variable-mu pentode
20	Triode hexode

followed by a letter and a serial number:

A	4 V heater
B	2 V heater
D	Indirectly heated, other than 2 V or 4 V cathode.

If special valves, the code consists of a number, such as:

2	Diode
3	Triode
4	Tetrode
5	Pentode
22 etc.	Double diode etc.

followed by a letter giving maximum anode dissipation and type, followed by a serial number, followed by one or two letters indicating base type and special features.

BRITAIN: BVA NUMBERS

BVA (British Valve Association) numbers were assigned to valves used for civilian replacements in WW II.

BRITAIN: ERICSSON VALVE CODE

The code consists of a group of letters followed by one or two numbers (separated by a solidus), and then another letter. Some devices have a suffix consisting of another letter, preceded by a solidus. Examples are GC10B, GC12/4B, GTR150M/S.

First letter gives the construction of the device:

G	Gas-filled
V	Hard vacuum

Remaining letters in the first group give the type of device:

C	Counter (stepping tube where not all electrodes are available on separate pins)

D	Diode (stabilizer etc.)
TE	Tetrode (trigger tubes etc.)
T	Triode (trigger tubes etc.)
TR	Triode (trigger tubes etc.)
S	Selector (stepping tube with all outputs available on separate pins)
R	Register display tube (e.g. digitrons, Nixie tube-like devices)

First or only number gives the most important characteristic of the device:

Counters, selectors	– number of states
Register displays	– number of display electrodes
Diodes/stabilizers	– operating voltage
Trigger tubes	– striking voltage on trigger electrode

Second number is used for counters only and specifies the number of available output pins (if more than one).

Final letter gives the base:

A,B,C,D,E	Plastic base shell (Octal, B5, duodecal, solder tags etc.)
G,H	26-pin B26A, B27A
M,P,Q	B7G
T	B9A
W,X,Y	Wire-ended

Suffixes:

/S	Tested to military (services) specs
/R	Tested for resistance to vibration

FRANCE: MAZDA

The code comprises two letters followed by three or four figures.

First letter (heater voltage):

A	1 V
B	2 V
D	4 V
E	5 V
F	6 V
G	7 V

Second letter (heater current)

Z	0.05 A
Y	0.1–0.14 A
X	0.15 A
W	0.2 A and above

First two figures:
Amplification factor (one digit only if less than 10)
Last two figures
Internal resistance (in kilo-ohms)

FRANCE: MINIWATT

This is a subset of the European system. Either two or three letters are used, followed by a sequential number.

First letter (heater):

A	4 V AC
B	180 mA DC
C	200 mA AC/DC
D	1.4 V battery
E	AC and 6.3 V car battery
F	13 V car battery
H	4 V battery
K	2 V battery
U	100 mA AC/DC

Second and optional third letter (type of valve):

A	Diode
B	Double diode
C	Triode (oscillator, detector, amplifier)
D	Audio power triode
E	Pentode
F	RF pentode
H	Hexode or heptode
K	Octode
L	Audio power pentode
M	Visual tuning indicator
X	Dual-anode rectifier (gas-filled)
Y	Single anode rectifier (vacuum)
Z	Dual-anode rectifier (vacuum)

FRANCE: VISSEAUX

The code comprises two letters followed by four figures.

Letters

RO	Directly heated
RS	Indirectly heated

First figure (heater voltage):
4	4 V

Second figure:
No particular significance

Last two figures:
Amplification factor

GERMANY: TELEFUNKEN

This original code had just two series, RE and RS, standing for Röhren-Empfangen (receiving valves) and Röhren-Senden (transmitting valves). The three-digit number following indicates first the heater current (first two digits, in hundredths of an amp) then the heater voltage (4, 3, 2 or 1 for 4 V, 3 V, 2 V or 1.5 V). The final letter indicates the valve base pattern, t for Telefunken, d for Europa-base with side contact for a space grid, and no letter at all for Europa-base.

The later V series valves used the first letter V to signify 0.05 amp heater current, followed by one or two letters denoting the function of the valve (following the standard European Pro-Electron/Mullard system). The numeric suffix indicates base type; 11 is an 8-pin Telefunken pattern. Thus VCL11 indicates a combined triode and output pentode with a Telefunken-style base.

NETHERLANDS: PHILIPS

The early Philips sets used tubes (not rectifiers) coded by a letter plus three or four digits, such as A428, E434 etc. The coding was used before the introduction of the European (Mullard) codes.
The letter gives current as follows.

A	0.06–0.10 amp
B	0.10–0.20 amp
C	0.20–0.40 amp
D	0.40–0.70 amp
E	0.70–1.25 amp
F	1.25 amp and above.

First digit(s): heater voltage, from 1 to 20 V.

Last two digits: give the gain if the tube is a triode; 41 is a tetrode; 42, 52 and 62 are HF tetrodes; 43,53 and 63 are output pentodes; 46 is an RF pentode.

Examples: E443 is an output pentode, A415 is a triode with a gain of 15.

USA: PRE-RMA CODE

Until about 1932, tube designations included two letters and three numbers, as in 'UV199'.

The first letter gives the brand, e.g.:

C Cunningham
D Duovac
N Philmore
V RCA

The second letter gives the base type, e.g.:

X Four pins
Y Five pins

The first digit shows the licensee while the last two figures show the type number.

Some designations appear not to follow this code. Another authority asserts that the two letters indicate base configuration and the first digit denotes the designer or manufacturer: 'The "UV" denotes four-pin stub base and the 1xx is Western Electric. The "PA 250" does not follow this completely – it denotes the RCA design, but has PA instead of UX for the base designator.'

USA: RADIO MANUFACTURERS ASSOCIATION AND RADIO TELEVISION MANUFACTURERS ASSOCIATION

There are two codes, which is confusing to some. The more commonly encountered code (Standard ET-110) consists of digits, a group of letters, then digits plus an optional suffix. A variant is used by the Russians.

The first digits give the approximate filament or heater voltage rating in bands, a point misunderstood by many. Digit 0 denotes cold cathode, 1 between 0.1 and 2.1, 2 between 2.1 and 2.9, 3 between 3.0 and 3.9, 5 between 5.0 and 5.9, 6 between 6.0 and 6.9, and so on. Another interpretation states that 6 indicates a voltage between 5.6 and 6.6 V, while 5 indicates a voltage between 4.6 and 5.6 V; 1 indicates a voltage in excess of 0, including 1.6 V, whilst 0 indicates a cold cathode.

The exception is that codes 7 and 14 are 'nominal' voltages and are used to indicate 6.3 V and 12.6 V valves with a loctal base. In the case of tapped filaments or heaters, the first figure indicates the total voltage with both sections in series.

The second symbol is a letter which is allotted in sequence commencing with A, except that I and O are not used; rectifiers follow the sequence backwards commencing with Z. When all the single letters of a group are exhausted, the system then proceeds with two letters commencing with AB; combinations of identical letters are not normally used. The single-ended (no top-cap) alternating current has a first letter S while the second letter may be that of the nearest equivalent in the double-ended range, e.g. type 6SK7 is the nearest single-ended equivalent to type 6K7. Another special case is the first letter L which is used for lock-in (loctal) types in the battery range. P as a second letter indicates a cathode ray tube. The rectifiers all start at the end of the alphabet such as 5Z3, 35Z5, 5X4 and 5U4.

Second digits give either the number of active electrodes or elements, or else the number of external connections. In metal tubes the shell counts as one element. Shielding by or in the base does not count as an element.

Whenever possible the 12 V equivalent of a 6 V tube retains the same letters, thus 6SK7GT and 12SK7GT are similar except for the heater rating.

The suffix letters specify the type of envelope or base. The code is:

G	Large glass envelope on octal base
GC	Glass compact: a tubular glass envelope
GT	Glass tubular: the smallest glass envelope
GT/G	Glass tubular: interchangeable with G and GT types
LM	Loctal-metal
LT	Locking base
M	Metal-coated glass envelope on octal base
MG	Metal-glass
ML	Metal-loctal
S	Spray-shielded (zinc-clad)
W	Military type assigned on behalf of armed forces, ruggedized
X	Low-loss base for HF use (ceramic)
Y	Intermediate-loss base (mica-filled phenolic)
A,B, C, D, E, F	Modified version of a type without other suffix (note 2)

Note 1: Valves used specifically for industrial purposes have a sequential all-numeric designation. RCA used 1600 numbers for special-purpose/industrial tubes, and Sylvania used 1200 numbers.

Note 2: Devices calling for the A versions of the tube may really need them. The canonical rule is that higher suffixes replace lower suffixes (or no suffix at all), but not vice-versa. Then again, that might just have been the current version at the time of design. In some A tubes, the A indicates things like higher anode dissipation, particularly in the case of power tubes. In other cases, the A means controlled heater warm-up time for series-string filaments.

USA: RMA CODE FOR SPECIAL-PURPOSE TUBES (DEVISED ABOUT 1942)

This is the code which covers the tubes that appear not to follow the rules; for instance 2C22 which has 6.3 V, not 2 V, heaters.

First number (heater power in watts):

1 None (no filament)
2 Up to 10 watts
3 10–20 watts
4 >20–50 watts
5 >50–100 watts
6 >100–200 watts
7 >200–500 watts
8 >500–1000 watts
9 >1000 watts

Letter (type of device):

A Single element (ballast etc.)
B Diode (e.g. 1B24, 2B21)
C Triode (e.g. 2C39, 2C51)
D Tetrode (e.g. 4D21)
E Pentode (e.g. 2E22, 2E26)
F Hexode (e.g. 2F21)
G Heptode (e.g. 2G21)
H Octode
J Magnetically controlled (e.g. 4J30)
K Electrostatically controlled (e.g. 2K26)
L Vacuum capacitor (e.g. 1L21)
N Crystal rectifier (e.g. 1N21, 2N706)
P Photoemissive (e.g. 1P25)
Q Cavities (e.g. 1Q21)
R Ignitrons (e.g. 1R21)
S Switches (e.g. 1S22)
T Storage
V Photoflash
W Travelling-wave
X X-ray
Y Thermionic converters

It may be that letters T–Y were never used. Some others were only used once.

MILITARY VALVES

BRITISH ARMY

Letters followed by digits. The letter part gives the valve type as follows:

ACR	Army CRT
AR	Army receiving valve (generally triode)
ARD	Army receiving diode
ARDD	Army receiving double diode
ARH	Army receiving hexode
ARP	Army receiving pentode
ARS	Army receiving screen grid (tetrode)
ARTH	Army receiving triode hexode
ARTP	Army receiving triode pentode
AT	Army transmitting valve (generally power triode)
ATP	Army transmitting pentode
ATS	Army transmitting screen grid (tetrode)
AU	Army rectifier
AW	Army stabilizer

BRITISH ROYAL AIR FORCE

Again the code consists of letters followed by digits. The letters have the following meanings :

VCR	Valve cathode ray (CRT)
VGT	Valve gas triode (thyratron)
VI	Valve indicator (tuning eye)
VR	Valve receiving (general receiving valve)
VS	Valve stabilizing (gas stabilizer)
VT	Valve transmitting (general transmitting valve)
VU	Rectifier

BRITISH ROYAL NAVY

Another letters followed by digits code. The letters give the type of valve, as follows:

NC	Navy CRT
NGT	Navy gas triode (thyratron)
NR	Navy receiving valve
NS	Navy stabilizer

NT Navy transmitting (or power) valve
NU Navy rectifier

OTHER BRITISH SERIES

CV (common valve) numbers replaced the above three codes during WW II. There is no way to decode these by simple inspection of their number. ZA and 10E are also used as prefixes to catalogue numbers. For example, UHF 'horned' triode 10E/392 is also equivalent to ZA 3055, VR 135, NR 80, CV 1135 and E 1148. On British Post Office components, VT means either valve, thermionic or valve, transistor (i.e. a transistor, not a valve) according to context.

FRENCH ARMY

Originally numbered in the TM (Télégraphie Militaire) series with the letters TM followed by the type number. There was also an '11' series of all-metal construction with octal base, equivalent to comparable American types but using 11-V heaters. Thus 11L6 was similar to 6L6 but using 11-V heaters. Valves marked ECMR indicate French Army origin (Établissement Centrale du Matériel Radio-télégraphique Militaire).

GERMAN DEFENCE (REICHSWEHR)

Most German Army valves use a special base and bear codes in the format: two letters, two digits, one letter and two digits.

First letter:
R Reichswehr (Defence); includes both Wehrmacht (Army) and Kriegsmarine (Navy)

Second letter:
D Dekametric wavebands
G Rectifier (Gleichrichter) or diode
K Cathode ray tube
L Transmitting or power
V Amplifier (Verstärker)

First number:
Heater voltage

Third letter:
A Cathode ray indicator

D	Dual anode
G	Rectifier or diode
H	Hexode
L	Speed modulation
M	Magnetron
P	Pentode
T	Triode

Final number:
This indicates the maximum power output in watts if the valve is a power tube, the coefficient of amplification if an amplifier or maximum rectified current if a rectifier.
A different scheme was used by the Air Force, employing two letters and a number. The number is a simple type number, issued progressively as each new valve was registered.

First letter:

L	Luftwaffe (German Air Force)

Second letter:

B	Cathode ray tube
D	Dekametric waves
F	Special type
G	Rectifier or diode
K	Stabilizer
M	Magnetron
S	Transmitting, wavelengths above 1 metre
V	Amplifier, wavelengths above 1 metre

Exceptions to this scheme include SA (rectifier), SD (triode) and SF (pentode), all using 1.9 V heaters.

Other markings commonly found include 'Eigentum der RLM' (Air Ministry property; RLM stood for Reichsluftfahrtsministerium which was the Government agency responsible for research and development of aircraft and avionic systems), 'Kriegsmarine' (Navy) and 'Wehrmachtseigentum' (Army property). BAL means Bau Aufsicht Luftfartsministerium', i.e. Air Ministry inspection.

US ARMY: SIGNAL CORPS

VT numbers are USA military valves, with no way of decoding their pattern without look-up tables, although many valves were dual-marked with VT and civilian part numbers.

US NAVY

Before the Joint Army Navy (JAN) plan was implemented in 1942, the Navy used two different numbering systems. The first scheme was placed in service just before the First World War, and was fairly simple – the first letters described who built the device (SE for the Bureau of Steam Engineering and C* for a civilian contractor, for example CG for General Electric), and the number was simply assigned sequentially. About 1932, a new scheme was introduced that would describe the parts better, using some of the old scheme to help. The 'C' contractor code was kept and greatly expanded and, of the digits following, the first two designated the class and the following three the specific type number. This system stayed in service until the outbreak of the Second World War, when it was dropped due to the huge influx of new tube types. The Navy did not immediately jump completely to the AN nomenclature system in 1942, however, placing some systems under the old Navy model letter/type number system and others under the new AN system (Ray Mote's two-part article in the June and July 1995 issues of *Electric Radio* magazine covered this topic).

Standard military marking requirements for parts are now (normally) as follows: part number (usually the NSN = National Stock Number), cage code (= prime contractor), date code (or lot date code), contract number and other sundry markings per specific contract. There is a good search engine for the cage code (and other goodies):

http://www.dlsc.dla.mil/

CATHODE RAY TUBES

AMERICAN RETMA SYSTEM

The code comprises a group of figures, a group of letters including the letter P, plus another figure or group of figures.

First figure or group of figures (screen size):
For round screens – diameter in inches
For rectangular screens – screen diagonal in inches

Letter(s) preceding letter P:
The letter(s) is (are) a serial code applying to tubes of the same diameter of diagonal and indicates the particular design of tube (A, B, C . . . Z, AB, AC etc.)

Letter P followed by a number (screen characteristics):
The suffix letter (A, B, C or F etc.), when used, indicates a later and modified version which can be substituted for any previous version but not vice versa.

The letter W indicates a military type and, when used, precedes any other suffix letters.

P1	Green	Medium persistence (pre-war 'scope tubes)
P2	Blue–green	Medium persistence
P3	Yellow	Medium persistence
P4	White	Medium persistence (standard TV tubes)
P5	Blue	Very Short persistence
P7	Blue–white	Short persistence,
	Yellow	Long persistence (*radar indicators)
P10	Dark trace	Long persistence
P11	Blue	Short persistence, ultraviolet
P12	Orange	Long persistence (*radar indicators)
P14	Purple–orange	Short persistence
P15	Blue–Green + near UV	Short persistence
P16	Near UV	Short persistence
P22	Colour TV	Medium persistence
P31	Green	Fast writing, used for oscilloscopes

EUROPEAN TYPE DESIGNATION (OLD SYSTEM)

The type number consists of two capital letters followed by two sets of figures, e.g. DG13-2, MW31-16.

First letter (method of focusing and deflection):

A	Electrostatic focusing and electromagnetic deflection
D	Electrostatic focusing and electrostatic deflection in two directions
M	Electromagnetic focusing and electromagnetic deflection

Second letter (properties of the screen):

P Phosphor type

First group of figures (dimensions of the screen):

Round tubes – screen diameter in cm
Rectangular tubes – screen diagonal in cm

Second group of figures:

Serial number

EUROPEAN TYPE DESIGNATION (NEW PRO-ELECTRON SYSTEM)

This codes applies to all patterns of cathode ray tubes. It consists of one letter followed by two groups of figures joined by a hyphen and one or two letters.

First letter (application and/or construction of tube):

A TV display tube for domestic applications
D Oscilloscope tube, single trace
E Oscilloscope tube, multiple trace
F Radar display tube, direct view
L Display storage tube
M TV display tube for professional use, direct view
P Display tube for professional applications projection
Q Flying spot scanner

First figure or group of figures (screen dimensions):
Round tubes – screen diameter in cm
Rectangular tubes – screen diagonal in cm

Second figure or group of figures:
Serial number

Final letters (screen properties):
The first letter denotes the colour of the fluorescence (or phosphorescence in the case of long or very long persistence screens) according to the regions of the Kelly chart of colour designations for lights, where applicable.

A Reddish-purple, purple, bluish purple
B Purplish-blue, blue, greenish-blue
D Blue-green
G Bluish-green, green, yellowish-green
K Yellow-green
L Orange, orange-pink
R Reddish-orange, red, pink, purplish-pink, purplish-red, red-purple
W Standard white television display tube phosphor
X Tricolour screens
Y Greenish-yellow, yellow, yellowish-orange

The second letter is a serial letter to denote other specific differences in screen properties.

MAZDA (FRANCE)

Television tubes use the code letter C (for cathode ray tube), a number (indicating screen size in mm), then two letters followed by a number. The first indicates the method of deflection (M = magnetic, S = electrostatic) and the second the screen colour (B = blue, G = green, R = radar, long persistence, W = white). The numbers were allocated sequentially.

MINIWATT (FRANCE)

Television tubes use two letters followed by a number. The first indicates the method of deflection (D = electrostatic, M = magnetic) and the second the screen colour (B = blue, G = green, R = radar, long persistence, W = white). The numbers were allocated sequentially.

PHOTOTUBES AND PHOTOMULTIPLIERS (OLD SYSTEM)

The type number consists of two figures followed by two letters (e.g. 90AV).

First figure (tube base):
2 Loctal eight-pin base
3 Octal eight-pin base
5 Special base
8 Noval nine-pin base
9 Miniature seven-pin base

Second figure:
Serial number

First letter (type of cathode):
A Caesium–antimony cathode (blue-sensitive)
C Caesium on oxidized silver cathode (red-sensitive)
T Tialkali cathode
U Caesium–antimony cathode with quartz window

Second letter (class of phototube):
G Gas-filled
V High vacuum

Third letter:
Always P for photomultiplier

VOLTAGE STABILIZERS (OLD SYSTEM)

The type number consists of a number followed by a capital letter, a figure and, in some cases, by a second capital letter (e.g. 85A2, 150C1K).

Number:
Average burning voltage in volts

First letter (current range):
A Max. 10 mA
B Max. 22 mA
C Max. 40 mA
D Max. 100 mA
E Max. 200 mA

Figure:
Serial number

Second letter (tube base):
E Edison
K Octal eight-pin base
P P-base

TRANSISTORS

EUROPEAN PRO-ELECTRON SYSTEM

Transistors, diodes and similar devices were originally treated as zero-filament voltage thermionic devices and given numbers prefixed by OA and OC. Today they are defined by two letters followed by a serial number (three figures or one letter and two figures). The serial number is manufacturer-dependent. The initial two letters define the device function as follows.

The first letter indicates the semiconductor material used:

A Germanium
B Silicon
C Compound materials such as gallium arsenide
D Compound materials such as indium antimonide
R Compound materials such as cadmium sulphide

The second letter indicates the general device function:

A Detection, high-speed or mixer diode
B Variable capacitance diode
C Transistor for AF (not power) applications

D	Transistor for AF power applications
E	Tunnel diode
F	Transistor for RF (not power) applications
G	Miscellaneous devices
L	Transistor for RF power applications
N	Photocoupler
P	Radiation-sensitive device (such as photo diode etc.)
Q	Radiation generating device (such as light emitting diode)
R	Controlling or switching device (not power) such as thyristor
S	Transistor for switching (not power) applications
T	Controlling or switching power device
U	Transistor for power switching
X	Multiple diode (such as varactor)
Y	Rectifier, booster or efficiency diode
Z	Voltage reference, voltage regulator or transient suppressor diode

NORTH AMERICAN (JEDEC) SYSTEM

The Joint Electronic Device Engineering Council devised a system using a numeral, the letter N and a registration number of two, three or four digits. The first numeral indicates the number of diode junctions; 1 = diode, 2 = transistor, 3= FET. The prefix JAN (joint Army/Navy) indicates a military specification variant of the transistor.

JAPANESE SYSTEM

Japanese transistors are numbered in series commencing 2S and one more letter, then two, three or four figures. The series are 2SA, 2SB, 2SC and 2SD (including both germanium and silicon devices). A = p-n-p, radio frequency applications; B = p-n-p, audio frequency, C = n-p-n, RF and D = p-n-p, AF. The 2SJ and 2SK series are all FETs.

Note that the first two characters (e.g. 2S) are normally omitted on device markings, so a 2SC438 transistor will normally be marked C438.

SOVIET SYSTEM

A code of one letter and three digits is used; the letter D denotes a diode and P denotes a transistor.

For Germanium diodes, 1–100 are point-contact types, 301–400 are junction types, 401–600 are mixers, 601–800 are video types and 1001 and above are rectifier stacks. For silicon diodes, 101–200 are point-contact types,

201–300 are junction types and 801–900 are zener diodes.

Transistors are numbered P1–100 germanium LP/LF, P101–200 silicon LP/LF, P201–300 germanium HP/LF, P301–400 silicon HP/LF, P401–500 germanium LP/HF, P501–600 silicon LP/HF, P601–700 germanium HP/HF and P701–800 silicon HP/HF. LP is low power, below 0.25 W, HP is high power greater than 0.25 W; LF is low frequency up to 5 MHz; HF is high frequency, above 5 MHz.

BRITISH GOVERNMENT TRANSISTORS

Transistors used by the armed services and the Post Office were also numbered in the CV (common valve) series. Some Post Office types were numbered PO1, PO2 etc.

US Army/Navy equipment designations: Letter/symbols used in making basic indicators

First letter: Installation

- A Airborne (installed assemblies and operated in aircraft) operating
- C Air transportable
- D Pilotless carrier
- F Ground, fixed
- G Ground, general use
- K Amphibious control
- M Ground, mobile
- P Pack or portable
- S Shipboard
- T Ground, transportable
- U General utility
- V Ground, vehicular
- W Underwater, fixed

Second letter: Type of equipment

- B Pigeon
- C Carrier (wire)
- F Photographic
- G Telegraph or teletype (wire)
- I Interphone or public address
- K Telemetering
- L Countermeasures
- M Meteorological
- N Sound in air
- P Radar
- Q Underwater sound
- R Radio
- S Special types
- T Telephone (wire)
- V Visual and light
- X Facsimile and television

Third letter: Purpose

- A Auxiliary (not complete sets)
- C Communications
- D Direction finder
- G Gun directing
- L Searchlight
- M Maintenance and test assemblies, also tools
- N Navigational aids
- P Reproducing photographic or sound
- Q Special types
- R Receiving
- S Search and/or detecting
- T Transmitting
- W Remote control
- X Identification and recognition

US Army/Navy component designations

Component indicator	*Description*
AB	Antenna base
AM	Amplifier
AS	Antenna system
AT	Antenna
BA	Battery, dry
BB	Battery storage
C	Control box
CG	Radio-frequency cable transmission lines and waveguides with terminals
CM	Comparator
CN	Compensators, regulators
CP	Computer
CR	Crystal units
CU	Coupling units
CV	Converter (electronic)
CW	Cover
CX	Cord
CY	Case
DY	Detector (Non-electronic; see CV)
DY	Dynamotor unit
F	Filters
G	Generator; see PU
GO	Goniometer
H	Headsets, handsets, head and chest sets
ID	Indicator
J	Junction, jack and terminal boxes
KY	Keyers, coders and interrupters
LS	Loudspeaker
M	Microphone
MD	Modulator
MK	Maintenance kit
MT	Meteorological apparatus
MT	Mountings
MX	Miscellaneous
O	Oscillator
PG	Pigeon articles
PH	Photographic articles
PP	Power pack
PU	Power units and motors
R	Radio receiver
RD	Recorder
RE	Relay assembly
RF	Radio-frequency unit
RG	Radio-frequency cables, transmission lines and waveguides
RL	Reel assembly
RT	Receiver and transmitter
S	Shelter
SA	Switching assembly
SB	Switchboard
SN	Synchronizer
T	Radio transmitter
TA	Telephone apparatus
TD	Timing device
TH	Telegraph apparatus
TK	Tool kit
TN	Tuning unit
TS	Test and measuring apparatus
TT	Teletypewriter and wire facsimile apparatus
U	Unions, connectors, plugs, sockets etc.
UG	Unions, connectors, plugs, sockets etc. designed for radio frequencies
V	Vehicles
VS	Visual signalling equipment

Index

References to illustrations shown *in italics*.